Gels for Oil and Gas Industry Applications

Gels for Oil and Gas Industry Applications

Editors

Qing You
Guang Zhao
Xindi Sun

MDPI • Basel • Beijing • Wuhan • Barcelona • Belgrade • Manchester • Tokyo • Cluj • Tianjin

Editors

Qing You
School of Energy Resources
China University of Geosciences
Beijing
China

Guang Zhao
School of Petroleum Engineering
China University of Petroleum
Qingdao
China

Xindi Sun
Physics and Engineering Department, College of Health, Engineering and Science
Slippery Rock University of Pennsylvania
Slippery Rock, PA 16057
United States

Editorial Office
MDPI
St. Alban-Anlage 66
4052 Basel, Switzerland

This is a reprint of articles from the Special Issue published online in the open access journal *Gels* (ISSN 2310-2861) (available at: www.mdpi.com/journal/gels/special_issues/gels_oil_gas).

For citation purposes, cite each article independently as indicated on the article page online and as indicated below:

LastName, A.A.; LastName, B.B.; LastName, C.C. Article Title. *Journal Name* **Year**, *Volume Number*, Page Range.

ISBN 978-3-0365-5196-8 (Hbk)
ISBN 978-3-0365-5195-1 (PDF)

Contents

gels

MDPI

Editorial

Editorial on Special Issue "Gels for Oil and Gas Industry Applications"

Qing You [1,*], Guang Zhao [2] and Xindi Sun [3]

1 School of Energy Resources, China University of Geosciences (Beijing), Beijing 100083, China
2 School of Petroleum Engineering, China University of Petroleum (East China), Qingdao 266580, China
3 Physics and Engineering Department, College of Health, Engineering and Science, Slippery Rock University of Pennsylvania, Slippery Rock, PA 16057, USA
* Correspondence: youqing@cugb.edu.cn

Citation: You, Q.; Zhao, G.; Sun, X. Editorial on Special Issue "Gels for Oil and Gas Industry Applications". *Gels* **2022**, *8*, 513. https://doi.org/10.3390/gels8080513

Received: 12 August 2022
Accepted: 16 August 2022
Published: 18 August 2022

Publisher's Note: MDPI stays neutral with regard to jurisdictional claims in published maps and institutional affiliations.

This Special Issue includes many advanced high-quality papers that focus on gel applications in the oil and gas industry. The papers in this Special Issue present the new development of gels that can be used as conformance control agents, drilling fluid additives, and hydraulic fracturing agents.

As common conformance control agents, gels have been applied in conventional reservoirs for decades. Recently, more and more research has focused on gel application in harsh environments, including high salinity and/or high temperature. Ding et al. [1] introduced the gelling behavior of PAM/Phenolic crosslinked gel and its profile control in a low-temperature and high-salinity reservoir. The results indicated that this gel could form a strong "stem-leaf"-shaped 3D network structure in deionized water. In addition, this structure remains stable in high-concentration salt solution. Therefore, this novel gel is suitable for conformance control of high-salinity and low-temperature reservoirs. Wang et al. [2] reported the enhanced oil recovery mechanism and technical boundary of gel foam for Changqing oilfield. In this study, the composite gel system and gel foam system were compared from the perspective of plugging performance and oil displacement performance. The results revealed that the composite gel system is stronger in terms of plugging performance, however, the gel-enhanced foam showed high oil displacement efficiency due to the "plug-flooding-integrated" feature of the foam. Qu et al. [3] studied the performance of soft movable polymer gel for water coning control of horizontal wells in offshore heavy oil cold production. The results indicated that this gel has a compact network structure and excellent creep property. The field application results revealed that the oil rate increased from 9.2 m^3/d to 20.0 m^3/d, the average water cut was reduced to 60–70%, and the cumulative oil production was predicted to increase almost 3-fold. Cheng et al. [4] used experiments and numerical simulation to investigate the oil displacement mechanism of weak gel in waterflood reservoirs. They concluded that weak gel selectively enters and blocks large channels and diverts subsequent water flow to the unswept area. In addition, the oil droplets converge to form an oil stream due to the negative pressure oil absorption created by the weak gel's viscoelastic bulk motion. In addition to conformance control agents, gels can also be used as additives in drilling fluids. Tang et al. [5] reported novel nanoparticles (acrylamide/2-acrylamide-2-3 methylpropanesulfonic acid/styrene/maleic anhydride polymer-based nanoparticles), which can improve the filtration of water-based drilling fluids at a high temperature. The rheological properties of the drilling fluids that were treated with these nanoparticles were stable before and after aging at 200 °C/16 h and the filtration control was improved. As a consequence, this novel nanoparticle can improve colloidal stability and mud cake quality at a high temperature. Tang et al. [6] also studied the effect of novel catechol-chitosan biopolymer encapsulator-based drilling mud on wellbore stability. The rheology properties of this drilling fluid were stable before and after 130 °C/16 h hot rolling and the shale inhibition behavior is good. Additionally, the novel drilling fluid could chelate with metal ions and form a stable covalent bond, which improved the adhesiveness, inhibition, and blockage. Wei et al. [7] presented a salt resistance copolymer and its application in oil well drilling fluids. Compared with other small molecular copolymer filtrate reducers, this new copolymer has better resistance to complex salts, better filtration-loss-controlling performance, and better resistance to saturated brine. Long et al. [8] introduced a new drilling fluid filtrate reducer and reported its preparation and hydrogelling performance. Compared with commercial filtrate reducers, this novel filtration reducer shows good thermostability and salinity resistance. The filtration loss performance is comparable with a low viscosity sodium carboxymethyl cellulose and carboxymethyl starch. Gels are also an important component in hydraulic fracturing. Shibaev et al. [9] summarized the novel trends in the development of surfactant-based hydraulic fracturing fluids. This review paper described the novel concepts and advances of viscoelastic surfactant-based (VES) fracturing fluids published in the last few years. This paper covered the use of oligomeric surfactants, surfactant mixtures, hybrid nanoparticles, or polymer/VES fluids. The advantages and limitations of different VES fluids were systematically discussed in this paper. Wang et al. [10] presented a new compound staged gelling acid fracturing method for ultra-deep horizontal wells. Sichuan carbonate gas reservoirs are used as models in a simulator to predict the performance of this new staged gelling acid fracturing method. The simulation results indicated that the crosslinked authigenic acid and gelling acid in the Sichuan carbonate gas reservoir are injected alternatively in three stages and the total production was 2.1 times higher than the conventional acid fracturing.

Funding: This research received no external funding.

Conflicts of Interest: The authors declare no conflict of interest.

References

1. Ding, F.; Dai, C.; Sun, Y.; Zhao, G.; You, Q.; Liu, Y. Gelling Behavior of PAM/Phenolic Crosslinked Gel and Its Profile Control in a Low-Temperature and High-Salinity Reservoir. *Gels* **2022**, *8*, 433. [CrossRef] [PubMed]
2. Wang, L.-L.; Wang, T.-F.; Wang, J.-X.; Tian, H.-T.; Chen, Y.; Song, W. Enhanced Oil Recovery Mechanism and Technical Boundary of Gel Foam Profile Control System for Heterogeneous Reservoirs in Changqing. *Gels* **2022**, *8*, 371. [CrossRef] [PubMed]
3. Qu, J.; Wang, P.; You, Q.; Zhao, G.; Sun, Y.; Liu, Y. Soft Movable Polymer Gel for Controlling Water Coning of Horizontal Well in Offshore Heavy Oil Cold Production. *Gels* **2022**, *8*, 352. [CrossRef] [PubMed]
4. Cheng, H.; Zheng, X.; Wu, Y.; Zhang, J.; Zhao, X.; Li, C. Experimental and Numerical Investigation on Oil Displacement Mechanism of Weak Gel in Waterflood Reservoirs. *Gels* **2022**, *8*, 309. [CrossRef] [PubMed]
5. Tang, Z.; Qiu, Z.; Zhong, H.; Mao, H.; Shan, K.; Kang, Y. Novel Acrylamide/2-Acrylamide-2-3 Methylpropanesulfonic Acid/Styrene/Maleic Anhydride Polymer-Based $CaCO_3$ Nanoparticles to Improve the Filtration of Water-Based Drilling Fluids at High Temperature. *Gels* **2022**, *8*, 322. [CrossRef] [PubMed]
6. Tang, Z.; Qiu, Z.; Zhong, H.; Kang, Y.; Guo, B. Wellbore Stability through Novel Catechol-Chitosan Biopolymer Encapsulator-Based Drilling Mud. *Gels* **2022**, *8*, 307. [CrossRef] [PubMed]
7. Wei, Z.; Zhou, F.; Chen, S.; Long, W. Synthesis and Weak Hydrogelling Properties of a Salt Resistance Copolymer Based on Fumaric Acid Sludge and Its Application in Oil Well Drilling Fluids. *Gels* **2022**, *8*, 251. [CrossRef] [PubMed]
8. Long, W.; Zhu, X.; Zhou, F.; Yan, Z.; Evelina, A.; Liu, J.; Wei, Z.; Ma, L. Preparation and Hydrogelling Performances of a New Drilling Fluid Filtrate Reducer from Plant Press Slag. *Gels* **2022**, *8*, 201. [CrossRef] [PubMed]
9. Shibaev, A.V.; Osiptsov, A.A.; Philippova, O.E. Novel Trends in the Development of Surfactant-Based Hydraulic Fracturing Fluids: A Review. *Gels* **2021**, *7*, 258. [CrossRef] [PubMed]
10. Wang, Y.; Fan, Y.; Wang, T.; Ye, J.; Luo, Z. A New Compound Staged Gelling Acid Fracturing Method for Ultra-Deep Horizontal Wells. *Gels* **2022**, *8*, 449. [CrossRef] [PubMed]

gels

MDPI

Article

A New Compound Staged Gelling Acid Fracturing Method for Ultra-Deep Horizontal Wells

Yang Wang [1,*], Yu Fan [1], Tianyu Wang [2], Jiexiao Ye [1] and Zhifeng Luo [3]

1 Engineering Technology Research Institute of Southwest Oil & Gas Field Company, PetroChina, Chengdu 610017, China; fanyu@petrochina.com.cn (Y.F.); yejiexiao@petrochina.com.cn (J.Y.)
2 Southwest Oil & Gas Field Company, Petrochina, Chengdu 610017, China; wty01@petrochina.com.cn
3 State Key Laboratory of Oil and Gas Reservoir Geology and Exploitation, Southwest Petroleum University, Chengdu 610500, China; lzfswpu2011@163.com
* Correspondence: wangyang0996@petrochina.com.cn

Abstract: Carbonate gas reservoirs in Sichuan are deeply buried, high temperature and strong heterogeneity. Staged acid fracturing is an effective means to improve production. Staged acidizing fracturing of ultra-deep horizontal wells faces the following problems: 1. Strong reservoir heterogeneity leads to the difficulty of fine segmentation; 2. The horizontal well section is long and running too many packers increases the completion risk; 3. Under high temperatures, the reaction speed between acid and rock is rapid and the acid action distance is short; and 4. The fracture conductivity is low under high-closure stress. In view of the above problems, the optimal fracture spacing is determined through productivity simulation. The composite temporary plugging of fibers and particles can increase the plugging layer pressure to 17.9 MPa, which can meet the requirements of the staged acid fracturing of horizontal wells. Through the gelling acid finger characteristic simulation and conductivity test, it is clear that the crosslinked authigenic acid and gelling acid in the Sichuan carbonate gas reservoir are injected alternately in three stages. When the proportion of gelling acid injected into a single section is 75% and the acid strength is 1.6 m^3/m, the length and conductivity of acid corrosion fracture are the best. A total of 12 staged acid fracturing horizontal wells have been completed in the Sichuan carbonate gas reservoir, and the production is 2.1 times that of ordinary acid fracturing horizontal wells.

Keywords: compound; staged; gelling acid fracturing; temporary plugging

Citation: Wang, Y.; Fan, Y.; Wang, T.; Ye, J.; Luo, Z. A New Compound Staged Gelling Acid Fracturing Method for Ultra-Deep Horizontal Wells. *Gels* **2022**, *8*, 449. https://doi.org/10.3390/gels8070449

Academic Editor: Mario Grassi

Received: 22 March 2022
Accepted: 13 May 2022
Published: 18 July 2022

1. Introduction

Sichuan carbonate gas reservoir is rich in natural gas resources. The reservoir has the characteristics of high temperature, low porosity, and low permeability [1–3]. The formation temperature can reach 160 °C, and the reservoir lithology is mainly dolomite. The gas production of vertical wells after fracturing is low, which cannot support the economic development of gas reservoirs [4,5]. Staged acid fracturing of horizontal wells is a favorable means to improve the development effect of gas reservoirs [6–8]. Staged fracturing technologies mainly comprise double-pack single-slip multistage fracturing [9], packer-sliding sleeve fracturing [10], and hydraulic jetting multi-stage fracturing [11]. Staged acid fracturing of ultra-deep horizontal wells faces many problems, such as fracture parameter optimization, staged fracturing process optimization, and acid corrosion fracture length improvement.

Productivity simulation and fracture parameter optimization of horizontal wells are the basis of staged acid fracturing of horizontal wells [12–14]. Simple analytical solutions are derived on the basis of the assumption of infinite drain hole conductivity. These models are widely adopted because they are easy to use. However, they are not able to produce an accurate prediction of productivity because they ignore the pressure drop in the wellbore [15–18]. Hemanta Mukherjee et al. [19] conducted a series of studies about the productivity of fractured horizontal wells in a uniform sandstone reservoir. Yang He et al. [20]

established the productivity-prediction model, which was suitable for the acid fracturing horizontal well in the fracture-cavity reservoir, by comprehensively considering primary factors, wellbore pressure drop, cracks for supporting liquid, and connected fracture system.

The reaction rate of acid liquid rises sharply at high temperatures, which seriously affects the acid fracturing effect of ultra-deep reservoirs [21–23]. Settari [24] developed a comprehensive acid fracturing model that solved for 2D/pseudo-3D fracture geometry, leakoff, heat transfer, and acid transport. Rencheng Dong et al. [25] developed a 3D acid fracturing model to compute the rough acid fracture geometry induced by multi-stage alternating injection of pad and acid fluids, they investigated the effects of viscous fingering, perforation design, and stage period on the acid etching process. S.J. Jackson et al. [26] investigated the stability of immiscible viscous fingering in Hele-Shaw cells with spatially varying permeability, providing an efficient, high accuracy scheme to track the interfacial displacement of immiscible fluids. Tu Nguyen et al. [27] developed an improved VOF interface-sharpening method on the general curvilinear grid for two-phase flows. Haoran Xu et al. [28] developed a modeling framework for the coupled thermal-hydro-mechanical chemical processes during acid fracturing in carbonatite geothermal reservoirs. Jiahui You et al. [29] developed a pore–scale numerical model to analyze the impact of acid–rock interaction on multiphase flow behavior, and the pore–scale model in this research provides the fundamental knowledge of the physical and chemical phenomena of acid–rock interactions and their impact on acid transport. Hossein Mehrjoo et al. [30] studied the final fracture conductivity during acid fracturing. Xu P et al. [31] studied the influences of rock microstructure on acid dissolution on a dolomite surface. Daobing Wang et al. [32] used a cohesive zone method (CZM)-based finite element model to obtain the fracture closure pressure and minimum horizontal principal stress of the major fracture and the branch fracture based on pressure fall-off analysis.

Deep high-temperature vuggy naturally fractured formations where tools are not reliable and costive [33], and diverted fracturing with temporary plugging has become an option [34]. Degradable fiber is another commonly used diverter. It can bridge on the fracture, or fill the gap between particulate diverters to form a tight barrier. A combination of particulate diverter and fiber was often applied in the fields [35,36]. Jianye Mouet et al. [37] designed a multi-stage triaxial fracturing system and experimental procedures to satisfy the requirements of diverted fracturing in horizontal wells. Among the various existing chemical diverters, temporary plugging agents have the greatest potential for being used in HPHT reservoirs. Temporary plugging agents are made from degradable polymers, typically polylactic acid (PLA), which can degrade at the reservoir temperature and leave no residue [38]. All reported field applications with this type of diverter are in reservoirs less than 100 °C, and only used for plugging the perforation holes during the refracturing [39–41]. Zhou et al. [42] investigated the filtration characteristic of temporary plugging and the length of filter cake under the condition of different fracture widths. Daobing Wang et al. [43] presented a comprehensive workflow to model hydraulic fracture by accounting for interactions with numerous crosscutting natural fracture or joint sets, as well as the effect of temporary plugging in opened fractures. This model is a fully coupled seepage flow in porous media, fluid flow in fractures, and rock deformation finite element model with adaptive insertion of cohesive elements as a crosscutting natural fracture or joint sets. Wang Mingxing et al. [44] designed a high-pressure evaluation system for fracture temporary plugging and investigated the key factors and their influencing pattern. Chen Yang et al. [45] conducted a series of experiments to investigate the plug formation time, diverter consumption, and plugging zone characteristics with different fracture widths and diverter concentrations. Zhu Baiyu et al. [46] established the visualization model of a microfluidic chip and CFD–DEM numerical simulation to evaluate the effect of particle plugging in a fracture-vuggy reservoir.

According to the previous studies, it can be concluded that there are still many technical problems in the staged acid fracturing of ultra-deep horizontal wells. Therefore, this paper introduces a new compound staged gelling acid fracturing method for ultra-deep horizontal wells. Using productivity simulation technology, we optimize the optimal fracture spacing and fracture length of segmented acid fracturing in horizontal wells. Through dissolution and plugging pressure tests of the temporary plugging agent, the combination of high-strength temporary plugging agent is optimized, and the key parameters of alternating injection acid fracturing process are optimized by the VOF model. This study deepens the understanding of the staged acid fracturing mechanism in an ultra-deep horizontal well, and provides great help for acid fracturing stimulation design. The compound staged gelling acid fracturing technology is widely used in ultra-deep horizontal wells in the Sichuan Basin, which has achieved a good stimulation effect and improved the development effect of gas reservoirs.

2. Results and Discussion

The buried depth of the carbonate gas reservoir in the Sichuan Basin is 6 to 7 km and the temperature is 140 to 160 °C. It is mainly composed of biological micrite ash and algal agglomerate dolomite. There are various types of gas reservoir space, including intergranular pores, intragranular pores, and intergranular pores (Figure 1), as well as karst caves, solution fractures, and structural fractures. The average porosity of the reservoir is 3.4% and the average permeability is 0.4 mD. In terms of natural fractures, they are mainly structural fractures. The section of structural fractures is generally straight, mostly with high-angle fractures. The wall of dissolution fractures is uneven and harbor shaped, and the structural dissolution fractures are generally filled with asphalt or dolomite. The core observation shows that the fractures are relatively developed, the fracture density is 2.86 pieces/m, and the connectivity between the fractures and various effective pores is good. According to the comparison of reservoir vertical development characteristics of multiple wells, the thickness of a single well reservoir is thin, and the average thickness is only 11.8 m.

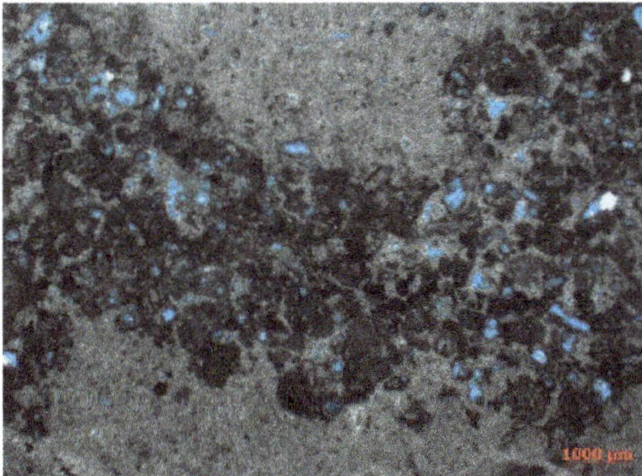

Figure 1. Thin section of well QY1.

According to the core test, the average Young's modulus of the carbonate reservoir in the Sichuan Basin is as high as 60 GPa and the average compressive strength is 510 Mpa. The high Young's modulus and compressive strength lead to the narrow width of acid fracturing fracture. The Poisson's ratio of the reservoir is 0.29. We used a GCTS RTR-2000 tester to evaluate the rock mechanical properties of cores in the Sichuan Basin, and it can be observed from Figure 2 that, with the increase in axial stress, there is no inflection point in the axial strain and volumetric strain curves, indicating that the reservoir cores in the Sichuan Basin have typical plastic characteristics, so it is difficult to maintain the conductivity of acid corrosion fractures under high closure pressure. Carbonate reservoirs in the Sichuan Basin have obvious plastic characteristics. Because the plastic deformation of rocks needs to consume additional energy, the fracture extension needs high pressure. Under the condition of the same injection displacement, the fracturing fracture formed in the elastic–plastic formation is relatively short and wide.

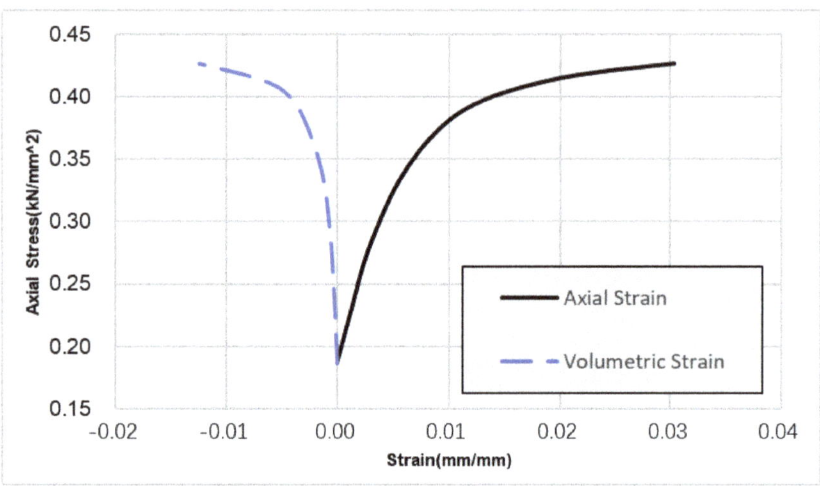

Figure 2. Stress—strain curve of well QY32.

The differential strain test results show that the maximum horizontal principal stress of the reservoir is 170.2 MPa, the minimum horizontal principal stress is 148.1 MPa, the vertical principal stress is 200.8 MPa, and the horizontal two-way stress difference is 22.1 MPa, and the horizontal stress difference coefficient is 0.15. According to the distribution law of three-way principal stress, the fractures formed by acid fracturing of carbonate rock reservoir in the Sichuan Basin are mainly vertical fractures.

Gelling acid is the most commonly used acid system for acid fracturing of gelling acid carbonate gas reservoir. Gelling acid has the characteristics of low friction and low acid-rock reaction speed. The gelling agent of acid is a linear high-molecular polymer. After being fully dissolved in acid solution, the molecular chains stretch and entangle with each other to form a spatial network structure and produce structural viscosity. The mechanical and thermal properties of gelled acid were evaluated by HAAKE rheometer, its density was 1.1 g/cm^3 and its viscosity was 24 mPa·s after shearing for 60 min at 160 °C (Figure 3). Gelling acid can maintain good viscosity at high temperatures, which is conducive to acid fracturing stimulation.

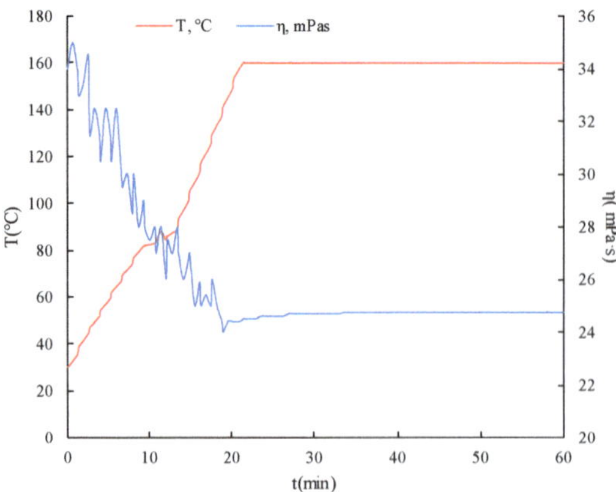

Figure 3. Stress–strain curve of well QY32.

2.1. Optimal Fracture Parameters of Acid Fracturing for Horizontal Wells

The optimization of fracture spacing is the key to the optimization of acid fracturing design [47–49]. Reasonable fracture spacing can reduce unnecessary packers and mutual interference between acid fracturing fractures. Reducing the fracture spacing is conducive to expanding the drainage area of gas wells, but the research shows that, when the fracture spacing is shortened to a certain extent, the productivity of gas wells will not be significantly improved, and there is an economic optimal fracture spacing for staged acid fracturing of horizontal wells [50–53]. According to the reservoir porosity, permeability, gas saturation, and other physical parameters in the Sichuan Basin, we used Eclipse to establish the numerical simulation model of staged acid fracturing horizontal wells in carbonate gas reservoirs, and the relevant parameters selected in the model are shown in Table 1.

Table 1. Numerical simulation-model parameters.

Grid side length (m)	0.1	Reservoir permeability (mD)	0.308
Model I-axis length (m)	3000	Gas saturation (%)	86.2
Model J-axis length (m)	600	Fracture conductivity (D·cm)	14
Model K-axis length (m)	12	Rock compressibility	3.9×10^{-7}
Reservoir porosity (%)	3.12	Formation pressure (MPa)	98.2

It can be observed from Figure 4 that the gas production of horizontal wells increases significantly with the decrease in fracture spacing. However, when the fracture spacing is reduced to 50 m, the gas production of horizontal wells is only 5.3% higher than that when the fracture spacing is 70 m, and the gas production does not significantly increase. Therefore, the optimal fracture spacing for the segmented acid fracturing of horizontal wells in carbonate gas reservoirs in the Sichuan Basin is 70 m.

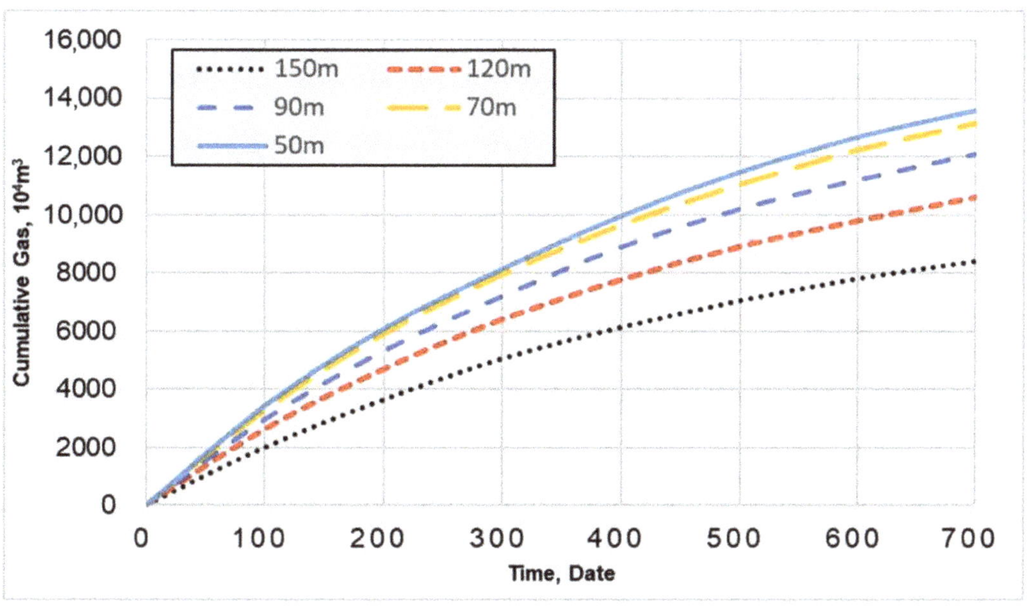

Figure 4. Productivity simulation under different fracture spacing.

2.2. Packer + Temporary Plugging

Staged acid fracturing technology of packer is an important measure to increase the production of the carbonate reservoir [54]. Its process is to run multiple packers in an open-hole horizontal well at one time, and divide the horizontal well section into several independent well sections for acid fracturing through the packer. The length of the horizontal section of the horizontal wells in carbonate gas reservoirs in the Sichuan Basin can reach 1.2 km, and the optimal fracture spacing according to productivity simulation is 70 m. If we only use packers for staged acid fracturing, we need to run at least 18 packers in the well to fully stimulate the formation, but running too many packers not only increases the risk of well completion [55], it also leads to greatly increases the throttling pressure, thus affecting the injection rate [56]. Through calculations, the carbonate gas reservoir in the Sichuan Basin only adopts the packer, which increases the throttling pressure by at least 30 MPa, and the acid injection rate is only 2 m^3/min. From such a low injection rate it is difficult to obtain long acid corrosion fractures at high temperature. In this paper, a new compound staged acid fracturing method was proposed to solve the problem of difficult stimulation of long interval horizontal wells. The goal of this method is to achieve tight cutting stimulation while reducing packers.

Aiming at the problem of difficult stimulation in the long well section, a new compound staged acid fracturing method is proposed in this paper. The goal of this method is to stimulate the reservoir while reducing the packer. The process principle is to use the temporary plugging agent to replace some packer functions. After throwing the ball to open the sliding sleeve, we first injected liquid to create the first fracture in the formation, then added the temporary plugging agent to block the end of the first fracture, and finally injected liquid for the second fracturing. This stimulation method can not only ensure the optimal fracture spacing, but also reduce the packer by half (Figure 5).

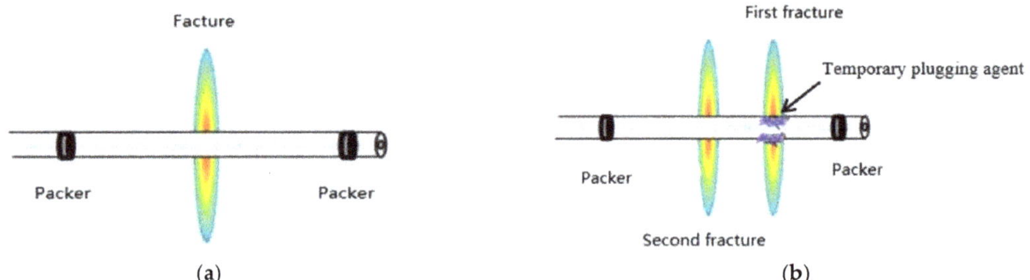

Figure 5. Schematic diagram of the staged acid fracturing technology. (**a**) Packer staged acid fracturing; (**b**) "Packer + temporary plugging" staged acid fracturing.

The reservoir temperature in the Sichuan Basin is high. The first requirement of acid fracturing for the temporary plugging agent is that the dissolution rate of the temporary plugging agent in a high-temperature acid liquid cannot be too high, because if the temporary plugging agent dissolves quickly, there is no way to form a stable plugging layer in the formation. This paper evaluated the dissolution experiments of temporary plugging agents of different materials at 150 °C, and the experiment shows that the dissolution rate of the polyemulsion-modified polyvinyl alcohol resin is low at 150 °C (Table 2), but the dissolution rate of the temporary plugging agent of other materials is more than 50%. The polyemulsion-modified polyvinyl alcohol resin with a low dissolution rate was selected for the temporary plugging acid fracturing in the Sichuan Basin. The polyemulsion-modified polyvinyl alcohol resin used in the experiment came from the Chengdu LEPS technology company; its molecular weight was 180,000–240,000 and it was completely dissolved in water after 3 h at 150 °C. Urea methyl ester is the urea formaldehyde resin. It is the condensation polymerization of urea and formaldehyde under the action of a catalyst to form the initial urea formaldehyde resin, and then form the thermosetting resin under the action of the curing agent. The molecular formula of polyethylene glycol is $HO(CH_2CH_2O)_nH$.

Table 2. Experiment on the solubility of the temporary plugging agents with different materials.

Experimental Sample	Dissolution Rate of Test Sample in 20% Hydrochloric Acid at 150 °C
Polyemulsion-modified polyvinyl alcohol resin	3.54%
Urea methyl ester	92.04%
Modified polyethylene glycol	96.65%

The reservoir in the Sichuan Basin is highly heterogeneous. According to statistics, the maximum fracture pressure difference between points in the horizontal section is 15 MPa. Therefore, in order to achieve an effective temporary plugging acid fracturing, the plugging layer formed by the temporary plugging agent must bear a pressure of at least 15 MPa. Fibers and particles are commonly used temporary plugging materials [57]. The plugging principle is to achieve a high-strength plugging through particle bridging and fiber winding. The polyemulsion-modified polyvinyl alcohol resin was processed into fibers with a length of 6–8 mm and particles with different particle sizes. In this paper, the plugging ability of fibers and particles was evaluated using a three-dimensional fracture dynamic plugging experimental equipment. By placing fibers and particles into clean water in a certain proportion and driving them into the simulated acid fracturing fracture at a uniform speed with a precision pressure pump, the displacement pressure change of the temporary plugging material after entering the simulated fracture was observed with an electronic pressure gauge.

To carry out the fracture plugging test with the temporary plugging agent, the first step was to determine the fracture width of the temporary plugging agent. Through the simulation of fracturing software, it was found that the average width of acid corrosion fracture in the Sichuan carbonate gas reservoir is 6 mm. Therefore, the fracture width of this test was determined as 2 mm and the test temperature was 150 °C.

It can be seen from Figure 6 that, after adding 40/70 mesh particles to 100/140 mesh particles, the pressure bearing capacity of the plugging layer is significantly improved. Especially when the concentration of 40/70 mesh particles increases to 1.5%, the pressure bearing capacity of the plugging layer can be increased to 13.7 MPa. Therefore, the addition of large particles is helpful to quickly bridge and form a sealing plugging layer. The pressure of the plugging layer is required to be higher than 15 MPa for the temporary plugging acid pressure of the Sichuan carbonate gas reservoir. Therefore, we considered adding an appropriate concentration of fibers into the particles and filling the gap between the particles of different sizes with fiber, so as to further improve the pressure bearing capacity of the plugging layer.

It can be seen from Figure 7 that, after adding fiber to particles, the pressure bearing capacity of the plugging layer is significantly improved. Especially when the fiber concentration reaches 1.0%, the pressure bearing capacity of the plugging layer can be increased to 17.9 MPa, which is 4.2 MPa higher than that of the plugging layer with only particles. The fiber + particle composite temporary plugging method meets the requirements of the staged acid fracturing of the Sichuan carbonate gas reservoir. It can be seen from Figure 8 that fibers and particles almost fill the whole simulated fracture, 40/70 mesh particles are distributed in a scattered manner, fibers and 100/140 mesh particles are intertwined to form a dense layer, and they fill the gap between the 40/70 mesh particles.

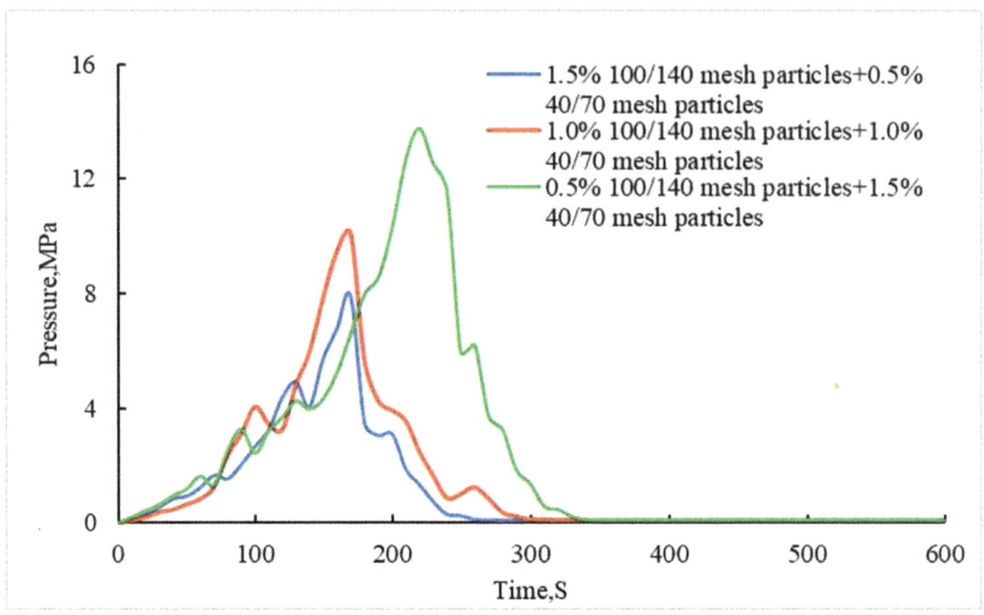

Figure 6. Comparison of the fracture sealing ability of the combination of differently sized particles.

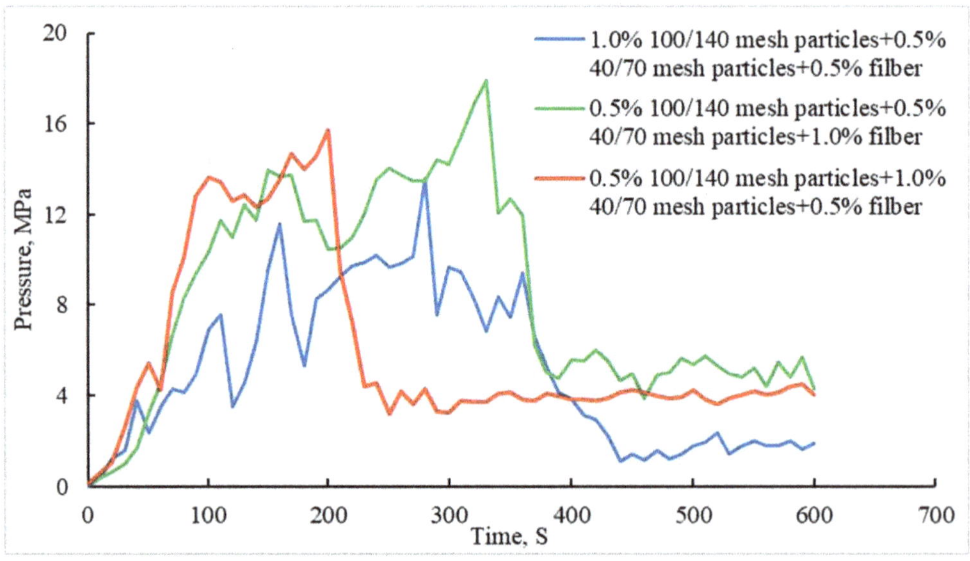

Figure 7. Comparison of the fracture sealing ability of the particle and fiber combination.

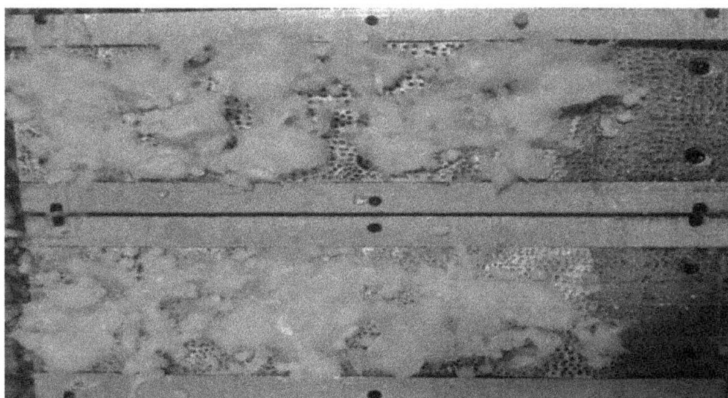

Figure 8. Distribution of the temporary plugging materials in a simulated fracture.

The sealing layer is composed of fibers and particles. The particles quickly accumulate in the fracture to form a sealing layer. At the same time, the fibers are intertwined to block the gap between the particles, so as to improve the pressure bearing capacity of the sealing layer. As can be seen from Figure 7, with the increase in the fiber dosage, the peak pressure bearing capacity of the plugging layer increases significantly.

2.3. Acid Fracturing Technology for the Ultra-Deep Reservoir

A high-temperature carbonate reservoir has a fast acid rock reaction speed and short acid corrosion fracture length [58]. Alternating acid fracturing injection is an effective means to improve the length of the acid fracturing fracture. The fingering phenomenon is produced by injecting liquids with different viscosities, which not only improves the effective action distance of acid, but also forms non-uniform etching on the fracture wall to further improve the conductivity [59]. Based on the multiphase flow VOF model and the test results of acid corrosion fracture conductivity, the acid fracturing process of alternating injection in the Sichuan carbonate gas reservoir was studied, and the key parameters, such as the optimal injection liquid viscosity ratio, injection displacement, and scale, were determined.

It can be seen from Figure 9 that, with the increase in the viscosity ratio of the two liquids, the fingering characteristics of acid liquid in the fracture become more and more obvious, and the flow pattern of acid liquid evolves from uniform piston propulsion to narrow fingering propulsion. When the viscosity ratio of fracturing fluid (blue part) to acid (red part) increases to 8:1, the fingering pattern of acid liquid tends to be stable. The non-uniform distribution of acid solution is conducive to strengthening the differential acid corrosion of rock and plays a positive role in maintaining and improving the fracture conductivity in the later production stage [59]. The viscosity of crosslinked autogenous acid is 150 mPa·s and the viscosity of gelling acid is 18 mPa·s. The viscosity ratio of the above two liquids is slightly greater than 8:1. The crosslinked authigenic acid and gelling acid meet the requirements of the best viscosity ratio of the alternating injection.

We selected two carbonate rock plates with the same mineral composition for the experiments and compared the surface morphology of the fractures by injecting acids with different viscosities. It can be seen from Figure 10 that the differential corrosion ability of acid is stronger after increasing the viscosity ratio of the acid solution. The fracture corrosion effect observed in the acid etching experiment is the same as that of the numerical simulation, which further proves the accuracy of the numerical simulation in the simulation of the acid fingering behavior.

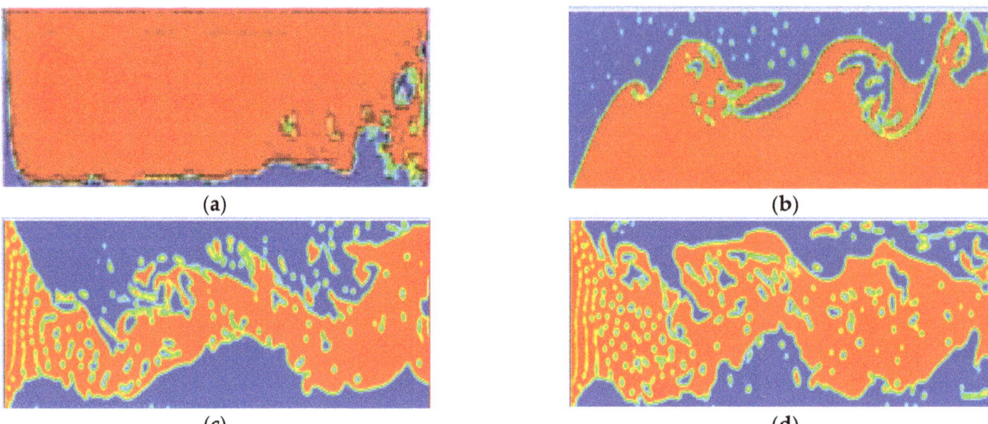

Figure 9. Fingering form of the acid solution under different viscosity ratios (red represents gelling acid and blue represents crosslinked authigenic acid). (**a**) Viscosity ratio 1:1; (**b**) Viscosity ratio 3:1; (**c**) Viscosity ratio 8:1; and (**d**) Viscosity ratio 12:1.

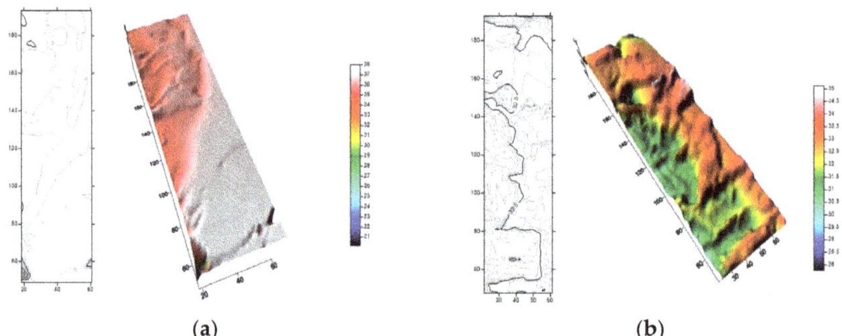

Figure 10. Acid corrosion fracture wall after injecting acid with different viscosities. (**a**) Viscosity ratio 3:1; and (**b**) Viscosity ratio 8:1.

We Kept the displacement and scale of the acid injection unchanged, and simulated the length of the acid corrosion fracture and dimensionless conductivity under the condition of series 1 to 5 injection. As can be seen from Figure 11, with the increase in the injection series, the length of acid corrosion fracture and dimensionless conductivity rise synchronously. When the alternating injection series is 3, the growth range of acid corrosion fracture length and dimensionless conductivity slows down. Compared with a three-stage alternate injection, the length of acid corrosion fracture of the five-stage alternate injection increases by only 6 m, with an increased rate of 6.12%. The best series of alternating injection acid fracturing in the Sichuan carbonate gas reservoir is 3, and the length of acid corrosion fracture can reach 99.6 m.

We used the acid fracturing software to simulate the fracture shape of alternating injection acid fracturing. The physical model was established according to the actual geological and mechanical parameters of carbonate rocks in the Sichuan Basin. The acid corrosion fracture shape was simulated according to the three-stage alternating injection of authigenic acid and gelling acid with an injection displacement of 7 m^3/min. Figure 12 is the simulation diagram of the acid corrosion fracture width. The color represents the width of the acid corrosion fracture, and the dark color shows that the width of the acid corrosion fracture is large. It can be seen from Figure 13 that the width of the acid corrosion fracture changes greatly after a three-stage alternating injection of acid fracturing. The multiple alternating injection of the crosslinked authigenic acid and gelling acid intensifies the interface effect between liquids and promotes the generation of the acid fingering phenomenon.

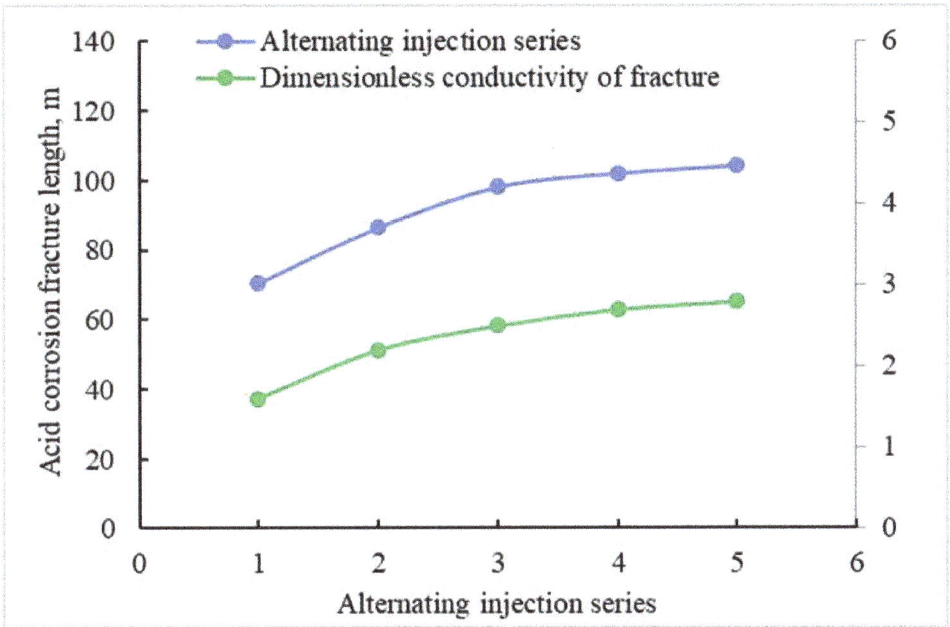

Figure 11. Acid corrosion fracture parameters under different injection series.

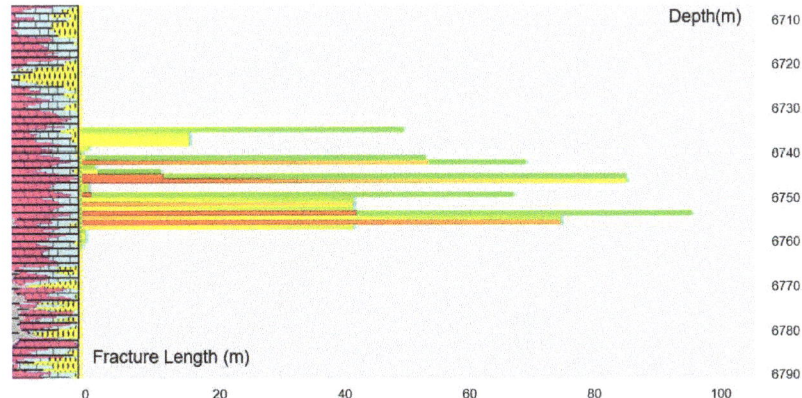

Figure 12. Simulation diagram of the acid corrosion fracture width under the three-stage alternating injection. (Red represents fractures with a width of 7–9 mm, yellow represents fractures with a width of 5–7 mm, and green represents fractures with a width of 3–5 mm).

Under the condition of keeping the injection displacement, injection stages, and total amount of liquid unchanged, the fracturing software was used to simulate the acid fracturing fracture extension under different liquid injection ratios. It can be seen from Figure 13 that, with the decrease in the amount of crosslinked autogenous acid, the length of acid fracturing fracture decreases, and the conductivity of acid etching fracture increases steadily. When the volume ratio of the crosslinked authigenic acid and gelling acid is 1:3, the length of acid etching fracture is close to 100 m, and the dimensionless conductivity of fracture is close to the maximum. The acid fracturing design of the Sichuan carbonate gas reservoir takes into account the two parameters of fracture length and conductivity, and the volume ratio of the injected crosslinked authigenic acid and gelling acid is preferably 1:3.

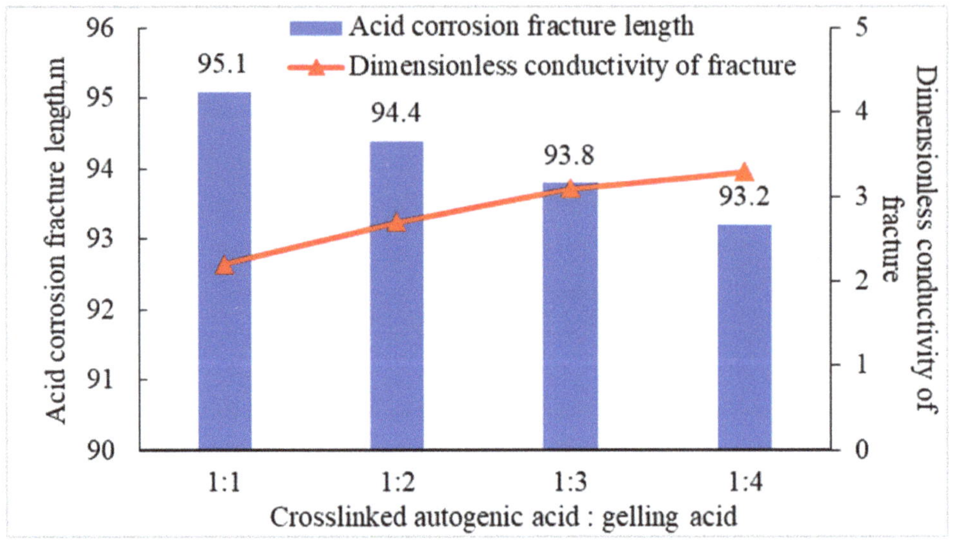

Figure 13. Acid corrosion fracture parameters under different liquid injection ratios.

As can be seen from Figure 14, when the injection volume ratio of the crosslinked autogenous acid and gelling acid is 1:3, the acid liquid creates non-uniform corrosion on the rock wall, the acid liquid creates a groove corrosion with a certain height difference on the rock wall, and the non-corrosion rock creates a support resembling a pier, which improves the effectiveness of the crack under high closure stress.

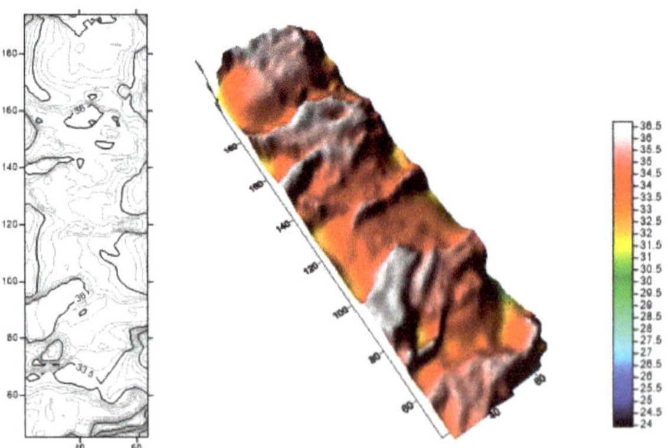

Figure 14. Wall morphology of the acid corrosion fracture.

Acid strength refers to the amount of acid used per meter of the reservoir during acid fracturing. The acid corrosion fracture length and dimensionless conductivity when the acid strength is from 0.7 to 1.9 m^3/m were simulated by fracturing software. It can be seen from Figure 15 that, with the increase in acid strength, the acid corrosion fracture length and dimensionless conductivity increase rapidly because increasing the amount of acid can ensure that more acid liquid reacts with the rock, so as to improve the acid fracturing effect. When the acid strength increases to 1.6 m^3/m, the growth rate of the acid corrosion fracture length and dimensionless conductivity slows down. When the acid strength is increased to 2.0 m^3/m, the length of acid corrosion fracture is 2.8 m longer than that of 1.6 m^3/m. The optimal acid strength of the Sichuan carbonate gas reservoir is 1.6 m^3/m, which can not only ensure a better fracture length and conductivity, but also reduce the excessive consumption of acid liquid and improve the economy of the acid fracturing stimulation.

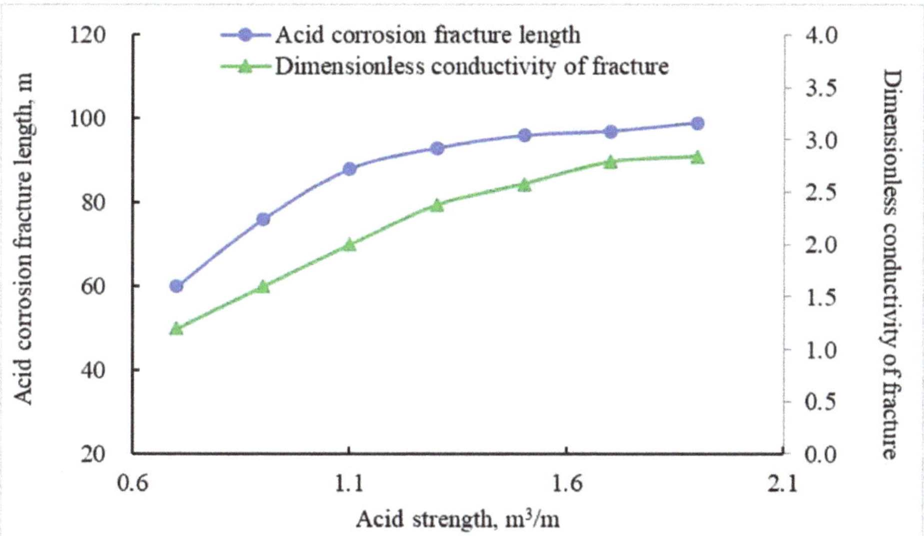

Figure 15. Simulation results of the acid corrosion fracture parameters under different acid strengths.

2.4. Field Application

In the Sichuan carbonate gas reservoir, a total of 12 horizontal wells have been stimulated by staged acid fracturing, and the production is 2.1 times higher than that of general stimulation. Among them, the drilling depth of well QY1019 is 6823 m, the horizontal section length is 1185 m, the reservoir section length is 901 m, the average porosity of the reservoir is 3.2%, and the average permeability is 0.31 mD. In 2021, the QY1019 well completed the compound staged acid fracturing stimulation through packer + temporary plugging. A total of 480 m^3 authigenic acid, 1440 m^3 gelling acid, 800 kg fiber, and 800 kg temporary plugging particles were injected into the formation. In order to verify the effectiveness of staged acid fracturing for the temporary plugging, two kinds of chemical tracers were added before and after the injection of the temporary plugging agent. The chemical tracer interpretation results showed that new fractures were formed after the addition of a temporary plugging agent. Net pressure fitting shows that, compared with conventional acid fracturing, the fracture length and conductivity of alternating injection acid fracturing increased by 78.3% and 52.5% respectively. The daily gas production is 1.58×10^4 m^3/d, which is 2.4 times of the test daily gas production of the adjacent wells.

3. Conclusions

In this study, we proposed a new staged acid fracturing method for ultra-deep horizontal wells, which uses the packer and the temporary plugging agent to realize the uniform stimulation of long well sections. We used numerical simulation software to determine that the optimal fracture spacing of horizontal wells in the carbonate gas reservoirs in the

Sichuan Basin is 70 m. Through the pressure test of the plugging layer, the composite temporary plugging mode of fiber and particle was proposed. The fiber and particle can form a plugging layer with a pressure bearing capacity of more than 17 MPa, which meets the stimulation requirements. Through the simulation of gelling acid fingering form and the test of conductivity, it was found that the best acid fracturing effect can be obtained by injecting crosslinked autogenous acid and gelling acid alternately in three stages and increasing the amount of acid. This study has great guiding significance for the stimulation of ultra-deep carbonate gas reservoirs.

Author Contributions: Conceptualization, Y.W.; methodology, Y.W.; software, Y.F.; validation, J.Y.; formal analysis, Z.L.; investigation, Z.L.; resources, T.W.; data curation, Y.F.; writing—original draft preparation, Y.W.; writing—review and editing, Y.W.; visualization, T.W.; project administration, Y.W.; funding acquisition, Y.W. All authors have read and agreed to the published version of the manuscript.

Funding: This research was funded by PetroChina Southwest Oil and Gas Field Company grant number grant no. 20200302-14, 20210302-19. And The APC was funded by PetroChina Southwest Oil and Gas Field Company.

Data Availability Statement: Not applicable.

Conflicts of Interest: The authors declare no conflict of interest.

References

1. Ma, X.; Yan, H.; Chen, J. Development patterns and constraints of superimposed karst reservoirs in Sinian Dengying Formation, Anyue gas field, Sichuan Basin. *Oil Gas Geol.* **2021**, *42*, 1281–1294.
2. Wang, K.; Teng, J.; Deng, H.; Fu, M.; Lu, H. Classification of Void Space Types in Fractured-Vuggy Carbonate Reservoir Using Geophysical Logging: A Case Study on the Sinian Dengying Formation of the Sichuan Basin, Southwest China. *Energies* **2021**, *14*, 5087. [CrossRef]
3. Xie, J. Innovation and practice of key technologies for the efficient development of the supergiant Anyue Gas Field. *Nat. Gas Ind. B* **2020**, *7*, 337–347. [CrossRef]
4. Li, S.; Ma, H.; Zhang, H.; Ye, J.; Han, H. Study on the acid fracturing technology for high-inclination wells and horizontal wells of the Sinian system gas reservoir in the Sichuan Basin. *J. Southwest Pet. Univ.* **2018**, *40*, 146.
5. Hu, Y.; Peng, X.; Li, Q.; Li, L.; Hu, D. Progress and development direction of technologies for deep marine carbonate gas reservoirs in the Sichuan Basin. *Nat. Gas Ind. B* **2020**, *7*, 149–159. [CrossRef]
6. Mi, Q.; Yi, X.; Luo, P.; Ren, L. Horizontal well staged fracturing technology of tight sandstone reservoirs with super depth in Tazhong Area. *J. Southwest Pet. Univ.* **2015**, *37*, 114.
7. Chen, Z.; Wang, Z.; Zeng, H. Status quo and prospect of staged fracturing technique in horizontal wells. *Nat. Gas Ind.* **2007**, *27*, 78.
8. Zhang, H.; Liu, Z.; Cheng, Z.; Tian, W.; Zhang, P.; Han, Y. Research on open hole packer of staged fracturing technique in horizontal wells and its application. *Oil Drill. Prod. Technol.* **2011**, *33*, 123–125.
9. Haishan, Y.; Jian, W.; Cuo, W. Horizontal Well Fracturing Technology Progress of Putaohua Reservoir in Daqing Peripheral Oilfield. *Reserv. Eval. Dev.* **2013**, *3*, 51–56.
10. Zou, G.; Che, M.; Ji, X. Multi-stage fracturing technology and application for ultra-high pressure fractured gas reservoir. *Nat. Gas Geosci.* **2012**, *23*, 365–369.
11. Chu, X.; Wu, J.; Duan, P.; Li, W.; Liang, C.; Liu, L. Application of fracturing technique combining hydraulic Jet with small-diameter packer on horizontal wells in Changqing low-permeability Oilfield. *Oil Drill. Prod. Technol.* **2013**, *34*, 73–76.
12. Liu, J.; Lian, Z.; Lin, T. Production forecasting formulas of horizontal wells under multiform completions. *Spec. Oil Gas Reserv.* **2006**, *13*, 61–63.
13. Hui, X.; Jianchun, G.; Jun, Z. Technical study on staged acid fracturing of horizontal well in fractured-cavernous carbonate reservoir. *Fault-Block Oil Gas Field* **2011**, *18*, 119–122.
14. Chunming, H.; Feng, H.; Zhe, L. Research on Acid Fracturing Technology of Horizontal Section of the Deep Marine Carbonate Reservoir. *Well Test.* **2013**, *22*, 5–9.
15. Babu, D.K.; Odeh, A.S. Productivity of a horizontal well appendices A and B. In Proceedings of the SPE Annual Technical Conference and Exhibition, Houston, TX, USA, 2–5 October 1988; Society of Petroleum Engineers (SPE): Aberdeen, UK, 1988.
16. Hegre, T.M.; Larsen, L. Productivity of multifractured horizontal wells. In Proceedings of the European Petroleum Conference, Aberdeen, UK, 15–17 March 1994; Society of Petroleum Engineers (SPE): Aberdeen, UK, 1994.
17. Elgaghah, S.A.; Osisanya, S.O.; Tiab, D. A simple productivity equation for horizontal wells based on drainage area concept. In Proceedings of the SPE Western Regional Meeting, Anchorage, AK, USA, 22 May 1996; Society of Petroleum Engineers (SPE): Aberdeen, UK, 1996.
18. Giger, F.M. Horizontal wells production techniques in heterogeneous reservoirs. In Proceedings of the Middle East Oil Technical Conference and Exhibition, Manama, Bahrain, 11 March 1985; Society of Petroleum Engineers (SPE): Aberdeen, UK, 1985.

19. Mukherjee, H.; Economides, M.J. A parametric comparison of horizontal and vertical performance. *SPE Form. Eval.* **1996**, *6*, 209–216. [CrossRef]
20. He, Y.; Li, Y.M.; Zhao, J.Z.; Zhang, Y.; Lin, T.; Mi, Q.B. Productivity-Prediction Model of Acid Fracturing Horizontal Well in Fracture-Cavity Reservoir. In *Applied Mechanics and Materials*; Trans Tech Publications Ltd.: Freienbach, Switzerland, 2013; Volume 321, pp. 872–876.
21. Wang, Y.; Zhou, C.; Yi, X.; Li, L.; Chen, W.; Han, X. Technology and Application of Segmented Temporary Plugging Acid Fracturing in Highly Deviated Wells in Ultradeep Carbonate Reservoirs in Southwest China. *ACS Omega* **2020**, *5*, 25009–25015. [CrossRef]
22. Wang, Y.; Zhou, C.; Yi, X.; Li, L.; Zhou, J.; Han, X.; Gao, Y. Research and Evaluation of a New Autogenic Acid System Suitable for Acid Fracturing of a High-Temperature Reservoir. *ACS Omega* **2020**, *5*, 20734–20738. [CrossRef]
23. Li, N.; Dai, J.; Liu, P.; Luo, Z.; Zhao, L. Experimental study on influencing factors of acid-fracturing effect for carbonate reservoirs. *Petroleum* **2015**, *1*, 146–153. [CrossRef]
24. Settari, A. Modeling of acid-fracturing treatments. *SPE Prod. Facil.* **1993**, *8*, 30–38. [CrossRef]
25. Rencheng, D.; Mary, F.W. Modeling Acid Fracturing Treatments with Multi-Stage Alternating Injection of Pad and Acid Fluids. In Proceedings of the 2021 SPE Reservoir Simulation Conference, online, 19 October 2021; SPE-203985-MS; Society of Petroleum Engineers: Aberdeen, UK, 2021. [CrossRef]
26. Jackson, S.J.; Power, H.; Giddings, D. Immiscible Thermo-Viscous Fingering in Hele-Shaw cells. *Comput. Fluids* **2017**, *156*, 621–641. [CrossRef]
27. Nguyen, V.T.; Park, W.G. A volume-of-fluid (VOF) interface-sharpening method for two-phase incompressible flows. *Comput. Fluids* **2017**, *152*, 104–119. [CrossRef]
28. Xu, H.; Cheng, J.; Zhao, Z.; Lin, T.; Liu, G.; Chen, S. Coupled thermo-hydro-mechanical-chemical modeling on acid fracturing in carbonatite geothermal reservoirs containing a heterogeneous fracture. *Renew. Energy* **2021**, *172*, 145–157. [CrossRef]
29. You, J.; Lee, K.J. Analyzing the Dynamics of Mineral Dissolution during Acid Fracturing by Pore-Scale Modeling of Acid-Rock Interaction. *SPE J.* **2021**, *26*, 639–652. [CrossRef]
30. Mehrjoo, H.; Norouzi-Apourvari, S.; Jalalifar, H.; Shajari, M. Experimental study and modeling of final fracture conductivity during acid fracturing. *J. Pet. Sci. Eng.* **2022**, *208*, 109192. [CrossRef]
31. Xu, P.; Sheng, M.; Lin, T.; Liu, Q.; Wang, X.; Khan, W.A.; Xu, Q. Influences of rock microstructure on acid dissolution at a dolomite surface. *Geothermics* **2022**, *100*, 102324. [CrossRef]
32. Wang, D.; Ge, H.; Wang, X.; Wang, Y.; Sun, D.; Yu, B. Complex Fracture Closure Pressure Analysis During Shut-in: A Numerical Study. *Energy Explor. Exploit.* **2022**, 01445987221077311. [CrossRef]
33. Wang, Y.; Fan, Y.; Zhou, C.; Luo, Z.; Chen, W.; He, T.; Fang, H.; Fu, Y. Research and Application of Segmented Acid Fracturing by Temporary Plugging in Ultradeep Carbonate Reservoirs. *ACS Omega* **2021**, *6*, 28620–28629. [CrossRef]
34. McNeil, F.; van Gijtenbeek, K.; Van Domelen, M. New hydraulic fracturing process enables far-field diversion in unconventional reservoirs. In Proceedings of the SPE/EAGE European Unconventional Resources Conference & Exhibition-From Potential to Production, Vienna, Austria, 20–22 March 2012; cp-285-00049. [CrossRef]
35. Bulova, M.N.; Nosova, K.E.; Willberg, D.M.; Lassek, J.T. Benefits of the novel fiber-laden low-viscosity fluid system in fracturing low-permeability tight gas formations. In Proceedings of the SPE Annual Technical Conference and Exhibition, Vienna, Austria, 20–22 March 2012; Society of Petroleum Engineers (SPE): Aberdeen, UK, 2006.
36. Yuan, L.; Zhou, F.; Li, B.; Gao, J.; Yang, X.; Cheng, J.; Wang, J. Experimental study on the effect of fracture surface morphology on plugging efficiency during temporary plugging and diverting fracturing. *J. Nat. Gas Sci. Eng.* **2020**, *81*, 103459. [CrossRef]
37. Zhang, L.; Zhou, F.; Mou, J.; Pournik, M.; Tao, S.; Wang, D.; Wang, Y. Large-scale true tri-axial fracturing experimental investigation on diversion behavior of fiber using 3D printing model of rock formation. *J. Pet. Sci. Eng.* **2019**, *181*, 106171. [CrossRef]
38. Anup, V.; Hunter, W.; Jennifer, R.; Andrew, C.; Brian, S. Sequenced fracture treatment diversion enhances horizontal well completions in the Eagle Ford Shale. In Proceedings of the SPE/CSUR Unconventional Resources Conference, Calgary, AB, Canada, 30 September–2 October 2014; Society of Petroleum Engineers (SPE): Aberdeen, UK, 2014.
39. Fry, J.; Roach, E.; Kreyche, B.; Yenne, T.; Geoffrey, G.; Jespersen, M. Improving Hydrocarbon Recovery in Sliding Sleeve Completions Utilizing Diverters in the Wattenberg Field. In Proceedings of the SPE Hydraulic Fracturing Technology Conference, The Woodlands, TX, USA, 9–11 February 2016; Society of Petroleum Engineers (SPE): Aberdeen, UK, 2016.
40. Barraza, J.; Capderou, C.; Jones, M.C.; Lannen, C.T.; Singh, A.K.; Shahri, M.P.; Babey, A.G.; Koop, C.D.; Rahuma, A.M. Increased cluster efficiency and fracture network complexity using degradable diverter particulates to increase production: Permian Basin Wolfcamp shale case study. In Proceedings of the SPE Annual Technical Conference and Exhibition, San Antonio, TX, USA, 9–11 October 2017; Society of Petroleum Engineers (SPE): Aberdeen, UK, 2017.
41. Senters, C.W.; Leonard, R.S.; Ramos, C.R.; Wood, T.M.; Woodroof, R.A. Diversion-Be Careful What You Ask For. In Proceedings of the SPE Annual Technical Conference and Exhibition, San Antonio, TX, USA, 9–11 October 2017; Society of Petroleum Engineers (SPE): Aberdeen, UK, 2017.
42. Zhou, F.; Yang, X.; Xiong, C.; Zhang, S.; Zong, Y.; Sun, L.; Guan, Z. Application and study of fine-silty sand control technique for unconsolidation quaternary sand gas reservoir, Sebei Qinghai. In Proceedings of the SPE International Symposium and Exhibition on Formation Damage Control, Lafayette, LA, USA, 18–20 February 2004; Society of Petroleum Engineers (SPE): Aberdeen, UK, 2004.

43. Wang, D.; Dong, Y.; Sun, D.; Yu, B. A three-dimensional numerical study of hydraulic fracturing with degradable diverting materials via CZM-based FEM. *Eng. Fract. Mech.* **2020**, *237*, 107251. [CrossRef]
44. Wang, M.; Wang, J.; Cheng, F.; Chen, X.; Yang, X.; Lv, W.; Wang, B. Diverter plugging pattern at the fracture mouth during temporary plugging and diverting fracturing: An experimental study. *Energy Rep.* **2022**, *8*, 3549–3558. [CrossRef]
45. Yang, C.; Feng, W.; Zhou, F. Formation of temporary plugging in acid-etched fracture with degradable diverters. *J. Pet. Sci. Eng.* **2020**, *194*, 107535. [CrossRef]
46. Baiyu, Z.; Hongming, T.; Senlin, Y.; Gongyang, C.; Feng, Z.; Ling, L. Experimental and numerical investigations of particle plugging in fracture-vuggy reservoir: A case study. *J. Pet. Sci. Eng.* **2022**, *208*, 109610. [CrossRef]
47. Wang, Y.; Zhou, C.; Zhang, H.; He, T.; Tang, X.; Peng, H.; Fang, H.; Luo, L.; Gang, S. Research and Application of Segmented Acid Fracturing Technology in Horizontal Wells of Ultra-deep Carbonate Gas Reservoirs in Southwest China. In Proceedings of the International Petroleum Technology Conference, online, 23 March–1 April 2021; Society of Petroleum Engineers (SPE): Aberdeen, UK, 2021.
48. Yue, J.; Duan, Y.; Qing, S.; Qing, P.; Liu, X. Study on productivity performances of horizontal fractured gas well with many vertical fractures. *Pet. Geol. Oilfield Dev. Daqing* **2004**, *23*, 46–48.
49. Yan, X.Y.; Hu, Y.Q.; Zhao, J.Z.; Shen, B.B. Production Simulate Analysis of Multiple-Fractured Horizontal Gas Wells. In *Advanced Materials Research*; Trans Tech Publications Ltd.: Freienbach, Switzerland, 2012; Volume 402, pp. 728–733.
50. Guo, J.; Gou, B.; Qin, N.; Zhao, J.; Wu, L.; Wang, K.; Ren, J. An innovative concept on deep carbonate reservoir stimulation: Three-dimensional acid fracturing technology. *Nat. Gas Ind. B* **2020**, *7*, 484–497. [CrossRef]
51. Miao, W.J.; Zhao, L. Development Status and Prospect of Staged Fracturing Technology in Horizontal Wells. *Electron. J. Geotech. Eng.* **2017**, *22*.
52. Al-Qahtani, M.Y.; Zillur, R. Optimization of acid fracturing program in the Khuff gas condensate reservoir of south Ghawar field Saudi Arabia by managing uncertainties using state-of-the-art technology. In Proceedings of the SPE Annual Technical Conference and Exhibition, New Orleans, LA, USA, 30 September–2 October 2001; Society of Petroleum Engineers (SPE): Aberdeen, UK, 2001.
53. Lei, Q.; Xu, Y.; Yang, Z.; Cai, B.; Wang, X.; Zhou, L.; Liu, H.; Xu, M.; Wang, L.; Li, S. Progress and development directions of stimulation techniques for ultra-deep oil and gas reservoirs. *Pet. Explor. Dev.* **2021**, *48*, 221–231. [CrossRef]
54. Li, G.; Sheng, M.; Tian, S.; Huang, Z.; Li, Y.; Yuan, X. Multistage hydraulic jet acid fracturing technique for horizontal wells. *Pet. Explor. Dev.* **2012**, *39*, 107–112.
55. Yang, W.; Bing, Z.; Qingyun, Y. Integrated techniques in tight reservoir development for horizontal wells in Block Shun 9. *Pet. Drill. Tech.* **2015**, *43*, 48–52.
56. Wang, T.; Zhao, B.; Qu, Z. Temporarily plugging and acid-fracturing technology in weak channels near wellbore region in Tahe Oilfield. *Fault-Block Oil Gas Field* **2019**, *26*, 794–799.
57. Qu, Z.; Lin, Q.; Guo, T. Experimental study on acid fracture conductivity of carbonate rock in Shunbei Oilfield. *Fault-Block Oil Gas Field* **2019**, *26*, 533–536.
58. Li, X.; Yang, Z.; Chen, R. Study on the physical model and fractal of the acid solution in the pre-liquid acid fracturing seam. *J. Southwest Pet. Univ.* **2007**, *29*, 105–108.
59. Wu, H.; Kong, C.; Yi, X.; Zhang, H.; Liu, H.; Wu, Y. Numerical analysis of acid-corroded wormholes in fractured carbonate rocks. *Drill. Fluid Completion Fluid* **2016**, *33*, 105–108.

 gels

MDPI

Article

Gelling Behavior of PAM/Phenolic Crosslinked Gel and Its Profile Control in a Low-Temperature and High-Salinity Reservoir

Fei Ding [1], Caili Dai [1,*], Yongpeng Sun [1], Guang Zhao [1], Qing You [2] and Yifei Liu [1]

1 School of Petroleum Engineering, China University of Petroleum (East China), No. 66 Changjiang West Road, Huangdao District, Qingdao 266580, China; b18020025@s.upc.edu.cn (F.D.); sunyongpeng@upc.edu.cn (Y.S.); zhaoguang@upc.edu.cn (G.Z.); liuyifei@upc.edu.cn (Y.L.)

2 School of Energy Resources, China University of Geosciences (Beijing), No. 29 Xueyuan Road, Haidian District, Beijing 100083, China; youqing@cugb.edu.cn

* Correspondence: daicl@upc.edu.cn

Abstract: Gel conformance control technology is widely used in moderate and high temperature reservoirs. However, there are few studies on shallow low-temperature and high-salinity reservoirs. The difficulties are that it is difficult to crosslink at low temperatures and with poor stability at high salt concentrations. Therefore, the PHRO gel was developed, which was composed of gelatinizing agent (polyacrylamide), crosslinking agents (hexamethylenetetramine and resorcinol) and crosslinking promoting agent (oxalic acid). The PHRO could form high-strength gels in both deionized water and high-concentration salinity solutions (NaCl, KCl, $CaCl_2$ and $MgCl_2$). The observation of the microstructure of PHRO gel shows that a strong "stem—leaf"-shaped three-dimensional network structure is formed in deionized water, and the network structure is still intact in high-concentration salt solution. The results show that PHRO has good salt resistance properties and is suitable for conformance control of low-temperature and high-salinity reservoirs.

Keywords: low temperature; high salinity; conformance control; salt resistance; phenolic; gel

Citation: Ding, F.; Dai, C.; Sun, Y.; Zhao, G.; You, Q.; Liu, Y. Gelling Behavior of PAM/Phenolic Crosslinked Gel and Its Profile Control in a Low-Temperature and High-Salinity Reservoir. *Gels* **2022**, *8*, 433. https://doi.org/10.3390/gels8070433

Academic Editor: Mario Grassi

Received: 31 May 2022
Accepted: 4 July 2022
Published: 11 July 2022

Publisher's Note: MDPI stays neutral with regard to jurisdictional claims in published maps and institutional affiliations.

1. Introduction

Waterflooding is a common method for the displacement of crude oil. It also has the function of a formation energy supplement. It is an effective way for oilfields to maintain stable production and is widely used worldwide. Nonetheless, with the development of water injection in the middle and late stages, the formation heterogeneity is aggravated by flushing with long-term water injection, and the injected water enters the production well along the large-pore or high-permeability zone. As a result, oilfields are facing problems of increased water cuts and declining oil production. There is still a considerable amount of remaining oil irregularly distributed in the unswept area of reservoirs by waterflooding [1,2]. Therefore, enhancing oil recovery in the middle and late stages of waterflooding development has become an important research topic for oilfield development researchers.

The North Buzachi oilfield in Kazakhstan is located on the Buzachi peninsula in the northeastern Caspian Sea [3]. The reservoir depth is less than 500 m. The average porosity is 32.4%. The average permeability is 1930 mD. The reservoir temperature is 29–33 °C, and the total salinity of formation water is $3.78 \times 10^4 - 6.09 \times 10^4$ mg/L. The North Buzachi oilfield is encountering a decline in the formation pressure coefficient, a high water–oil ratio and declining oil production in some wells after a long period of waterflooding. To improve waterflooding performance and ensure stable oil production, it is necessary to plug large channels or highly permeable zones to expand the swept volume of injected water.

The practice of oilfield development in the past 20 years shows that the most effective conformance control method is the application of chemical plugging agents [4]. The commonly used plugging agents are resin, foam, gel, etc., and gel is the most frequently applied among them [5–11]. Gel is a type of plugging agent with a three-dimensional network structure formed through a crosslinking reaction between gelation agents and crosslinking agents [12,13]. Gelation agents are water-soluble polymers, such as polyacrylamide and polyethylene imine, of which polyacrylamide and its derivatives are the most commonly used. For crosslinking agents, there are two main types [14–17]: (1) organic metal crosslinking agents, such as Cr^{3+}, Zr^{4+} and Al^{3+}, can be crosslinked with the carboxylic acid group in the polymer molecule; (2) organic crosslinking agents, such as formaldehyde-phenol and phenolic resin, can be crosslinked with amide groups in polymer molecules. At present, research on gel focuses on middle and deep reservoirs, whereas there are few studies on shallow reservoirs with low temperature and high salinity.

Due to the low activity of crosslinking agents at low temperatures, the crosslinking reaction of most gels is slow or does not occur. Chromium and phenolic gels are the most commonly used gels at low temperatures [18,19]. Chromium gels are composed of PAM and Cr^{3+} crosslinking agents, which have the characteristics of fast gelation and high strength at low temperatures [11,20–23]. However, they result in high water loss due to structural damage in high-salinity water over time. Then, the stability and strength of chromium gels will decline sharply, which will greatly reduce the plugging period. DiGiacomo [24] found that in the presence of calcium and magnesium ions, the dehydration of chromium gels was intensified and suggested that this was due to Ca^{2+} and Mg^{2+} ions reacting with the carboxyl group in PAM and producing carboxylates with low water solubility. Wang [25] investigated the impact of calcium chloride on the dehydration behavior and the microstructure of PAM/Cr^{3+} gel. The experimental results indicated that a high concentration of calcium chloride could promote the water loss of gels, while a low concentration of calcium chloride could inhibit the dehydration of gels. The reduction in water-holding holes in PAM/Cr^{3+} gel is the internal reason that a high concentration of calcium chloride aggravates the water loss of gels.

In this study, a phenolic crosslinking gel was prepared, referring to the temperature and salinity of formation water in the North Buzachi oilfield. The fluid consists of polyacrylamide, hexamethylenetetramine, resorcinol and oxalic acid (PHRO). The oxalic acid has dual functions of accelerating crosslinking and shielding saline ions, which can reduce the gelation time and strengthen the salt resistance of PHRO. First, the influence of different additive concentrations on gelling performance and fluid stability was studied. Secondly, the gelling performance and stability of the fluid in different concentrations of sodium chloride, potassium chloride, calcium chloride and magnesium chloride solutions were studied. Finally, the salt tolerance mechanism of the fluid was analyzed from the microscopic morphology. The results demonstrate that PHRO is suitable for conformance control in low-temperature and high-salinity conditions. This study provides basic theoretical support for the development of conformance control technology in shallow reservoirs with low temperatures and high salinity.

2. Results and Discussion

2.1. Preparation of PHRO Gel

2.1.1. Process of Crosslinking

The crosslinking reactions of PHRO gel are condensation reactions between polyacrylamide (PAM), hexamethylenetetramine (HMTA) and resorcinol (RO). There are three types of crosslinking reactions [17]:

1. HMTA is hydrolyzed to produce formaldehyde, which is crosslinked with PAM;
2. Formaldehyde reacts with RO to generate polyhydroxy methyl phenol, which is crosslinked with PAM;
3. Formaldehyde reacts with RO to phenolic resin, which is crosslinked with PAM.

The hydrolysis of HMTA to formaldehyde is the key step in the formation of the gel, and the reaction process is illustrated in Figure 1. The hydrolysis reaction is a reversible reaction, and the reaction rate is very slow at low temperatures. The addition of OA makes the solution acidic, which shortens the gelation time of PHRO.

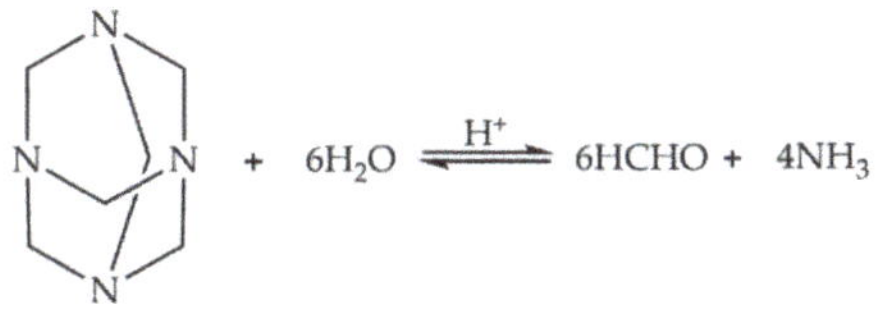

$$+ \ 6H_2O \xrightleftharpoons{H^+} 6HCHO + 4NH_3$$

Figure 1. Hydrolytic process of HMTA.

2.1.2. Macroscopic and Microscopic Morphology of Gel

The appearance of the PHRO gel in deionized water is shown in Figure 2. The gel formed in ionized water is dark orange, and its strength reaches the "G" level.

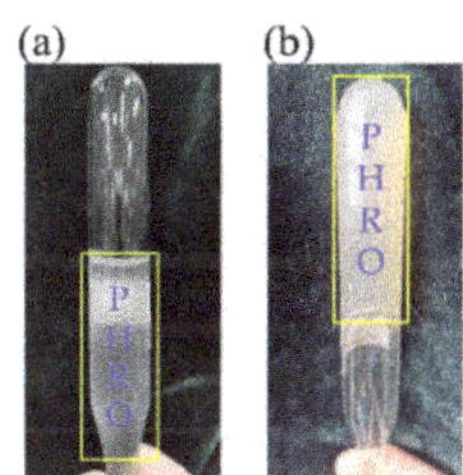

Figure 2. Picture of PHRO before and after crosslinking. (**a**) Before crosslinking; (**b**) after crosslinking (the formula is 0.4 wt% PAM + 0.5 wt% HMTA + 0.05 wt% RO + 0.3 wt% OA).

The microscopic morphology of the PHRO gel in deionized water is shown in Figure 3. The gel has a robust three-dimensional network structure like the "stem-leaf" of plants, and the network is embedded with granular substances of different sizes. It is speculated that the particle substance is phenolic resin with different molecular weights generated by the reaction of formaldehyde and RO. The particles could reinforce the strength of the gel structure.

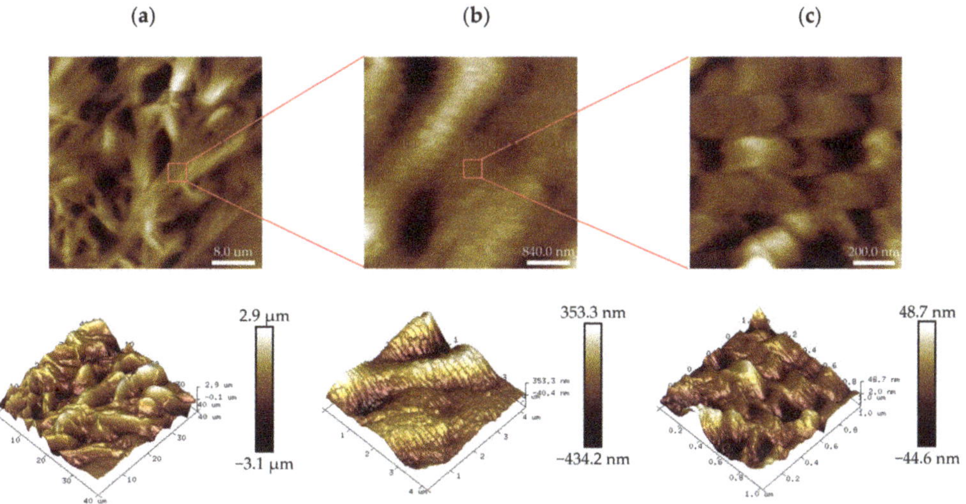

Figure 3. Two-dimensional (2D) and three dimensional (3D) images of the PHRO gel in deionized water with AFM. (**a**) Scanning area is 40.0 µm × 40.0 µm; (**b**) scanning area is 4.2 µm × 4.2 µm; (**c**) scanning area is 1.0 µm × 1.0 µm.

2.2. Effect of Additive Concentration on Gelling Performance of PHRO Gel

The gelation time, gel strength and dehydration rate of PHRO gels with different concentrations of PAM, HMTA, RO and OA were measured. The experimental temperature was 30 °C, and deionized water was used to prepare the liquid.

2.2.1. Effect of PAM Concentration on Gelling Performance

The gelation time, gel strength and dehydration rate of PHRO gels with different concentrations of PAM (0.1–0.8 wt%) were measured while keeping concentrations of HMTA, RO and OA constant, and the results are exhibited in Figure 4. With the increase in the concentration of PAM from 0.1 wt% to 0.8 wt%, the gel strength of PHRO remained at the "G" level. The gelation time was shortened from 173 h to 116 h, reduced by 32.9%. The dehydration rate declined by 14.9%, from 17.0% to 2.1%.

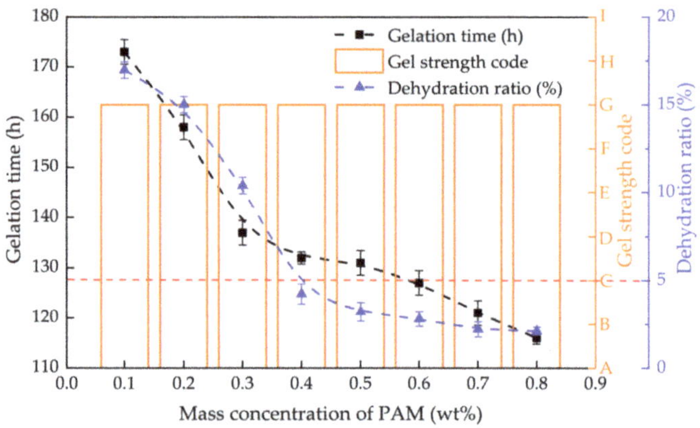

Figure 4. Gelling performance with different concentrations of PAM (0.5 wt% HMTA + 0.05 wt% RO + 0.3 wt% OA).

The results show that the increase in the concentration of PAM is beneficial for increasing the gelatinization rate and stability. This is because the number of "-$CONH_2$" groups involved in crosslinking increases with increasing PAM concentration, resulting in a faster reaction rate and stronger crosslinking density. With the dehydration rate < 5.0% as the standard, the optimal PAM dosage is $\geq$ 0.4 wt%.

2.2.2. Effect of HMTA Concentration on Gelling Performance

The gelation time, gel strength and dehydration rate of PHRO gels with different concentrations of HMTA (0.1–0.8 wt%) were measured while keeping concentrations of PAM, RO and OA constant, and the results are shown in Figure 5. When the concentration of HMTA was $\leq$0.2 wt%, the gel strength increased from "D" to "E", the gelation time increased from 161 h to 181 h, and the dehydration rate decreased from 8.2% to 6.2%. When the concentration of HMTA was >0.2 wt%, the gel strength increased from "E" to "G", the gelation time decreased from 181 h to 105 h, and the dehydration rate decreased from 6.2% to 3.5%.

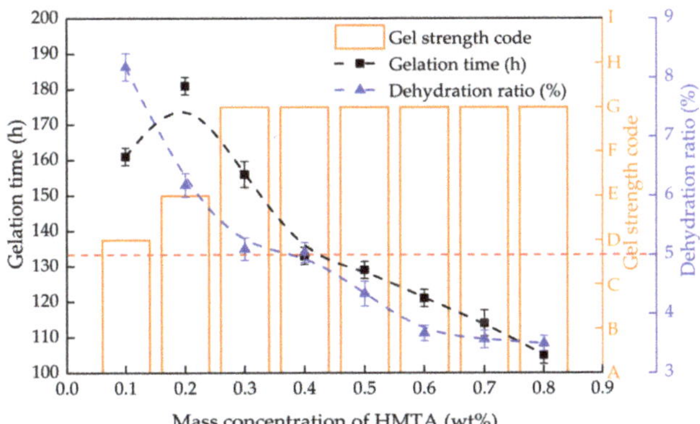

Figure 5. Gelling performance with different concentrations of HMTA (0.4 wt% PAM+ 0.05 wt% RO+ 0.3 wt% OA).

The results show that a low concentration of HMTA could not prepare high-strength gels, and the increase in the concentration of HMTA is beneficial for increasing the gelatinization rate and stability. This is because when the concentration of HMTA is low, the number of "-OH" groups involved in crosslinking is low. The crosslinking density is poor, and the gel strength is weak. With the increase in HMTA concentration, the number of "-OH" groups increased, the crosslinking reaction accelerated, and the crosslinking density increased. With the dehydration rate < 5.0% as the standard, the optimal HMTA dosage is ≥0.5 wt%.

2.2.3. Effect of RO Concentration on Gelling Performance

The gelation time, gel strength and dehydration rate of PHRO gels with different concentrations of RO (0.01–0.08 wt%) were measured while keeping concentrations of PAM, HMTA and OA constant, and the results are illustrated in Figure 6. When the RO concentration was ≤0.03 wt%, the gel strength increased from "D" to "G". The gelation time increased from 153 h to 181 h, and the dehydration rate increased from 4.7% to 4.8%. When the RO concentration was >0.03 wt%, the gel strength remained at the "G" level. The gelation time was shortened from 181 h to 97 h, and the dehydration rate was reduced from 4.8% to 4.6%.

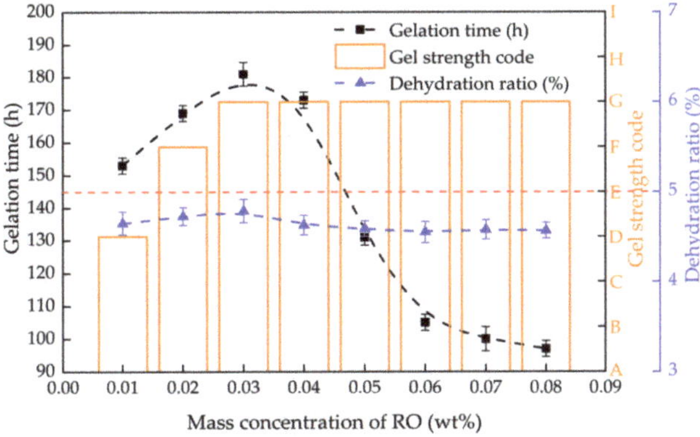

Figure 6. Gelling performance with different concentrations of RO (0.4 wt% PAM + 0.5 wt% HMTA + 0.3 wt% OA).

The results show that a low concentration of RO could not prepare high-strength gels. And increasing the concentration of RO is beneficial for increasing the gelatinization rate and stability. This is because with the increase in the concentration of RO, the contents of polyhydroxymethyl phenol and phenolic resin increased; that is, the number of crosslinking sites increased. Therefore, a low concentration of RO has low crosslinking density and weak gel strength, while a high RO concentration has a high crosslinking rate, high crosslinking density and high gel strength. The gelation time should be as short as possible in field applications, so the optimal RO dosage is $\geq$0.05 wt%.

2.2.4. Effect of OA Concentration on Gelling Performance

The gelation time, gel strength and dehydration rate of PHRO gels with different OA concentrations (0.1–0.8 wt%) were measured while keeping PAM, HMTA and RO concentrations constant, and the results are displayed in Figure 7. When the concentration of OA was $\leq$0.5 wt%, the gel strength remained at the "G" level, the gelation time was shortened from 157 h to 115 h, and the dehydration rate increased from 4.0% to 22.4%. When the concentration of was OA > 0.5 wt%, the gel strength remained unchanged at the "G" level, the gelatinization time was extended from 115 h to 164 h, and the dehydration rate increased from 22.4% to 41.6%.

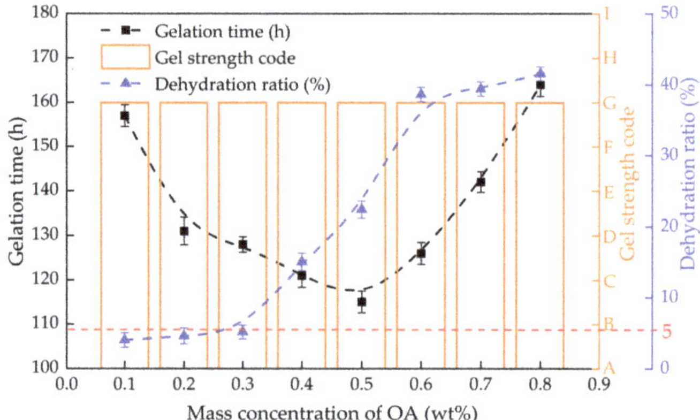

Figure 7. Gelling performance with different concentrations of OA (0.4 wt% PAM + 0.5 wt% HMTA + 0.05 wt% RO).

The results show that a low concentration of OA is beneficial for increasing the gelatinization rate and stability of gels, but a high concentration of OA is unfavorable. This is because the acidity of the solution increases with increasing OA concentration, which can accelerate the hydrolysis reaction of HMTA and thus accelerate the crosslinking reaction. When the concentration of OA is too high, the acid will destroy the structure of PAM, resulting in a slow crosslinking rate and a decrease in the stability of gels. With the dehydration rate < 5.0% as the standard, the optimal OA dosage is $\leq$0.3 wt%.

2.3. The Effect of Aqueous Salinity on Gelling Performance of PHRO Gel

The influence of Na^+, K^+, Ca^{2+} and Mg^{2+} on PHRO gelling performance was studied by measuring the gel strength, gelation time and dehydration rate of PHRO in different concentrations of NaCl, KCl, $CaCl_2$ and $MgCl_2$ solutions. The gel formula in the experiment was 0.4 wt% PAM + 0.5 wt% HMTA + 0.05 wt% RO + 0.3 wt% OA, and the experimental temperature was 30 °C.

2.3.1. Gelling Performance in Monovalent Salt Solution

The gel strength, gelation time and dehydration rate of PHRO gels in NaCl and KCl solutions with different concentrations (0–20.0 wt%) are shown in Figures 8 and 9, respectively. With the increase in the concentration of NaCl, the gel strength remained at the "G" level, and the gelation time was reduced from 130 h to 84 h. The dehydration rate increased from 4.1% to 5.5%. With the increase in the concentration of KCl, the gel strength was maintained at the "G" level. The gelation time was reduced from 131 h to 91 h, and the dehydration rate increased from 4.1% to 6.1%.

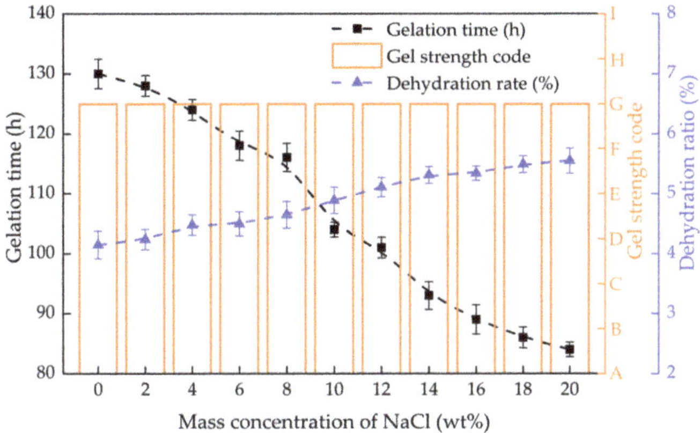

Figure 8. Gelling performance of PHRO in different concentrations of NaCl solution.

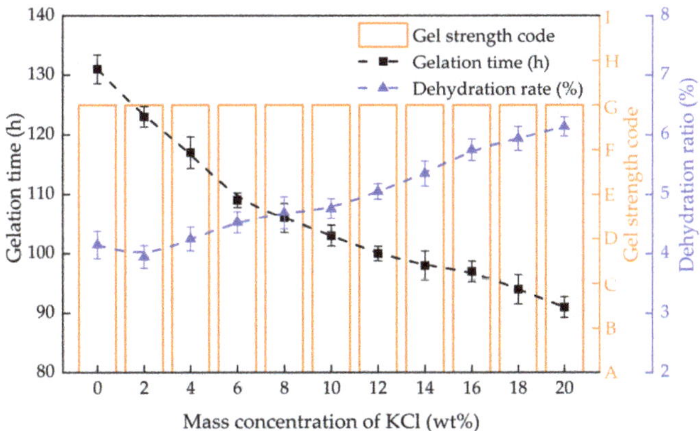

Figure 9. Gelling performance of PHRO in different concentrations of KCl solution.

The results indicate that PHRO can form high-strength gels and good stability in NaCl and KCl solutions with high concentrations; that is, the PHRO gels have good resistance to univalent salt ions. This is due to the electrostatic shielding effect of Na^+ and K^+, which can make the molecular chain of PAM curl up to a certain extent and shorten the distance between molecules [26]. However, this effect is slight and only accelerates the crosslinking reaction but has little negative influence on gel strength.

2.3.2. Gelling Performance in Divalent Salt Solutions

The gel strength, gelation time and dehydration rate of PHRO gels in $CaCl_2$ and $MgCl_2$ solutions with different concentrations (0–1.6 wt%) are shown in Figures 10 and 11, respectively. With the increase in the concentration of $CaCl_2$, the gel strength remained at the "G" level, the gelation time was first shortened from 131 h to 115 h and then increased to 135 h, and the dehydration rate increased from 4.1% to 8.6%. Moreover, with the increase in the concentration of $MgCl_2$, the gel strength remained at the "G" level, the gelation time was first shortened from 130 h to 117 h and then increased to 143 h, and the dehydration rate increased from 4.1% to 7.9%.

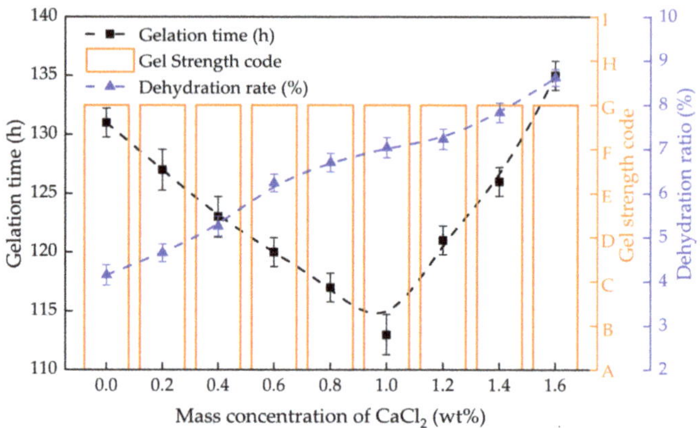

Figure 10. Gelling performance of PHRO in different concentrations of $CaCl_2$ solution.

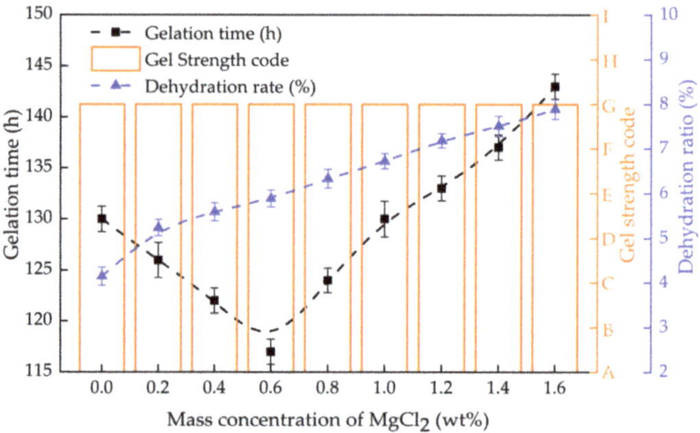

Figure 11. Dehydration rate of PHRO in different concentrations of $MgCl_2$ solution.

The results show that PHRO can form high-strength gels with good stability in high concentrations of $CaCl_2$ and $MgCl_2$ solutions; that is, PHRO gels have good resistance to bivalent saline ions. Ca^{2+} and Mg^{2+} have a stronger electrostatic shielding effect than Na^+ and K^+ [27,28]. When the concentration of $CaCl_2$ and $MgCl_2$ is low, they react with OA to form precipitation. As the concentration of Ca^{2+} and Mg^{2+} increases, the shielding effect of Ca^{2+} and Mg^{2+} on the PAM molecular chain is increased, which intensifies the curl of the PAM molecule. This tends to lead to internal crosslinking between "-$CONH_2$" and "-COO^-" on the PAM molecular chain, resulting in weakened inter-molecular crosslinking, and then the strength and stability of the gel are reduced.

2.4. Plugging Performance

The functional performance of PHRO gel in underground formation is shown in Figure 12. The main process is as follows: (1) select the well group with abundant remaining oil; (2) according to the construction design scheme, inject PHRO uncrosslinked gel solution into the injection well; and (3) waterflooding again. The experimental results of the plugging performance of PHRO gel are shown in Figure 13 and Table 1. In the process of injection, the injection pressure increased rapidly from 73.6–98.7 kPa due to the high viscosity of the uncrosslinked gel solution, which increased by 70.2–92.6 kPa compared with that of the simulated formation water. After PHRO was gelatinized completely, the initial pressure increased to the highest point in the subsequent waterflooding process and then decreased slightly. As the pressure increased, the simulated formation water broke through part of the core channel, and the pressure stabilized when the simulated formation water was injected. The calculated residual resistance coefficient is higher than 41.0, and the plugging rate of the gel is higher than 97.6%. Moreover, the residual resistance coefficient and plugging rate increased with the increase in core permeability.

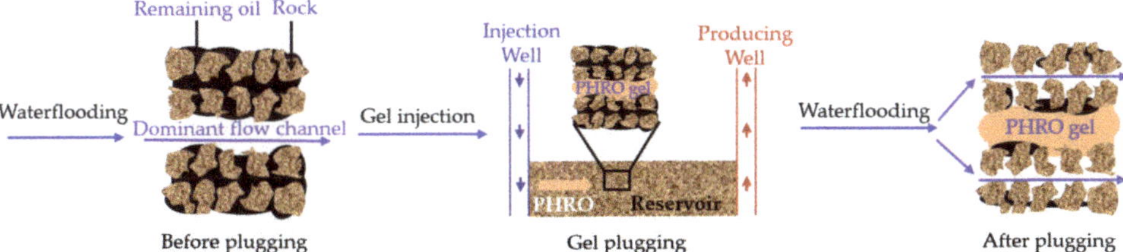

Figure 12. Schematic diagram of functional performance of PHRO gel in formation.

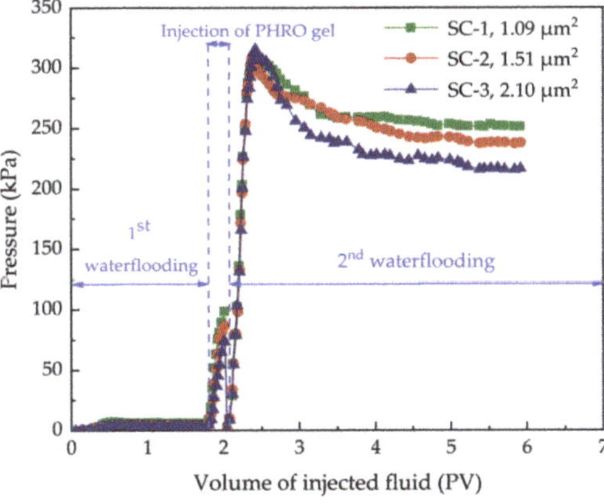

Figure 13. Pressure response of PHRO during conformance control experiment (the formula of gel is 0.4 wt% PAM + 0.5 wt% HMTA + 0.05 wt% RO + 0.3 wt% OA).

Table 1. Plugging ability of gels under different permeability conditions.

Number of Sandpack	P_1 (kPa)	P_2 (kPa)	F_{RR}	E (%)
SC-1	6.13	251.6	41.0	97.6
SC-2	4.23	238.2	56.3	98. 2
SC-3	3.31	216.8	62.5	98.5

The results show that PHRO gel has good plugging performance and is suitable for conformance control in low-temperature and high-salinity formation.

2.5. Discussion

The effect of saline ions on PHRO gels is mainly on the PAM molecular chain. After PAM hydrolysis, the molecular chain contains negative "-COO^-", and the molecular chain is extended due to electrostatic repulsion [26]. According to the "Lyotropic series" theory and osmotic pressure theory, the electrostatic shielding of Na^+, K^+, Ca^{2+} and Mg^{2+} can weaken the intermolecular and intramolecular electrostatic repulsion of PAM [29]. As a result, PAM molecules get closer to each other, and their chains curl up. The strength of electrostatic shielding is related to the valence state and concentration of saline ions. To explore the salt tolerance mechanism of PHRO gel, the microtopography of PHRO in NaCl and $CaCl_2$ solutions was observed.

Firstly, the microscopic morphology of PHRO in NaCl solution with a concentration of 20.0 wt% was investigated, and the results are shown in Figure 14. The PHRO gel has a complete network structure, and the mesh shrinks to form a denser network. The results show that the double-layer structure of PAM was not damaged in high-concentration sodium chloride solution, and there was still strong hydrogen bonding between PAM and water molecules. Therefore, the gel still has good stability.

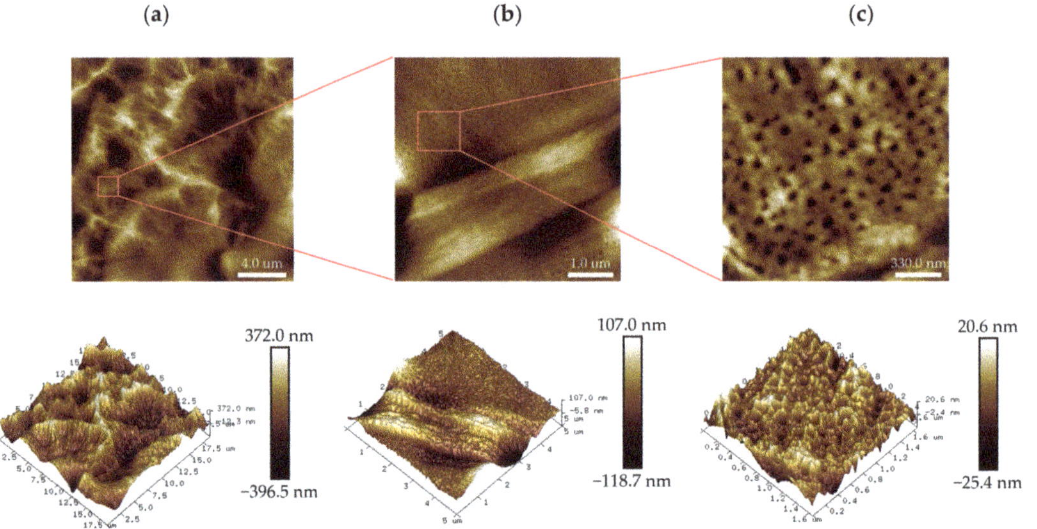

Figure 14. Two-dimensional (2D) and three-dimensional (3D) images of the PHRO gel in 20.0 wt% NaCl solution with AFM. (**a**) Scanning area is 18.8 μm × 18.8 μm; (**b**) scanning area is 5.0 μm × 5.0 μm; (**c**) scanning area is 1.7 μm × 1.7 μm.

Secondly, the microscopic morphology of PHRO in $CaCl_2$ solution with a concentration of 1.6 wt% was scanned, as shown in Figure 15. The PHRO gel has a complete network structure, with more mesh shrinkage and lower mesh numbers than those in NaCl solution. This is because the electrostatic shielding effect of Ca^{2+} is much stronger than that of Na^+, the molecular chain of PAM is more curled up, and the hydrogen bonding between PAM and water molecules is more damaged. However, due to the shielding effect of OA on Ca^{2+}, the gel still has good stability in saline water.

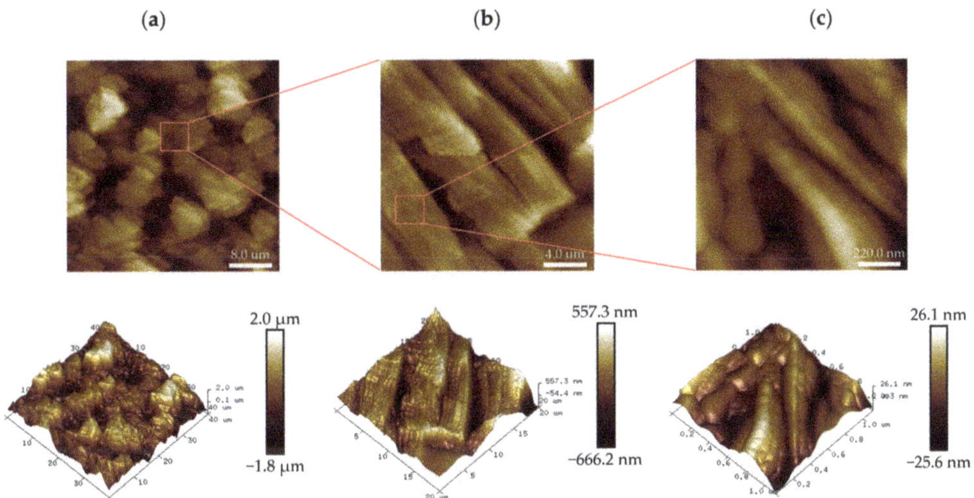

Figure 15. Two-dimensional (2D) and three-dimensional (3D) images of the PHRO gel in 1.6 wt% $CaCl_2$ solution with AFM. (**a**) Scanning area is 40.0 μm × 40.0 μm; (**b**) scanning area is 20.0 μm × 20.0 μm; (**c**) scanning area is 1.1 μm × 1.1 μm.

There are two reasons that PHRO gel has good salt resistance. On the one hand, formaldehyde and resorcinol react to produce phenolic resin particles of different molecular weights, which are evenly dispersed in the network structure of PHRO. These particles cover the PAM skeleton and attain a saline ion shielding effect. On the other hand, the addition of OA can partially eliminate the influence of Ca^{2+} and Mg^{2+} and improve the stability of gels.

In conclusion, the PHRO gel has good salt resistance, and appropriate salinity can accelerate the crosslinking reaction and improve the stability of gels. The results show that it is suitable for conformance control in high-salinity formation. In addition, gels of different strengths can be prepared by adjusting the concentration of the polymer and crosslinking agent. During the field application, the high-strength gel blocks the near-well zone and the weak gel blocks the in-depth zone to realize in-depth profile control.

3. Conclusions

A PHRO gel suitable for conformance control of low-temperature and high-salinity reservoirs was prepared, and the influence of the concentration of each component on the gelling performance of the gel was investigated. The gelling performance of PHRO in sodium chloride, potassium chloride, calcium chloride and magnesium chloride solutions were studied. The microstructure of PHRO in sodium chloride and calcium chloride solution was observed, and the salt resistance mechanism of PHRO gel was analyzed. Some conclusions may be drawn as follows:

(1) PHRO can form gels with high strength and good stability at a low temperature. The optimal concentrations of different additives were PAM $\geq$ 0.4 wt%, HMTA$\geq$ 0.5 wt%, RO $\geq$ 0.05 wt% and OA $\leq$ 0.3 wt%. The strength of the gel can reach the "G" grade. The gelation time is 105–173 h, and the dehydration rate is 2.1–5.2%.

(2) The strength of PHRO gels in different concentrations of NaCl and KCl solutions can reach the "G" level, and the dehydration rate is 4.1–6.1%. The gelatinization time is 84–131 h, and it decreases with the increase in saline ion concentration. Furthermore, the strength of PHRO gels in different concentrations of $CaCl_2$ and $MgCl_2$ solutions can be maintained at the "G" level, and the dehydration rate is 4.1–8.6%.

The gelatinization time is 115–143 h, and it decreases with the increase in saline ion concentration.

(3) PHRO gel has a robust "stem-leaf" three-dimensional network structure, which contains embedded phenolic resin with different sizes and different molecular weights. It can enhance the strength of the PAM skeleton and improve the stability of the gel. Moreover, the PHRO gel still has a complete network structure in a high-concentration salt solution, the network is more compact due to mesh contraction, and the gel structure still has high strength.

(4) The plugging rate of PHRO in simulated formation water is more than 97.6%, which is suitable for conformance control in low-temperature and high-salinity reservoirs.

(5) The salt-resistant performance of PHRO gel is mainly because the phenolic resin generated by the fluid can strengthen the gel network structure, and OA has the function of shielding salt ions. The appropriate number of saline ions is beneficial for accelerating the crosslinking reaction and improving the gel strength. The results show that PHRO gel is suitable for in-depth conformance control in low-temperature and high-salinity formation, which has broad application prospects.

4. Materials and Methods

4.1. Materials

4.1.1. Additives

The polymer used in this study was PAM with an average molecular weight of 1.0×10^7 g/mol and a hydrolysis degree of 3.6%, which was manufactured by Anhui Tianrun Chemical Industry Co., Ltd. (Bengbu, China). The crosslinking agents used were HMTA and RO, which were provided by Weifang Xingjia Chemical Co., Ltd. (Weifang, China). To encourage the fluid to crosslink at low temperatures, OA was used as a cross-promoting agent, purchased from Shandong Feishuo Chemical Technology Co., Ltd. (Jinan, China). The salts used for the salt resistance study fluid were NaCl, KCl, $CaCl_2$, $NaHCO_3$ and $MgCl_2{\cdot}6H_2O$, which were obtained from Pharmaceutical Group Chemical Reagents Co., Ltd. (Shanghai, China). The water used in the experiment was deionized water.

4.1.2. Simulated Formation Water

The simulated formation water used in the experiment in this paper was prepared according to the ionic composition of formation water from North Buzachi Oilfield. The total salinity of the simulated water was 63,246.2 mg/L, and the ionic composition is shown in Table 2.

Table 2. Ionic composition of simulated formation water.

Ionic Type	Cl^-	SO_4^{2-}	HCO_3^-	Mg^{2+}	Ca^{2+}	Na^+	K^+
Ionic content (mg/L)	36879.4	540.9	72.6	579.3	1985	17445.9	5743.1

4.2. Methods

4.2.1. Preparation of PHRO Gel

The PHRO gel was prepared at room temperature (25 °C). First, PAM was evenly dispersed in the solution according to the designed dosage and stirred with a mechanical stirrer (IKA RW20, IKA, Königswinter, Germany) at a rate of 300 r/min for 2 h. Then, HMTA, RO and OA were successively added according to the designed dosage and stirred at a rate of 300 r/min for 30 min. Finally, 15 mL of the well-stirred initial solution of the gel was put into an ampoule bottle, sealed with an alcohol burner and placed in the oven (UFP 500, Memmert Company, Schwabach, Germany) at 30 °C.

4.2.2. Determination of Gelation Time and Gel Strength

In this study, the GSC strength code method was used to determine the strength of PHRO [30,31]. Letters A-I were used to represent different strength levels of the gel, and grade "I" was the highest strength, as shown in Table 3. First, the initial solution of PHRO gel was prepared according to the method in Section 4.2.1. Secondly, the samples were taken out of the oven every hour to record the time and strength code. When the strength code no longer changes, the current code is the strength level of PHRO, and the time is the gelation time of PHRO. There were three parallel samples in each group, and the gelation time was the average value of the sample data.

Table 3. Gel strength code [30,31].

Gel Strength Code	Gel Category	Gel Description
A	No detectable gel	The same viscosity as the original polymer solution.
B	Highly flowing gel	Only slightly more viscous than the initial polymer solution.
C	Flowing gel	Most of the gel flows to the bottle cap by gravity upon inversion.
D	Moderately flowing gel	Only a small portion (5.0–10.0%) of the gel does not readily flow to the bottle cap by gravity upon inversion
E	Barely flowing gel	Significant portion (>15.0%) of the gel does not flow by gravity upon inversion.
F	Highly deformable non-flowing gel	The gel does not flow to the bottle cap by gravity upon inversion.
G	Moderately deformable non-flowing gel	The gel deforms about halfway down the bottle by gravity upon inversion.
H	Slightly deformable non-flowing gel	Only the gel surface slightly deforms by gravity upon inversion.
I	Rigid gel	There is no gel surface deformation by gravity upon inversion.

4.2.3. Stability of PHRO Gel

In this study, the dehydration rate (D_w) was used to characterize the stability of PHRO gel. The lower the D_w of PHRO gel in the same aging time, the better its stability. First, the initial solution of PHRO gel was prepared according to the method in Section 4.2.1., and the mass of gel-forming solution was recorded as m_0. Secondly, the prepared samples were aged for 90 days in the oven at 30 °C. Finally, the mass of the solid phase in the ampoule bottle aged for 90 days was weighed at room temperature (25 °C) and denoted as m_1. There were three parallel samples in each group, and the dehydration rate was the average value of the sample data. The calculation formula of D_w is:

$$D_w = \frac{m_0 - m_1}{m_0} \times 100\% \tag{1}$$

4.2.4. Plugging Performance of PHRO Gel

The plugging performance of PHRO gels is characterized by the plugging rate (E) and residual resistance coefficient (F_{RR}). The higher the E and F_{RR}, the better the plugging performance.

The plugging performance of PHRO gels was evaluated by using a sandpack with a diameter of 25 mm and a length of 200 mm. The flow chart of the evaluation device is shown in Figure 16. Sandpacks with different permeabilities were obtained by filling them with quartz sand with different sizes, and 300 mesh nets for sand leaking prevention were placed at the inlet and outlet of the sandpack to prevent sand production. The porosity and permeability parameters of the sandpack are shown in Table 4. The plugging evaluation experiments of gels were carried out in the oven at 30 °C, and the fluid injection rate was 1.0 mL/min.

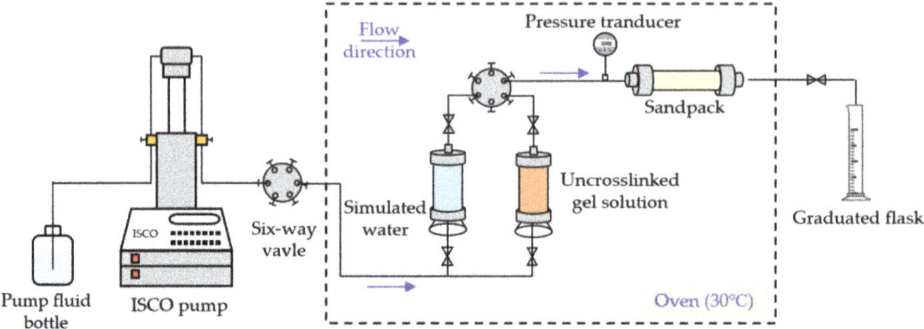

Figure 16. Flow chart of sandpack model.

Table 4. Porosity and permeability of sandpack.

Number of Sandpack	Permeability (μm^2)	Porosity (%)
SC-1	1.09	25.4
SC-2	1.51	29.3
SC-3	2.10	31.0

First, the simulated formation water was injected, and the stable pressure P_1 was recorded after the pressure stabilized. Secondly, the uncrosslinked gel solution of PHRO gel was prepared according to the method in Section 4.2.1. Thirdly, 0.3 PV (pore volume) of the initial solution of the gel was injected, and the sandpack was placed in the oven and aged for 7 days. Finally, simulated formation water was injected, and the stable pressure P_2 was recorded after the pressure stabilized.

The calculation formula of the residual resistance factor is:

$$F_{RR} = \frac{P_2}{P_1} \tag{2}$$

The calculation formula of the plugging rate is:

$$E = \left(1 - \frac{P_1}{P_2}\right) \times 100\% \tag{3}$$

4.2.5. The Microscopic Morphology of PHRO Gel

Atomic force microscopy (AFM) (Multimode 8, Bruker Company, Billerica, MA, USA) was used to observe the microstructure of PHRO gels. First, the initial solution of PHRO gel was prepared as described in Section 4.2.1. Secondly, an appropriate amount of gel sample was cut and placed on a copper sheet with scissors to make it spread into a film as thin as possible and dried in a freeze dryer (SCIENTZ-10N, Ningbo Scientz Biotechnology Co., Ltd., Ningbo, China) for 24 h. Finally, the microstructure of PHRO was observed in air in AFM tapping mode, and the related parameters were determined. The experimental temperature was 30 °C, and the room humidity was 30%.

Author Contributions: Data curation, Y.S.; funding acquisition, C.D.; investigation, F.D. and G.Z.; methodology, C.D. and Q.Y.; writing—original draft, F.D.; writing—review and editing, C.D. and Y.L. All authors have read and agreed to the published version of the manuscript.

Funding: The authors sincerely appreciate the financial support from the National Natural Science Foundation of China (Nos. 51834010 and 52120105007).

Institutional Review Board Statement: Not applicable.

Informed Consent Statement: Not applicable.

Data Availability Statement: Data are contained within the article.

Conflicts of Interest: The authors declare no conflict of interest.

References

1. Cheng, H.; Zheng, X.; Wu, Y.; Zhang, J.; Zhao, X.; Li, C. Experimental and numerical investigation on oil displacement mechanism of weak gel in waterflood reservoirs. *Gels* **2022**, *8*, 309. [CrossRef] [PubMed]
2. Xiong, C.; Tang, X. Technologies of water shut-off and profile control: An overview. *Pet. Explor. Dev.* **2007**, *34*, 83–88.
3. Zhen, N.; Dong, B.; Xia, B.; Zheng, X.; Xie, C. New cementing technologies successfully solved the problems in shallow gas, low temperature and easy leakage formations. In Proceedings of the International Oil and Gas Conference and Exhibition in China, Beijing, China, 8–10 June 2010.
4. Aladasani, A.; Bai, B. Recent developments and updated screening criteria of enhanced oil recovery techniques. In Proceedings of the International Oil and Gas Conference and Exhibition in China, Beijing, China, 8–10 June 2010.
5. Zhao, G.; You, Q.; Tao, J.; Gu, C.; Aziz, H.; Ma, L.; Dai, C. Preparation and application of a novel phenolic resin dispersed particle gel for in-depth profile control in low permeability reservoirs. *J. Pet. Sci. Eng.* **2018**, *161*, 703–714. [CrossRef]
6. Bai, B.; Wei, M.; Liu, Y. Injecting large volumes of preformed particle gel for water conformance control. *Oil Gas Sci. Technol.* **2012**, *67*, 941–952. [CrossRef]
7. Yao, C.; Lei, G.; Hou, J.; Xu, X.; Wang, D.; Steenhuis, T.S. Enhanced oil recovery using micron-size polyacrylamide elastic microspheres: Underlying mechanisms and displacement experiments. *Ind. Eng. Chem. Res.* **2015**, *54*, 10925–10934. [CrossRef]
8. Imqam, A.; Bai, B. Optimizing the strength and size of preformed particle gels for better conformance control treatment. *Fuel* **2015**, *148*, 178–185. [CrossRef]
9. Goudarzi, A.; Zhang, H.; Varavei, A.; Taksaudom, P.; Hu, Y.; Delshad, M.; Bai, B.; Sepehrnoori, K. A laboratory and simulation study of preformed particle gels for water conformance control. *Fuel* **2015**, *140*, 502–513. [CrossRef]
10. Dai, C.; Zhao, G.; Zhao, M.; You, Q. Preparation of dispersed particle gel (DPG) through a simple high speed shearing method. *Molecules* **2012**, *17*, 14484–14489. [CrossRef]
11. Qing, Y.; Yefei, W.; Zhou, W.; Ziyuan, Q.; Fulin, Z. Study and application of gelled foam for in-depth water shutoff in a fractured oil reservoir. *J. Can. Pet. Technol.* **2009**, *48*, 51–55. [CrossRef]
12. Li, Y.; Li, Y.; Peng, Y.; Yu, Y. Water shutoff and profile control in china over 60 years. *Oil Drill. Prod. Technol.* **2019**, *41*, 773–787.
13. Liang, J.-T.; Sun, H.; Seright, R. Why Do Gels Reduce Water Permeability More Than Oil Permeability? *SPE Reserv. Eng.* **1995**, *10*, 282–286. [CrossRef]
14. Rozhkova, Y.A.; Burin, D.A.; Galkin, S.V.; Yang, H. Review of microgels for enhanced oil recovery: Properties and cases of application. *Gels* **2022**, *8*, 112. [CrossRef] [PubMed]
15. Yu, H.; Ma, Z.; Tang, L.; Li, Y.; Shao, X.; Tian, Y.; Qian, J.; Fu, J.; Li, D.; Wang, L.; et al. The effect of shear rate on dynamic gelation of phenol formaldehyde resin gel in porous media. *Gels* **2022**, *8*, 185. [CrossRef]
16. Ren, H.; Qu, N.; Wang, Y.; Qiu, X.; Liu, Q.; Chi, P. Effect of crosslinker concentration on the comprehensive properties of polyacrylamide-aluminium citrate/water glass gel. *IOP Conf. Ser. Earth Environ. Sci.* **2020**, *440*, 022021. [CrossRef]
17. Tokar, E.; Tutov, M.; Kozlov, P.; Slobodyuk, A.; Egorin, A. Effect of the resorcinol/formaldehyde ratio and the temperature of the Resorcinol–Formaldehyde gel solidification on the chemical stability and sorption characteristics of ion-exchange resins. *Gels* **2021**, *7*, 239. [CrossRef]
18. Zhang, C.; Qu, G.; Song, G. Formulation development of high strength gel system and evaluation on profile control performance for high salinity and low permeability fractured reservoir. *Int. J. Anal. Chem.* **2017**, *2017*, 2319457. [CrossRef]
19. Du, D.; Pu, W.; Tan, X.; Liu, R. Experimental study of secondary crosslinking core-shell hyperbranched associative polymer gel and its profile control performance in low-temperature fractured conglomerate reservoir. *J. Pet. Sci. Eng.* **2019**, *179*, 912–920. [CrossRef]
20. El-Karsani, K.S.; Al-Muntasheri, G.A.; Hussein, I.A. Polymer systems for water shutoff and profile modification: A review over the last decade. *SPE J.* **2013**, *19*, 135–149. [CrossRef]
21. Zhao, G.; Dai, C.; Chen, A.; Yan, Z.; Zhao, M. Experimental study and application of gels formed by nonionic polyacrylamide and phenolic resin for in-depth profile control. *J. Pet. Sci. Eng.* **2015**, *135*, 552–560. [CrossRef]
22. Fang, Y.; Yang, E.; Cui, X. Study on profile control and water shut-off performance of interpenetrating network polymer gel composite system in shallow low temperature fractured oil layer. *ChemistrySelect* **2019**, *4*, 8158–8164. [CrossRef]
23. Zhang, L.; Khan, N.; Pu, C. A new method of plugging the fracture to enhance oil production for fractured oil reservoir using gel particles and the HPAM/Cr$^{(3+)}$ system. *Polymers* **2019**, *11*, 446. [CrossRef] [PubMed]
24. DiGiacomo, P.M.; Schramm, C.M. Mechanism of polyacrylamide gel syneresis determined by C-13 NMR. In Proceedings of the SPE Oilfield and Geothermal Chemistry Symposium, Denver, CO, USA, 1–3 June 1983.
25. Wang, D. Study on mechanism of Ca^{2+} on the syneresis of HPAM gel. *Sci. Technol. Eng.* **2017**, *17*, 277–281, 287.
26. Kizilay, M.Y.; Okay, O. Effect of hydrolysis on spatial inhomogeneity in poly(acrylamide) gels of various crosslink densities. *Polymer* **2003**, *44*, 5239–5250. [CrossRef]
27. Bueno, V.B.; Bentini, R.; Catalani, L.H.; Petri, D.F.S. Synthesis and swelling behavior of xanthan-based hydrogels. *Carbohydr. Polym.* **2013**, *92*, 1091–1099. [CrossRef]

28. Muta, H.; Kojima, R.; Kawauchi, S.; Tachibana, A.; Satoh, M. Ion-specificity for hydrogen-bonding hydration of polymer: An approach by ab initio molecular orbital calculations. *J. Mol. Struct. THEOCHEM* **2001**, *536*, 219–226. [CrossRef]
29. Pu, J.; Geng, J.; Han, P.; Bai, B. Preparation and salt-insensitive behavior study of swellable, Cr^{3+}-embedded microgels for water management. *J. Mol. Liq.* **2019**, *273*, 551–558. [CrossRef]
30. Jia, H.; Pu, W.-F.; Zhao, J.-Z.; Liao, R. Experimental investigation of the novel phenol−formaldehyde cross-linking HPAM gel system: Based on the secondary cross-linking method of organic cross-linkers and its gelation performance study after flowing through porous media. *Energy Fuels* **2011**, *25*, 727–736. [CrossRef]
31. Sydansk, R.D. Delayed in Situ Crosslinking of Acrylamide Polymers for Oil Recovery Applications in High-Temperature Formations. U.S. Patent 4844168A, 4 July 1989.

 gels

MDPI

Article

Enhanced Oil Recovery Mechanism and Technical Boundary of Gel Foam Profile Control System for Heterogeneous Reservoirs in Changqing

Liang-Liang Wang , Teng-Fei Wang *, Jie-Xiang Wang, Hai-Tong Tian, Yi Chen and Wei Song

School of Petroleum Engineering, China University of Petroleum (East China), Qingdao 266580, China; llwang2017@163.com (L.-L.W.); wangjxupc@126.com (J.-X.W.); tian_haitong2022@163.com (H.-T.T.); chyi2020@163.com (Y.C.); song15954231226@163.com (W.S.)

* Correspondence: wangtengfei@upc.edu.cn; Tel.: +86-159-6691-6346

Abstract: The gel plugging and flooding system has a long history of being researched and applied, but the Changqing reservoir geological characteristics are complex, and the synergistic performance of the composite gel foam plugging system is not fully understood, resulting in poor field application. Additionally, the technique boundary chart of the heterogeneous reservoir plugging system has hardly appeared. In this work, reservoir models of porous, fracture, and pore-fracture were constructed, a composite gel foam plugging system was developed, and its static injection and dynamic profile control and oil displacement performance were evaluated. Finally, combined with the experimental studies, a technical boundary chart of plugging systems for heterogeneous reservoirs is proposed. The research results show that the adsorption effect of microspheres (WQ-100) on the surface of elastic gel particles-1 (PEG-1) is more potent than that of pre-crosslinked particle gel (PPG) and the deposition is mainly on the surface of PPG. The adsorption effect of PEG-1 on the surface of PPG is not apparent, primarily manifested as deposition stacking. The gel was synthesized with 0.2% hydrolyzed polyacrylamide (HPAM) + 0.2% organic chromium cross-linking agent, and the strength of enhanced gel with WQ-100 was higher than that of PEG-1 and PPG. The comprehensive value of WQ-100 reinforced foam is greater than that of PEG-1, and PPG reinforced foam, and the enhanced foam with gel has a thick liquid film and poor foaming effect. For the heterogeneous porous reservoir with the permeability of 5/100 mD, the enhanced foam with WQ-100 shows better performance in plugging control and flooding, and the recovery factor increases by 28.05%. The improved foam with gel enhances the fluid flow diversion ability and the recovery factor of fractured reservoirs with fracture widths of 50 μm and 180 μm increases by 29.41% and 24.39%, respectively. For pore-fractured reservoirs with a permeability of 52/167 mD, the PEG + WQ-100 microsphere and enhanced foam with WQ-100 systems show better plugging and recovering performance, and the recovery factor increases are 20.52% and 17.08%, 24.44%, and 21.43%, respectively. The smaller the particle size of the prefabricated gel, the more uniform the adsorption on the foam liquid film and the stronger the stability of the foam system. The plugging performance of the composite gel system is stronger than that of the enhanced gel with foam, but the oil displacement performance of the gel-enhanced foam is better than that of the composite gel system due to the "plug-flooding-integrated" feature of the foam. Combined with the plugging and flooding performance of each plugging system, a technique boundary chart for the plugging system was established for the coexisting porous, fracture, and pore-fracture heterogeneous reservoirs in Changqing Oilfield.

Keywords: enhanced gel system; enhanced foam system; heterogeneous reservoir; plugging adaptability; enhanced oil recovery

Citation: Wang, L.-L.; Wang, T.-F.; Wang, J.-X.; Tian, H.-T.; Chen, Y.; Song, W. Enhanced Oil Recovery Mechanism and Technical Boundary of Gel Foam Profile Control System for Heterogeneous Reservoirs in Changqing. *Gels* **2022**, *8*, 371. https://doi.org/10.3390/gels8060371

Academic Editors: Qing You, Guang Zhao and Xindi Sun

Received: 25 May 2022
Accepted: 11 June 2022
Published: 12 June 2022

1. Introduction

Due to the stimulation of low/ultra-low permeability reservoirs, high-permeability channeling channels are easily formed after artificial fractures are connected with natural fractures [1,2]. However, the high cost of fine water injection and the long shutdown period greatly affect the normal production of oilfields [3]. Given this, the development of plugging and flooding systems with strong applicability has received extensive attention.

Gel and foam systems, as the two most widely used plugging agents, have achieved positive progress in both laboratory experiments and field applications in recent decades [4,5]. Gel-based plugging systems have been used on a large scale in the 1990s, and various modified gels, micro-nano microspheres, polyethylene glycol gels (PEG), and chitosan-based gels have been developed, including preformed particle gel (PPG) and HPAM weak gels with various particle sizes and compositions [6–10].

Field tests show that a single plugging system cannot meet the selective plugging of highly heterogeneous reservoirs and cannot meet the needs of adjusting the water absorption profile and enhancing oil displacement. In recent years, scholars have carried out a series of composite plugging system applicability evaluation experiments to achieve the comprehensive effect of "deep control and flooding + near-well plugging". Jia et al., developed a novel nanocomposite gel with tunable gel formation time based on the in-situ polymerization method [11]. The nanocomposite gel system has good temperature resistance (75–105 °C), adding 5% nano-silica, and the compressive performance of the composite gel system is improved from 8.7 to 21 KPa; ammonium persulfate and hydrochloric acid can be caused to be effectively degraded. Given the development of cross-linking after polymer flooding, Haung et al., proposed a low initial viscosity gel plugging agent: 500–1000 mg/L polymer LH2500, 1000–2500 mg/L cross-linking agent CYJL, 200–500 mg /L citric acid, 100–150 mg/L sodium sulfite, and 100–200 mg/L sodium polyphosphate. For secondary polymer flooding reservoirs, the oil recovery can be enhanced by 13.6% after 0.1 PV gel plugging [12]. Liu et al., developed a gel system formed by a terpolymer (L-1) and a new cross-linking system (HB-1) for ultra-deep and high-temperature reservoirs [13]. This can create a stable continuous 3D network structure with good long-term thermal stability. The strength of the gel system can be adjusted by changing the concentrations of the terpolymer (0.05–1%) and the crosslinking agent (0.05–1%).

The commonly used foam sealing and channeling systems are mainly composed of a foaming agent, foam stabilizer, and gas phase [14,15]. Nitrogen comes from various sources and has become the most commonly used foam gas phase. Foam stabilizers include HPAM, worm-type surfactants, inorganic nanoparticles, and nano-microspheres. Using three crude oils with different properties, Lai et al., studied the stability and oil displacement capacity of the gel foam system [16]. For crude oil with more heavy components such as resins and asphaltenes, when the oil saturation is lower than 20%, the gel foam shows higher foam stability. The oil-containing gel foam generated from heavy oil has good stability, the plugging rate is 95.33%, and the enhanced oil recovery is 23.1%. Zhang et al., performed huff and puff experiments on core samples using different gases and foams at the temperature of unconventional liquid reservoirs, capturing time-lapse CT images with a computed tomography unit (CTU) [17]. Studies have shown that a combination of EOR techniques (foam or sequencing surfactant and gas injection) opens the possibility of achieving sound oil recovery. Natural fractures can easily lead to lost circulation and can easily limit the normal production of oil fields. Li et al., developed an oil-based pressure-bearing foam gel plugging agent for fracturing shale: 0.6% DSFA foaming agent + 0.3% BPMO foaming agent + 0.5% EPDM + 1.5% SBS + 0.05% rigid crosslinking agent DB + 0.02% modified SiO_2 nanoparticles [18]. The core plugging experiment shows that the plugging efficiency of the plugging agent is about 90%.

The concise literature shows that many studies have been carried out on the plugging control and flooding system and its mechanism in heterogeneous reservoirs, and good application results have been achieved in oilfields. However, in the face of complex and changeable reservoirs with strong heterogeneity, coupled with insufficient understanding of the synergistic mechanism of multi-component compound plugging systems, the existing plugging systems have poor water blocking and oil-increasing effects, and even cause secondary damage to the reservoir. In addition, the technical chart of the heterogeneous reservoir plugging system has hardly appeared. Therefore, in this work, combined with the geological characteristics of Changqing Oilfield, reservoir models of porous, fracture, and pore-fracture were constructed, the composite plugging system was developed based on the gel and foam system, and injection and dynamic performance were evaluated. The synergistic effect stimulates the new vitality of traditional plugging agents, optimizes and strengthens the gel and gel-enhanced foam plugging control and flooding systems, and achieves breakthroughs in the channeling channel of "injecting, plugging, long-lasting, and low-cost". Finally, based on the optimal plugging systems of different reservoir models, a technical boundary chart of plugging systems for heterogeneous reservoirs is proposed. The results of this study improve the composite gel foam plugging system and its action mechanism suitable for highly heterogeneous reservoirs; additionally, they provide crucial technical guidance for Changqing's low-permeability reservoirs to achieve stable production and increase production.

2. Experiments and Methods

2.1. Static Performance Evaluation Experiment of Plugging System

2.1.1. Gel Plugging Agent

1. Pre-crosslinking gel particle

The prefabricated gel particles include microspheres (WQ-100), PEG-1, PPG, HPAM, organochromium crosslinking agent, foaming agent, and formation water (60,000–100,000 mg/L), all provided by Changqing Oilfield. Microscopic tests of WQ-100, PEG-1, and PPG by Microscope First. A total of 0.5 g of prefabricated gel particle plugging agent and 100 mL of formation water/kerosene was sequentially added to 6 beakers and stirred with a magnetic stirrer for 10 min. Next, the magnetron was taken out and all the beakers were sealed and placed in a hot air drying oven with a constant temperature of 70 °C [19]. Finally, the microscopic morphology of WQ-100, PEG-1 and bulk particles was observed by the microscope, and the particle size changes were analyzed and counted.

2. Gel

A total of 500 mL of formation water was poured into a jar and stirred with a large torque stirrer. Slowly, 3 g of HPAM powder was added and was continued to be stirred until the HPAM was completely dissolved. The prepared mother liquor was diluted with formation water and stirred for 1 min to make HPAM evenly dispersed in the formation water. Next, 30 mL of the diluted HPAM solution was added to the cup, the organic chromium cross-linking agent was dropped into the beaker and stirred for 30 s (450 rpm), and 25 mL was poured into a stoppered measuring cylinder. All stoppered graduated cylinders were placed in a water bath preheated to 70 °C, the state of the mixture in the graduated cylinder was recorded every 1 h, and the time required for each group of formulations to reach grade G was recorded [20]. After the gel in the graduated cylinder reached the G level, the vacuum pump hose was connected to the inside of the forming gel, the vacuum pump was started, and the maximum reading of the vacuum gauge was recorded.

2.1.2. Foam

A 100 mL quantity of distilled water was put into the mud cup and different volume fractions of foaming agents were added. The high-speed stirrer was set to 6000 rpm and the foam was poured out after stirring for 3 min. The foaming volume and the time it took to produce 50 mL of liquid were recorded. The foam performance is quantitatively evaluated using the foam comprehensive value, and the calculation method is as follows [21]:

$$F_c = V_0 t_f$$

where F_c is foam comprehensive coefficient, mL·s; V_0 is initial foam volume, mL; t_f is foam half-life, s.

2.1.3. Binary/Ternary Complex System

1. Pre-crosslinking gel particles

The formula of the pre-crosslinking gel particle complex systems were 0.5% WQ-100 + 0.5% PEG-1, 0.5% WQ-100 + 0.5% PPG, and 0.5% PEG-1 + 0.5% PPG, respectively. The formation water was used as the dispersant for the compound system, totaling 100 mL. After the three groups of compound systems were sealed, they were aged in a constant temperature oven at 70 °C for 12 h. The pre-gel compound system was taken out and the poorly dispersed system was uniformly dispersed again by stirring with a glass rod.

2. Enhanced gels with pre-crosslinking gel particle

The HPAN dispersion method is the same as 2.1.1 (2). All dispersion systems were put into 25 mL graduated cylinders with a stopper and placed in a water bath with a constant temperature of 70 °C, and the gel formation time and the degree of vacuum breakthrough were recorded.

3. Enhanced foam with pre-crosslinking gel particles and gels

The formulation and interfacial tension of the gel-reinforced foam compound system is shown in Table 1. A 100 mL quantity of prefabricated gel particle-foaming agent dispersion system was put into a mud cup. After stirring for 3 min, the rotation was stopped and immediately poured into a graduated cylinder to measure the foaming volume and absorb the upper foam for microscopic observation. The time when the liquid in the graduated cylinder reaches 50 mL was recorded and the comprehensive foam value of the system was calculated. The 0.2% HPAM + 0.2% organic chromium cross-linking agent + 0.4% foaming agent was dispersed in formation water to obtain a jelly-reinforced foam system. The reinforced foam system sample was added to the sample cell of the interfacial tension tester, and the size of the bubbles at the outlet of the injector in the tester was controlled by software. The test was stopped when each sample is measured for 900 s, and the real-time gas–liquid interfacial tension values in the interval of 300–900 s were averaged as the test results of this group of samples [22].

Table 1. The formula of enhanced foam system with pre-crosslinking gel particles.

Enhanced Foam System	Interfacial Tension/(mN/m)	Note
0.4% foaming agent	35.2	/
0.4% foaming agent + 0.1% WQ-100	33.6	Reduce surface tension
0.4% foaming agent + 0.1% PEG-1	34.8	Not significantly
0.4% foaming agent + 0.1% PPG	/	Particle size is too large to measure
0.4% foaming agent + 0.2% HPAM + 0.1% organic chromium cross-linking agent	46.5	Liquid film gelation, foam stabilization

2.2. Physical Simulation Model Construction

To better simulate the reservoir conditions of strong heterogeneity with the development of micro-fractures, laboratory physical simulation experimental models of porous, fracture, and pore-fracture were constructed. For the porous reservoirs model, a one-dimensional sand filling model was filled by mixing quartz sands of different particle sizes, with a porosity of 26% and a permeability of 7×10^{-3} μm^2. Secondly, the artificial core (5×10^{-3} μm^2) was split along the axis direction by the Brazilian splitting device, and the split core adhered to the ceramsite. After the cores were merged, they were wrapped and tightened by the raw material belt to simulate the fractured reservoir after underground fracturing, and the opening of the fracture could be changed by adjusting the particle size of the ceramsite. Finally, the sand filter screen was cut into small strips and mixed with quartz sand (40–80 μm, 80–120 μm) evenly. The one-dimensional sand-packing model was filled to construct pores for simulating reservoirs with micro-fractures. The diagram and fundamental porosity and permeability characteristics of the fractured and pore-fractured experimental models are shown in Figure 1 and Table 2, respectively.

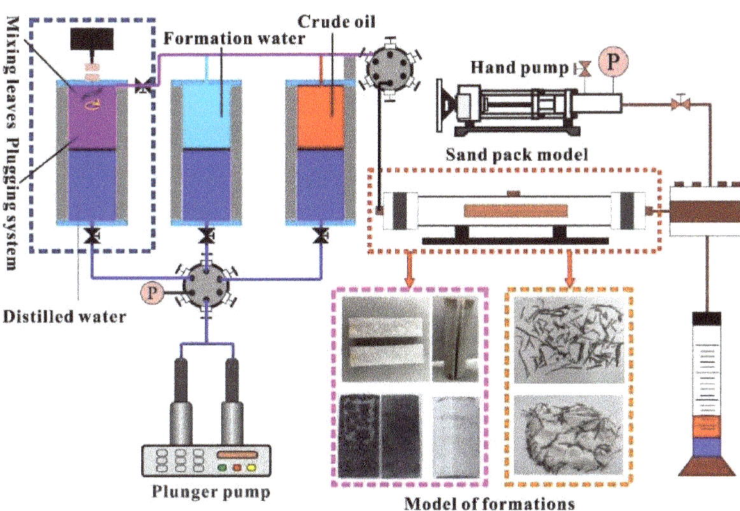

Figure 1. Flow chart of laboratory displacement experiment.

Table 2. Basic parameters of the fractured and pore-fractured experimental models.

Physical Models	Fracture Width /μm	Porosity before Splitting /%	Permeability before Splitting /$\times 10^{-3}$ μm^2	Permeability after Splitting /$\times 10^{-3}$ μm^2
Fractured models	50	19.5	5.3	90,000
	180	20.6	4.6	2,480,000
	340	21.2	6	31,200,000
	510	20.4	4.7	87,000,000
Pore-fractured models	Matrix volume /cm^3	Microfracture volume/cm^3	Porosity /%	Permeability /$\times 10^{-3}$ μm^2
	295	0.4	27.1	52.2
		0.8	28.4	91.2
		1.2	29.8	166.7
		1.6	31.2	252.7

The calculation formula of fracture permeability is as follows.

$$K = 1000 \times \frac{\phi b^2}{12}$$

where K is permeability, 10^{-3} μm^2, ϕ is porosity of fractured reservoir model, and b is fracture width, μm.

2.3. Plugging EOR Mechanisms and Performance

The prefabricated different reservoir models were loaded into the displacement system, and the experimental flow of plugging and enhanced oil recovery displacement is shown in Figure 1. The formation water was pumped into the model to measure the initial permeability (Ki), the fracture opening was adjusted by changing the particle size of ceramsite, and the number of screen bars was increased or decreased to alter the pore-fractured reservoir model. For each reservoir model, first, a plunger pump was used to saturate the model with water at 0.5 mL/min and record the steady pressure Pi. The injection rate was maintained and 0.5 PV of the plugging system was injected into the model to be tested. After the plugging system was injected, it was transferred to the formation water injection until the pressure was stable.

It is worth noting that for millimeter-centimeter PPG, it can be seen from the static performance experiments that it has evident stratification after adding liquid. During the injection and displacement process, the conventional intermediate container was used because a large amount of PPG is deposited or adheres to the top cover of the intermediate container. This can cause injection difficulties and even lead to the experiment's failure. As shown in Figure 1, the intermediate vessel injected with PPG was upgraded and improved in this study, mild shear was generated by stirring, and the pressure-resistant pipeline was thickened with an inner diameter of 3 mm. The slightly sheared bulk particles are further subjected to extrusion shearing while passing through the injection pipeline, which also improves the PPG injection performance. During the EOR performance evaluation experiment, after saturating the formation water and crude oil in sequence, firstly, the formation water was injected to stimulate the production. The produced fluid was injected into the plugging system after no oil phase flowed out, and then the formation water was injected again until no oil phase flowed out again. During the period, the cumulative oil production was recorded by the oil-water separation metering device, the cumulative oil production injection curve was drawn, and the enhanced oil recovery performance of the plugging system was evaluated. The plugging rate represents the plugging ability of the formation after a particular plugging agent is injected and forms stable plugging. Its calculation theory is as follows:

$$\text{Plugging efficiency} = 1 - \frac{K_1}{K_i}$$

where K_i is permeability after plugging, 10^{-3} μm^2, K_1 is initial permeability, 10^{-3} μm^2.

3. Results and Discussion

3.1. Static Performance of Plugging System

3.1.1. Gel Plugging Agent

1. Pre-crosslinking gel particle

Figure 2 shows the particle size test results of the microspheres (WQ-100) dispersed in the formation water and kerosene. WQ-100 delivers good dispersing performance in both formation water and kerosene, has good adaptability to both dispersing media, and has no precipitation occurs. WQ-100 still showed good expansion performance under high temperature and high salt environments, and the particle size distribution expanded from 1–5 to 10–20 μm within 24 h. Additionally, WQ-100 has good stability without swelling, cracking, and cross-linking polymerization within 120 h.

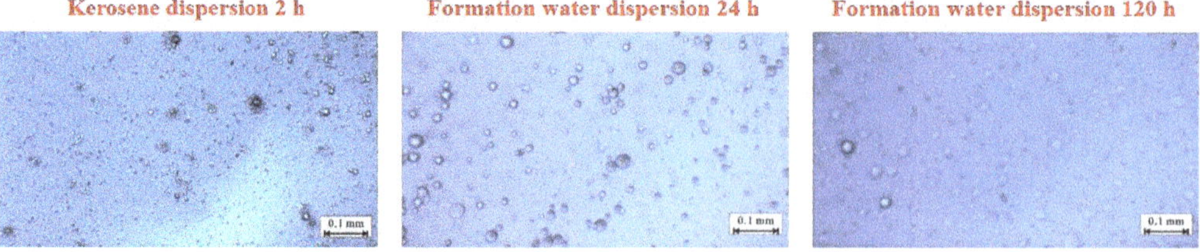

Figure 2. Micrograph of microspheres (WQ-100) after dispersion.

Figure 3 shows the particle size test results after PEG-1 formation water and kerosene are dispersed. When sampling the PEG-1 dispersion system, delamination can be observed, indicating that the dispersibility of the plugging agent in formation water is not excellent. PEG-1 continued to expand within 120 h and the expansion rate gradually decreased with time. The particle size was raised from 0.1 to 0.5 mm and the diameter expansion rate was

500%. The PEG-1 gel particles had good temperature and salt resistance. During the ageing process, the PEG-1 particles continued to absorb water, and the inner folded network of long chains gradually unfolded. At the macroscopic level, the particle volume expanded, the surface gradually became smooth, and the interior of the particles became transparent.

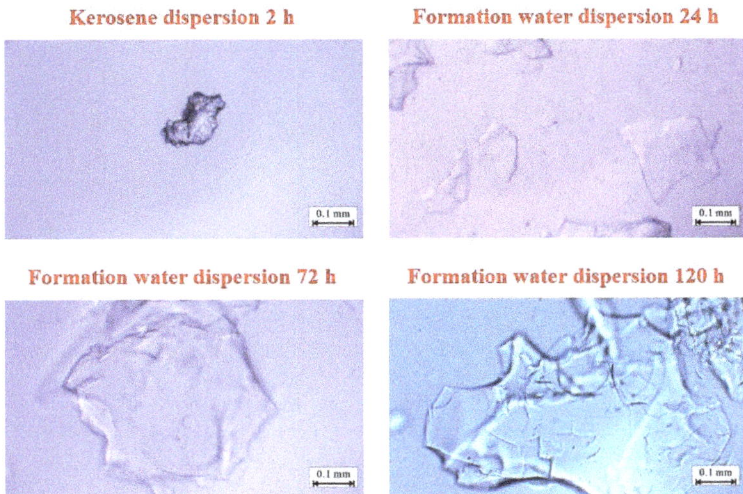

Figure 3. Micrograph of PEG-1 after dispersion.

The particle size test results of bulky particles dispersed in formation water and kerosene are shown in Figure 4. When sampling the bulky particle dispersion system, an apparent layering phenomenon was found, and reliable sampling could be carried out only after stirring. The swelling particle samples continued to expand within 120 h, the expansion rate decreased with time, and the particle size grew from 5 to 10 mm.

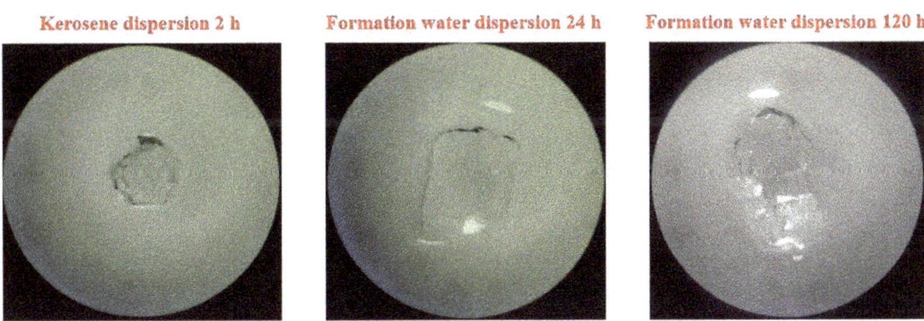

Figure 4. Micrograph of preformed particle gel after dispersion.

2. Gel

After HPAM was dispersed, it was a uniform viscous colloidal fluid. No residue was seen at the bottom of the bottle, indicating that it had good adaptability to the formation water with high salinity. The breakthrough vacuum degree and gel formation time of the gel is shown in Figure 5. With the increase of the mass fraction of HPAM and organic chromium cross-linking agent, the gelation time of the gel was shortened, and the strength of the gel was increased. When the mass fractions of HPAM and organic chromium crosslinking agent are 0.05–0.15% and 0.05–0.25%, respectively, the gelation time is 11–35 h and the gel strength is 0.015–0.027 MPa.

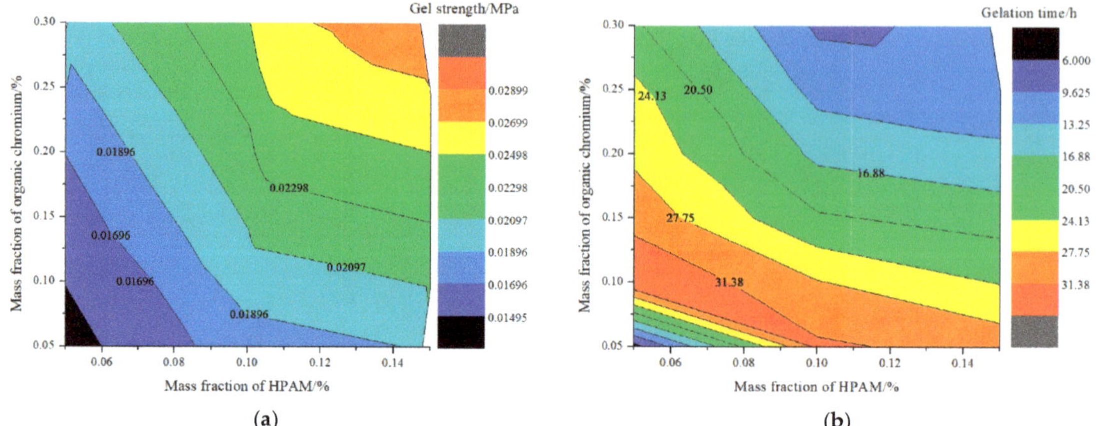

Figure 5. Contour diagrams of static performance of organic chromium gel: (**a**) gel strength, (**b**) gelation time.

The formulation adjustment of the gel system is more flexible. The concentration of each component can be adjusted according to the need to change the gelation time and the breakthrough vacuum degree to achieve the effect of deep displacement regulation and enhanced plugging. There is an obvious negative correlation between the breakthrough vacuum degree of the gel system and the gelation time, as shown in Figure 6. In order to meet the fluidity and blocking performance, the recommended gel system formulation is 0.2% HPAM + 0.2% organic chromium cross-linking agent.

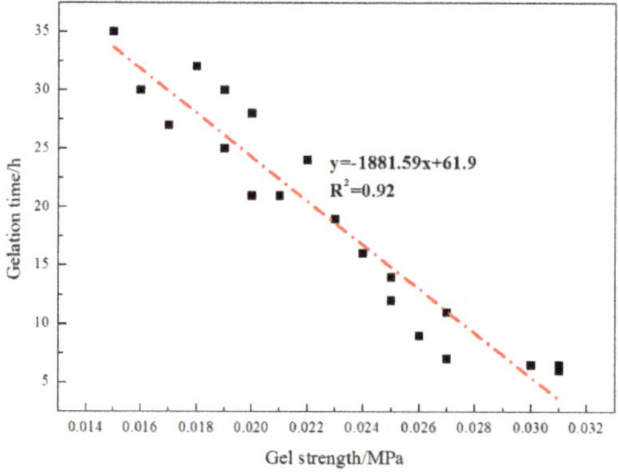

Figure 6. Fitting curves of breakthrough vacuum and gelation time.

3.1.2. Foam Comprehensive Coefficient at Varying Concentrations of Foaming Agent

As a surfactant, a foaming agent can be evenly distributed on the gas–liquid interface and slow down the time required for gas–liquid separation. The foaming performance and half-life results of different concentrations of foaming agents are shown in Table 3 and Figure 7. As the concentration of foaming agent (c) increased from 0.1% to 0.4%, the initial foaming volume increased from 160 mL to 450 mL. At c > 0.4%, the initial bubble volume and elution half-life changed gradually, and after c > 0.8%, the elution half-life decreased. With the increase of the foaming agent concentration, the density of the surfactant on the liquid film gradually increases after the foam is formed. After the monomolecular film is completely adsorbed on the surface of the liquid phase, the bimolecular film begins to form. The concentration of the foaming agent is defined as the critical concentration of the foaming agent. Above this concentration, the comprehensive performance of the foam begins to stabilize.

Table 3. Foam comprehensive coefficient under varying concentrations of foaming agent.

Concentration of Foaming Agent/%	Initial Foaming Volume/mL	Elution Half-Life/s	Foam Composite Value/(mL s)
0.1	160	83	13,280
0.2	230	262	60,260
0.4	450	467	210,150
0.6	480	495	237,600
0.8	500	520	260,000
1.0	520	515	267,800

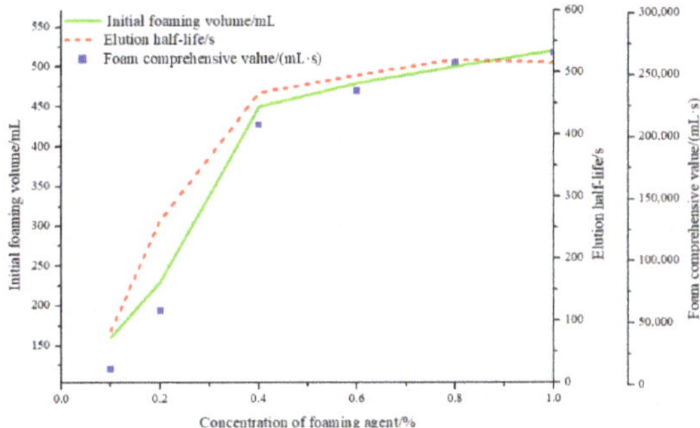

Figure 7. Concentration of foaming agent-initial foaming volume, foam half-life, and composite value.

3.1.3. Binary/Ternary Composite Plugging System

1. Pre-crosslinking gel particle

The microscopic test results of the gel particle compound system are shown in Figure 8. In the compound system of WQ-100 and PEG-1/PPG, the primary performance is adsorption. The adsorption effect of WQ-100 on the surface of PEG-1 particles is stronger than that on the surface of PPG particles. It is worth noting that the adsorption amount of WQ-100 on the PPG surface is much less than the amount deposited on the surface recesses, which indicates that the microspheres are mainly deposited on the surface of PPG particles. In addition, due to the large particle size of PEG-1 and PPG particles, the adsorption effect of PEG-1 on the surface of PPG particles is not obvious, mainly manifested as deposition stacking. WQ-100 can form secondary plugging on the formation after PEG plugging, improving the sweeping flow efficiency.

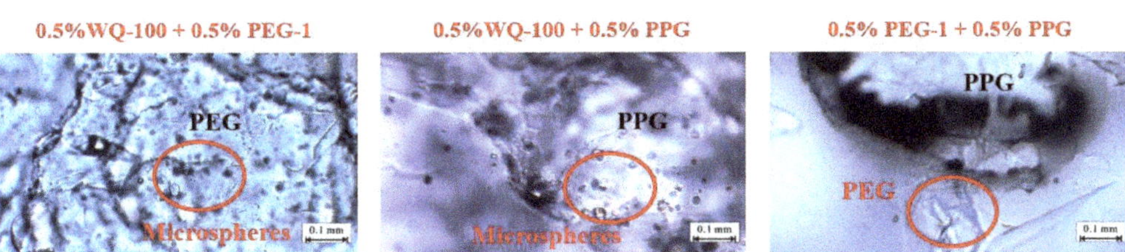

Figure 8. Pre-crosslinking gel particle absorptive effect.

2. Enhanced gels with pre-crosslinking gel particle

The gelation time and strength of the prefabricated gel particles and the polymer gel compound system are shown in Table 4. The relationship between the strength of the composite gel system and the gelation time is shown in Figure 9. Under a certain mass fraction of HPAM and organic chromium cross-linking agent, and with an increased particle size of the added prefabricated gel particle plugging agent, the breakthrough vacuum degree of different compound gel systems under the same gel formation time increases. In particular, the strength of the gel compound system increases more obviously after adding WQ-100.

Table 4. Strength and gelation time of gel compounds system.

HPAM/%	Gel Agent	Organic Chromium Crosslinking Agent/%	Breakthrough Vacuum /MPa	Gelation Time/h
0.2		0.1	0.024	21
		0.2	0.027	12
		0.3	0.031	7
	0.2% WQ-100	0.1	0.038	21.5
		0.2	0.040	12
		0.3	0.042	7.5
	0.2% PEG-1	0.1	0.032	22
		0.2	0.035	12
		0.3	0.037	8
	0.2% PPG	0.1	0.028	22
		0.2	0.032	13
		0.3	0.033	7

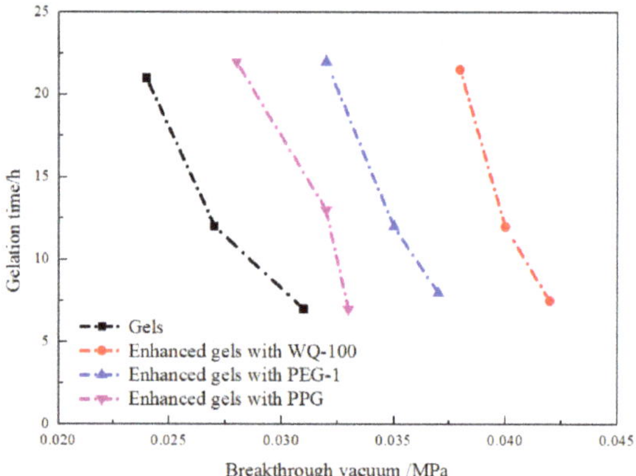

Figure 9. Gels strength and gelation of enhanced gels systems.

The microscopic test results of the enhanced gel systems are shown in Figure 10. The dispersion performance of the WQ-100 system gel is the best and it is evenly distributed. Compared with WQ-100, PEG-1 and PPG are both observed to have obvious granularity, indicating that the dispersion effect of PEG-1 and PPG is poor.

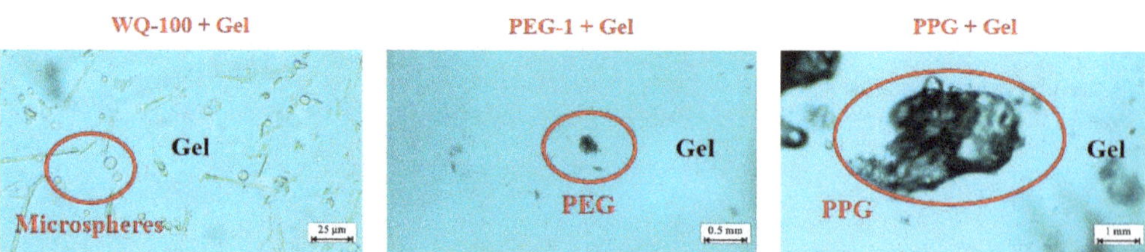

Figure 10. Micrograph of enhanced gels systems.

Generally, there are two ways to strengthen the gel. One is to add micro-nano-scale particles into the HPAM solution to strengthen the gel structure by using it as a skeleton. The second is that the HPAM solution and the added prefabricated gel particles strengthen the gel by forming additional chemical bonds. Figure 11 shows the mechanism of action of nanosphere-enhanced gel and PEG/PPG particle-enhanced gel. The red and yellow areas represent small particle plugging agents such as microspheres and large particle plugging agents such as PEG and PPG, respectively. For the micro-nano-enhanced bulk gel system, if nano-SiO_2 particles are added, it does not participate in the coordination reaction due to their stable chemical properties; it only exists in the form of a gel skeleton [23]. The essence of nano-microspheres is AM and its derivatives with a certain stable structure, so it can not only serve as the skeleton of the gel during the formation of the gel. In addition, the nano-microspheres can form coordination bonds with the colloidal HPAM dispersion system through the cross-linking agent, which leads to the largest improvement in the breakthrough vacuum degree of the gel system with the addition of microspheres [24].

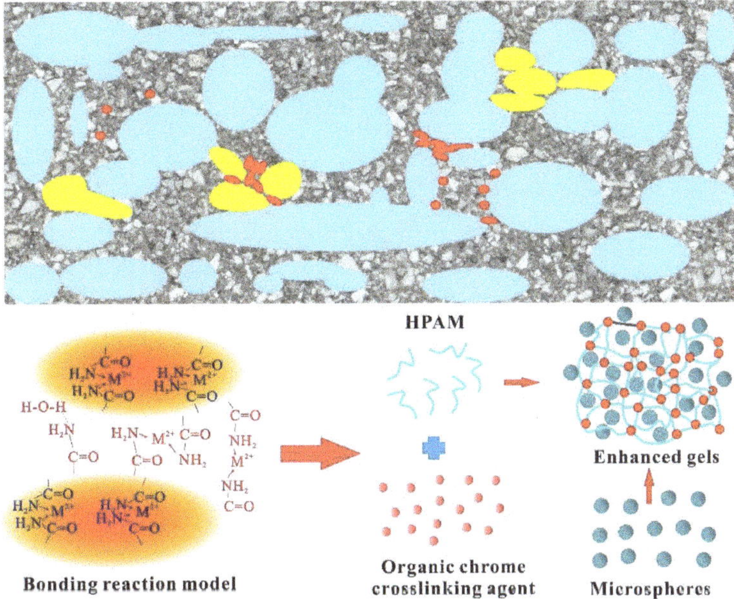

Figure 11. Formation mechanisms of enhanced gels with microspheres.

The structural properties and particle size of PEG enable it to form a reinforced structure in part of the gel. In contrast, the PPG particles can develop coordination with the HPAM dispersion system on its surface, which can both improve the gel strength to a certain extent. However, due to the poor dispersibility of PEG and PPG in the liquid phase, the gel can only be strengthened locally. Meanwhile, the larger particle size of PEG and PPG particles also aggravated the injection difficulty of the gel system.

3. Enhanced foam with pre-crosslinking gel particle and gels

The foaming performance of the reinforced foam system is shown in Figure 12 on the right. The foaming volume of microsphere-reinforced foam is more significant than that of single foam. The foaming volume of the reinforced foam system obtained by compounding other prefabricated particle plugging agents except microspheres is smaller than that of single foam. Combined with the results of the liquid elution half-life test, the foaming agent has good tolerance to the actual working environment with a small amount of white oil in the solution. However, since the specific surface area of PEG particles is larger than that of PPG, more white oil is attached to the surface of the PEG particles during sampling, resulting in a more obvious attenuation of the reduction of the foaming volume and half-life.

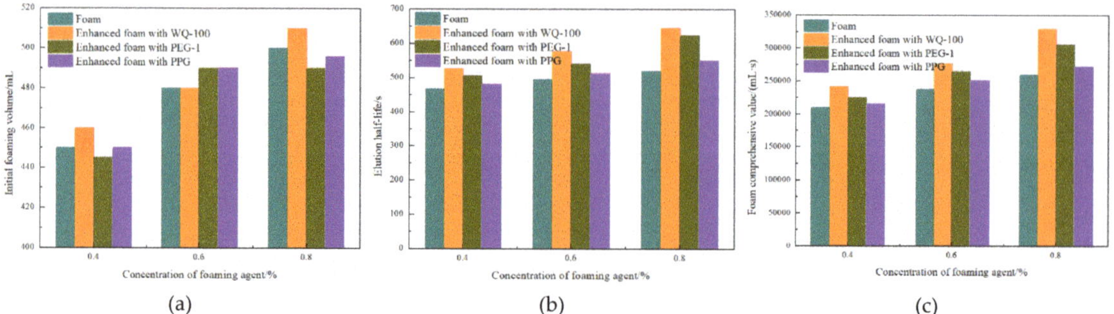

Figure 12. The basic performance of enhanced foam system and single foam: (**a**) foam volume, (**b**) foam liquid chromatography half-life, (**c**) foam composite value.

Sampling and microscopic observation of the reinforced foam system are carried out, and the results are shown in Figure 13. A single conventional foam liquid film is thinner, the particle size distribution of the bubble particles is relatively concentrated, and the liquid phase is clear and transparent. In the PEG and PPG reinforced foam system, it was observed that the particle size homogeneity of the bubble particles decreased significantly, no other particles were seen on the surface of the bubble particles, and the liquid phase was turbid. After PEG and PPG plugging agents were added to the foaming agent system, they did not participate in the foaming process. Only the surface white oil entered the liquid phase, resulting in the deterioration of the foaming performance of the foaming agent.

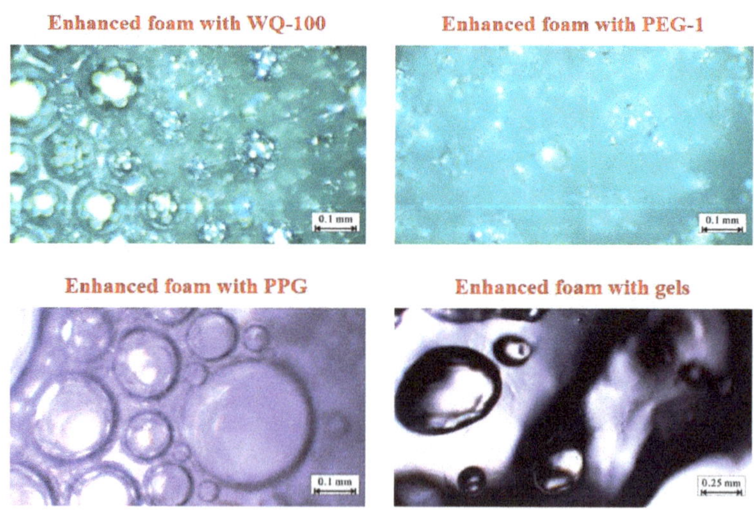

Figure 13. Enhanced foam system with pre-crosslinking gel particle and gels.

In the microsphere-reinforced foam system, it can be seen that a large number of microspheres are adsorbed on the surface of the bubble particles, the particle size of the bubble particles is concentrated, and the particle size is reduced. Since the microspheres are originally dispersed in the oil phase, the liquid phase is slightly turbid, but the weakening effect of the oil relative to the foaming agent is weak. For microsphere-reinforced foam, the microspheres are uniformly dispersed in the foaming agent solution and form a skeleton structure on the liquid film after foaming. During the foaming process, the microspheres absorb water and expand, which reduces the free water in the liquid film and strengthens the ability of the foaming agent to have interfacial tension. Additionally, the polymer surface has a large number of hydroxyl groups, negative ions appear in free water after dispersion, and the diffusion double electron layer appears, strengthening the repulsion between the liquid films and increasing the foam volume [25]. On the other hand, in the process of liquid film dehydration, the microspheres are dehydrated synchronously, resulting in the slowing down of the water loss rate of the foam on the macroscopic level, which is manifested as the prolongation of the half-life of the foam. The synergistic effect of the two mechanisms greatly increases the comprehensive value of the foam. The liquid film of the gel-reinforced foam system is thicker and there is basically no contact between the bubbles. After the foaming, the liquid phase of the gel is still the main body and the foaming effect of the foaming agent on the gel is extremely poor.

3.2. Plugging and EOR Performance of Gel System

3.2.1. Plugging Performance

1. Porous reservoir

Since the porosity reservoir model is a low-porosity and low-permeability reservoir, only the microspheres and PEG systems with smaller particle sizes are selected for the plugging performance evaluation experiment. The plugging effects of microspheres and PEG systems on porous reservoirs with different permeability are shown in Table 5. Compared with microspheres, the PEG system has a poor plugging effect in reservoirs with low permeability but has a stronger plugging effect on reservoirs with high permeability. As the permeability increased from 4.8×10^{-3} μm^2 to 156.7×10^{-3} μm^2, the plugging ability of the microsphere system decreased rapidly and the plugging rate decreased from 92.05% to 20.73%. The PEG system can play a good plugging role in the reservoir with relatively high permeability; the plugging rate decreases from 96.83% to 91.82%.

Table 5. Plugging efficiency of pre-crosslinking gel particle on the porous reservoir.

Plugging System	Initial Permeability /10^{-3} μm^2	Plugging Efficiency /%	Plugging Efficiency after Injection 10 PV Formation Water
0.5% WQ-100 0.5 PV	4.8	92.05	79.21
	15.9	90.52	71.93
	49.6	78.21	56.20
	99.8	52.34	21.54
	156.7	20.73	5.80
0.5% PEG-1 0.5 PV	4.8	End face seal	
	15.9	End face seal	
	49.6	96.83	90.24
	99.8	93.83	82.23
	156.7	91.82	77.53

Microscopic observation was performed on the end face of the model injection end after the PEG system was pressed, and the results are shown in Figure 14. It can be observed that PEG particles accumulate on the end face of the model and fail to enter the model to play a blocking role. Since the PEG system is mainly due to the hydration expansion after injection and the plugging of the hyperpermeable pores by a single particle, it is more inclined to reduce the permeability of the hyperpermeable layer using deep regulation and flooding. In addition, the microsphere-PEG compound system (0.5%, 0.25 PV) was used to simulate the heterogeneous reservoir formed by the interlayer permeability difference through the dual sand-packing tube model, and the effect of different injection sequences on the plugging performance was studied. Taking the reservoir model with a permeability difference of 30 as an example: when the microspheres were injected first, the injection pressure only increased from 0.21 MPa to 0.24 MPa, and when the PEG system continued to be injected, the pressure increased rapidly, reaching as high as 2.48 MPa. PEG was injected first, and the pressure in the model quickly rose to 2.30 MPa. Continued injection of microspheres mainly overcomes the resistance of the PEG particles. Since PEG is displaced into the formation more profoundly by the subsequent microspheres in the system with PEG injected first, it is more difficult for the injected water to flow around. Additionally, through the process of "broken-migration-re-blocking", the plugging of the hyperpermeable layer by PEG particles is more uniform, which further enhances the continuity of the plugging effect.

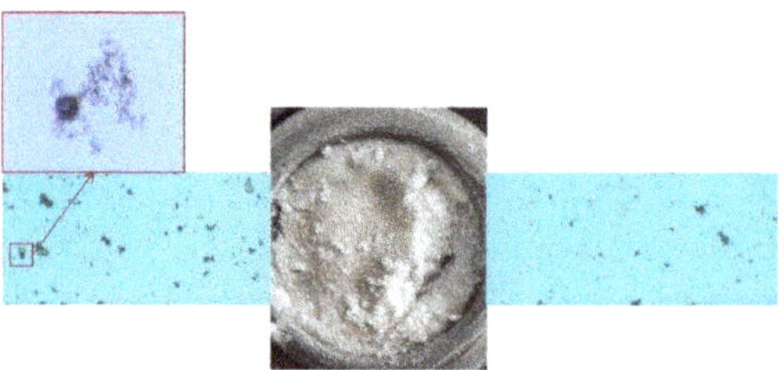

Figure 14. Micrograph of pore formation model's end face and internal after injecting PEG system.

2. Fractured reservoir

Table 6 shows the evaluation results of the plugging performance of a single prefabricated gel particle plugging agent. The injection pressure of the microspheres grows from 1.88×10^{-3} MPa to 2.5×10^{-3} MPa when the fracture opening is 50 μm. However, with the further increase of the fracture opening, its plugging ability disappears. The plugging ability of the PEG system began to decline when the permeability was greater than 180 μm, and the PEG particles quickly flowed out along the fracture, resulting in the final injection pressure of only 1.3 MPa. The plugging effect of microspheres and PEG decreased rapidly with the increase in fracture opening. The plugging rate of large-sized bulk particles and gel systems increased in the range of low fracture opening (50–180 μm).

Table 6. Plugging efficiency of pre-crosslinking gel particle on the fractured reservoir.

Plugging System	Fracture Width /μm	Plugging Efficiency /%	Residual Plugging Efficiency/%
0.5% WQ-100 (0.5 PV)	50	24.80	7.39
	180	16.67	5.17
	340	5.00	0.00
	510	1.23	0.00
0.5% PEG-1 (0.5 PV)	50	94.29	86.88
	180	82.98	38.46
	340	18.18	6.90
	510	13.64	0.72
0.5% PPG (0.5 PV)	50	98.76	91.83
	180	99.76	99.34
	340	99.53	98.49
	510	99.68	98.75
0.2% HPAM + 0.2% Cr^{3+} (0.5 PV)	50	97.11	90.91
	180	99.81	99.18
	340	99.88	99.67
	510	99.91	99.75

The plugging performance of the prefabricated gel particle composite plugging system is shown in Table 7. The microsphere-enhanced gel system showed stronger erosion resistance than the bulk particle-PEG system in formations with larger fracture openings. It showed a higher residual plugging rate at the end of the flooding. For large-scale fractures, the prefabricated gel particle plugging agent sheared and broken by the filler is more likely to be flushed out with the displacing fluid, thus showing insufficient plugging durability for large-scale fractures.

Table 7. Plugging efficiency of multiple pre-crosslinking gel particles on the fractured reservoir.

Plugging System	Fracture Width /μm	Plugging Efficiency /%	Residual Plugging Efficiency/%
0.5% PPG 0.25 PV + 0.5% PEG-1 0.25 PV	50	99.02	97.61
	180	99.85	99.50
	340	99.87	99.66
	510	99.90	99.81
0.5% WQ-100 + 0.2% HPAM + 0.2% Cr^{3+} 0.5 PV	50	98.93	92.85
	180	99.91	98.97
	340	99.90	99.05
	510	99.92	99.13

3. Pore-fractured reservoir

Pore-fractured reservoirs have the characteristics of both porosity and fractured reservoirs. Large-sized plugging agents have a poor injection effect, while small-sized plugging agents are prone to channeling. Figure 15 shows the plugging effect of microspheres, PEG, and their composite systems in the pore and fractured reservoirs. The microsphere-PEG compound system showed an excellent plugging effect on pore fractured reservoirs because the compound system appeared in the 52.2×10^{-3} μm^2, 91.2×10^{-3} μm^2, and 166.7×10^{-3} μm^2 models. The end face is blocked and only the experimental results are obtained for the reservoir with a permeability of 252.7×10^{-3} μm^2. The plugging effect of the granular system on pore and fractured reservoirs is unstable, and a reinforced foam system should be developed for plugging.

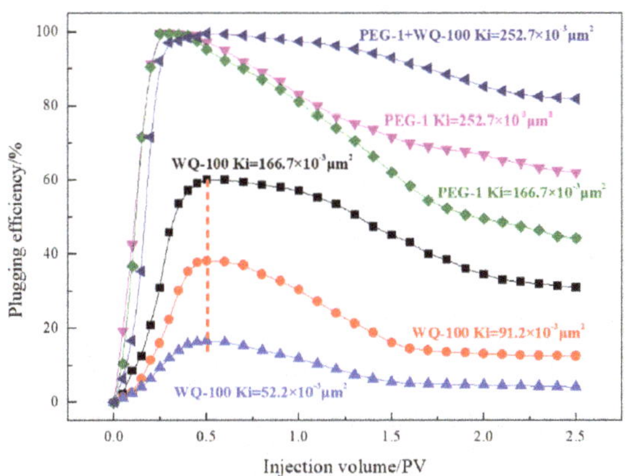

Figure 15. Plugging effects of pre-crosslinking gel particle on the pore-fractured reservoir.

3.2.2. EOR Performance

For different heterogeneous reservoir models, the enhanced oil recovery effect of each gel plugging system is shown in Table 8. Compared with the 2-segment plugging system, the recovery degree of the 4-segment plugging system after the first round of

PEG-microsphere injection was basically the same as that after the 2-segment plugging system was injected with PEG particles. In the first stage, the front-end PEG particles are mainly used to block and increase the swept area. After injecting the second round of PEG-microspheres, a new slug appeared in the 4-segment plugging system, which additionally blocked the inlet section of the high-permeability pipe and further forced the injected fluid to enter the low-permeability pipe for oil displacement. The level of output increased significantly. Subsequent microsphere systems are also mainly turned into low-permeability tubes, relying on the deformation and adsorption capacity of the microspheres to the oil phase to further utilize the residual oil. After plugging, the recovery factors of the three models were increased by 11.49%, 24.14%, and 25.89%, respectively. Increasing the number of slugs had a positive effect on the control and flooding performance. Compared with heterogeneous porosity reservoirs, the recovery degree of fractured reservoirs before plugging is low, only about 16%.

Table 8. Plugging efficiency of multiple pre-crosslinking gel particle on the fractured reservoir.

Reservoir Models		**Plugging System Formula**	**EOR before Plugging /%**	**EOR after Plugging /%**
Porous reservoirs	50/100 mD	0.5 PV WQ-100	39.19	50.68
		0.25 PV PEG + 0.25 PV WQ-100	38.82	64.71
		0.15 PV PEG + 0.15 PV WQ-100 + 0.1 PV PEG + 0.1 PV WQ-100	39.08	63.22
Fractured reservoirs	Fracture width /µm	Plugging system Formula	EOR before plugging /%	EOR after plugging /%
	50	0.5 PV WQ-100	14.63	18.29
	50	0.2 PV PEG + 0.3 PV WQ-100	15.63	37.5
	180		7.5	13.75
	50	0.2 PV PPG + 0.3 PV PEG	16.67	44.44
	180		7.32	29.27
Pore-fractured reservoirs	Permeability /mD	Plugging system Formula	EOR before plugging /%	EOR after plugging /%
	52	0.5 PV WQ-100	43.75	52.5
	167		40.79	44.74
	52	0.2 PV PEG + 0.3 PV WQ-100	46.15	66.67
	167		39.02	56.1

As shown in Figure 16, take the crack width of 50 µm and the PEG-microsphere system as an example; after all the PEG particles were injected, the recovery rate reached 31.56%. The injection of microspheres and the adsorption of PEG particles further blocked the hypertonic channel. In addition, the microspheres entered the low-permeability matrix together with the water phase and elastically deformed in the matrix pores, further producing the residual oil that was not used during the water flooding process. Continued water injection replaced the pre-microsphere slug in the low-permeability reservoir. Because the length of the slug formed by PEG particles failed to seal the entire fracture, the final recovery rate reached 37.50%. For the two reservoir models, the PEG-microsphere system enhanced oil recovery by 21.87% and 6.25%, respectively, and the PPG-PEG system enhanced oil recovery by 27.77% and 21.95%, respectively. The bulky particles have a good ability to control water and profile in fractured reservoirs, and the microsphere system also has a stronger oil washing effect than the single water phase after turning to the matrix.

For pore-fractured reservoirs, after the microsphere system is injected into a reservoir with high fracture density, due to the extremely small particle size and good fluidity, the amount of microspheres per unit fracture area decreases and the bridging instability between microsphere particles intensifies. The injected fluid can only perform secondary oil washing near the wall of the micro-cracks, so it enters the failure stage immediately after stopping the injection of microspheres, and the PEG-microsphere system can form a more reliable plugging. The microsphere system improves the oil recovery of porous and fractured reservoirs with permeability by 8.75% and 3.95%, respectively, and the

PEG-microsphere composite system improves the oil recovery by 20.52% and 17.08%, respectively, as shown in Figure 16. The addition of the PEG front slug effectively blocked the micro-fractures, and the subsequent microsphere system formed adsorption on the PEG surface and further strengthened the diversion of the injected water to the formation.

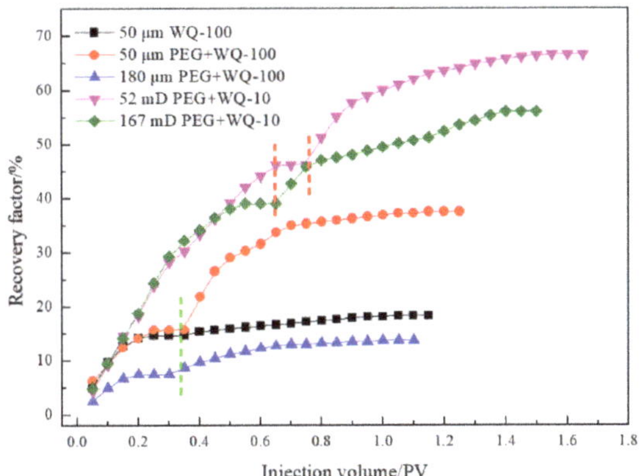

Figure 16. EOR effects of enhanced gels on fractured and pore-fractured reservoir.

3.3. Plugging and EOR Performance of Gel Enhanced Foam System

3.3.1. Plugging Performance

The plugging effects of the enhanced foam system on different reservoirs are shown in Table 9. Compared with the conventional foam system, the microsphere foam system has a higher comprehensive foam value, which means that its blocking performance is stronger than that of the conventional foam system in static experiments. In the microsphere foam system, after the foam fails, the microspheres—as foam stabilizers—continue to play a blocking role. Compared with the microsphere-enhanced foam system, the "break-migration-re-plugging" process of the PEG-microsphere system in the porous reservoir significantly slows down the decay rate of the plugging rate. After the foam system is injected into the fracture, it quickly flows back along the fracture, which is not suitable for formations with a fractured opening greater than 50 μm. Compared with the conventional jelly system, the structure of foam-modified jelly is more complex, and it is more difficult to break through the foam-modified jelly under the same injection amount. Microsphere-enhanced foam can be used for secondary blocking through microspheres to enhance the blocking ability.

3.3.2. EOR Performance

The enhanced oil recovery effect of the enhanced foam system on different heterogeneous reservoir models are shown in Table 10 and Figure 17. Compared with the 4 slug PEG-microsphere plugging system, the conventional foam and the microsphere-enhanced foam system can increase the recovery by 0.20% and 2.16%, respectively. Compared with the PEG-microsphere system, the foam system has a specific surfactant flooding ability while plugging, and the recovery degree is improved more greatly. For microsphere-enhanced foam, effective plugging was not formed in fractured formations with a fractured opening of 180 μm, only foaming agents and microspheres could be used to strengthen oil washing near the fracture surface, and a large amount of injected fluid was lost. The gel-reinforced foam system shows good plugging and profile control performance. Forming a gel system with worse rheology strengthens the ability to divert the injected water, allows the subsequent fluid to enter the depth of the matrix for oil washing, and increases the reserves in the low-permeability matrix [26]. After injection of the gel-enhanced foam system, the recovery factors of fractured reservoirs with fracture widths of 50 and 80 μm increased by 29.41% and 24.39%, respectively. The increase in fracture density leads to an aggravation of the ineffective circulation of injected fluid, which requires higher plugging, regulation, and displacement capability of the same plugging system. Compared with the PEG-microsphere composite system, the recovery degree of the microsphere-enhanced foam increased by 3.92% and 4.35% under the reservoir conditions of 52 mD and 167 mD.

Table 9. Plugging efficiency of enhanced foam system in varying reservoir models.

Reservoirs Models	Permeability/mD	0.25 PV PEG + 0.25 PV WQ-100	0.25 PV Foam + 0.25 PV WQ-100
Porous reservoirs	5/150	Section block 96.1%/97.6% 92.8%/94.5% 91.1%/92.8%	94.2%/97.8% 94.1%/96.3% 91.6%/92.5% 90.4%/90.9%
Reservoirs models	Fracture width/μm	Foam + WQ-100	Enhanced foam with gels + WQ-100
Fractured reservoirs	50 180 340 510	85.5/90.5 50.6/56.8 10.3/13.8 4.8/5.9	94.7/96.4 92.9/94.2 91.6/92.6 89.7/90.5
Reservoirs models	Microfracture/matrix volume ratio	Foam	WQ-100 + Foam
Pore-fractured reservoirs	0.4/295 0.8/295 1.2/295 1.6/295	91.4 89.7 86.1 80.3	93.6 92.8 91.2 90.5

Table 10. EOR efficiency of enhanced foam at varying reservoirs models.

Reservoir Models		Plugging System Formula	EOR before Plugging /%	EOR after Plugging /%
Porous reservoirs	5/100 mD	PEG-WQ-100	39.08	63.22
		Foam	39.13	65.22
		Enhanced foam with WQ-100	39.02	67.07
Fractured reservoirs	Fracture width/μm	Plugging system Formula	EOR before plugging /%	EOR after plugging /%
	50	PEG-WQ-100	15.63	37.5
		PPG-PEG-1	16.67	44.44
		WQ-100	14.63	18.29
		Enhanced foam with WQ-100	17.86	42.86
		Enhanced foam with gels	17.65	47.06
	180	PEG-WQ-100	7.5	13.75
		PPG-PEG-1	7.32	29.27
		Enhanced foam with WQ-100	7.89	16.58
		Enhanced foam with gels	8.54	32.93
Pore-fractured reservoirs	Permeability /mD	Plugging system Formula	EOR before plugging /%	EOR after plugging /%
	52	PEG-WQ-100	46.15	66.67
		Foam	44.44	65.56
		Enhanced foam with WQ-100	46.67	71.11
	167	PEG-WQ-100	39.02	56.1
		Foam	39.02	48.78
		Enhanced foam with WQ-100	42.86	64.29

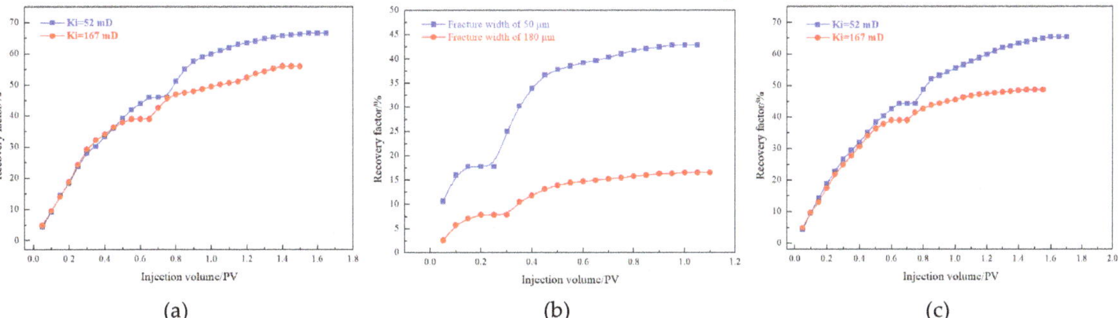

Figure 17. EOR effects of enhanced foam with WQ-100: (**a**) pore reservoirs, (**b**) fracture reservoirs, (**c**) pore-fractured reservoir.

3.4. Construction of Profile and Control System Technique Plate

Combined with the static performance, plugging mechanism, and enhanced oil recovery effect of the reinforced gel and reinforced foam systems, a technical chart of the plugging system was constructed according to the characteristics of different heterogeneous reservoirs, as shown in Figure 18. For pore-fractured reservoirs with a permeability of 50–200 mD, considering the stability of injection, the microsphere-reinforced foam system is still preferred to strengthen the plugging of micro-fractures and play the role of microspheres in regulating plugging. For pore-fractured reservoirs with a permeability higher than 200 mD, the reservoirs have begun to show obvious fractured reservoir characteristics, so the PEG-microsphere system with good injection performance should be prioritized to plug microfractures. For fractured reservoirs with fracture width less than 100 μm (permeability is about 400,000 × 10^{-3} μm^2 according to the empirical formula, which is affected by porosity in fractures), considering the effectiveness of plugging and reducing injection resistance, the use of microspheres-reinforced foam system or PEG-microsphere system. Considering the stronger mechanical strength of microsphere-enhanced gel, microsphere-enhanced gel can be used in fractured reservoirs with a fractured opening of more than 300 μm to enhance the plugging effect. Alternatively, the formation of bulk swelling particles and PEG particles can comprehensively strengthen the plugging of fractures and improve the fluid flow diversion ability, followed by the use of microsphere flooding to improve the oil washing efficiency of the remaining oil in the low-permeability matrix and comprehensively improve the recovery factor.

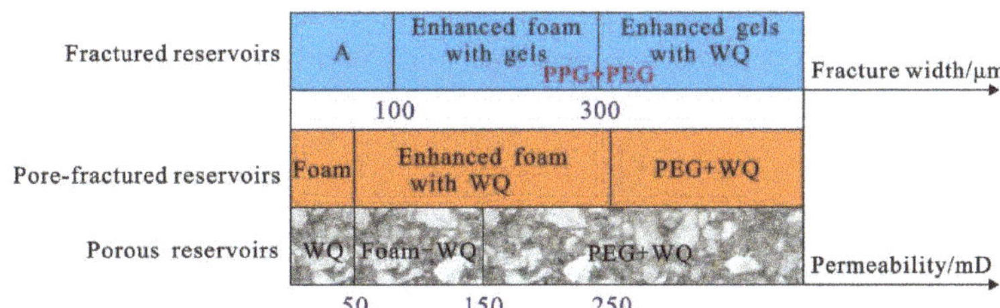

Figure 18. Technique boundary chart of plugging and flooding system in heterogeneous reservoirs.

4. Conclusions

The conclusions of this work are as follows.

(1) The adsorption effect of microspheres (WQ-100) on the surface of PEG-1 is more potent than that of PPG, and the deposition is mainly on the surface of PPG. The

adsorption effect of PEG-1 on the surface of PPG is not apparent, primarily manifested as deposition stacking.

(2) The gel was synthesized with 0.2% HPAM + 0.2% organic chromium cross-linking agent, and the strength of enhanced gel with WQ-100 was higher than that of PEG-1 and PPG. The comprehensive value of WQ-100 reinforced foam is greater than that of PEG-1, and PPG reinforced foam and the enhanced foam with gel has a thick liquid film and poor foaming effect.

(3) For the heterogeneous porous reservoir with the permeability of 5/100 mD, the enhanced foam with WQ-100 shows better performance in plugging control and flooding, and the recovery factor increases by 28.05%. The improved foam with gel enhances the fluid flow diversion ability and the recovery factor of fractured reservoirs with fracture widths of 50 μm and 180 μm increases by 29.41% and 24.39%.

(4) For pore-fractured reservoirs with a permeability of 52/167 mD, the PEG + WQ-100 microsphere and enhanced foam with WQ-100 systems show better plugging and recovering performance, and the recovery factor increases are 20.52% and 17.08%, 24.44% and 21.43%, respectively.

(5) The plugging performance of the composite gel system is stronger than that of the enhanced gel with foam. However, the oil displacement performance of the gel-enhanced foam is better than that of the composite gel system due to the "plug-flooding-integrated" feature of the foam.

Author Contributions: L.-L.W.: investigation, writing—original draft, formal analysis. T.-F.W.: methodology, data curation, validation, writing—review & editing, conceptualization. J.-X.W.: methodology, conceptualization. H.-T.T.: methodology, investigation, formal analysis. Y.C.: investigation, formal analysis, data curation, validation. W.S.: methodology, investigation. All authors have read and agreed to the published version of the manuscript.

Funding: This research was funded by the Natural Science Foundation of Shandong Province (ZR2021ME006), the Opening Fund of Shandong Key Laboratory of Oilfield Chemistry, and the Fundamental Research Funds for the Central Universities (19CX05006A). The authors also thank the anonymous reviewers for their valuable comments.

Institutional Review Board Statement: Not applicable.

Informed Consent Statement: Not applicable.

Data Availability Statement: Not applicable.

Acknowledgments: L. Wang and T. Wang acknowledge financial support from the the Natural Science Foundation of Shandong Province (ZR2021ME006). And J. Wang, H. Tian, Y. Chen and W. Song acknowledge financial support from the Fundamental Research Funds for the Central Universities (19CX05006A).

Conflicts of Interest: The authors declare no conflict of interest.

References

1. Wang, J.T.; Wang, Z.Y.; Sun, B.J.; Gao, Y.H.; Wang, X.; Fu, W.Q. Optimization design of hydraulic parameters for supercritical CO_2 fracturing in unconventional gas reservoir. *Fuel* **2019**, *235*, 795–809. [CrossRef]
2. Tian, Y.; Uzun, O.; Shen, Y.Z.; Lei, Z.D.; Yuan, J.R.; Chen, J.H.; Kazemi, H.; Wu, Y.S. Feasibility study of gas injection in low permeability reservoirs of Changqing oilfield. *Fuel* **2020**, *274*, 117831. [CrossRef]
3. Jia, D.L.; Liu, H.; Zhang, J.Q.; Gong, B.; Pei, X.H.; Wang, Q.B.; Yang, Q.H. Data-driven optimization for fine water injection in a mature oil field. *Petrol. Explor. Dev.* **2020**, *47*, 674–682. [CrossRef]
4. Yang, J.B.; Bai, Y.R.; Sun, J.S.; Lv, K.H.; Han, J.L.; Dai, L.Y. Experimental study on physicochemical properties of a shear thixotropic polymer gel for lost circulation control. *Gels* **2022**, *8*, 229. [CrossRef]
5. Zhu, D.Y.; Bai, B.J.; Hou, J.R. Polymer gel systems for water management in high-temperature petroleum reservoirs: A chemical review. *Energy Fuels* **2017**, *31*, 13063–13087. [CrossRef]
6. Xie, K.; Su, C.; Liu, C.L.; Cao, W.J.; He, X.; Ding, H.N.; Mei, J.; Yan, K.; Cheng, Q.; Lu, X.G. Synthesis and performance evaluation of an organic/inorganic composite gel plugging system for offshore oilfields. *ACS Omega* **2022**, *7*, 12870–12878. [CrossRef]
7. Jia, H.; Kang, Z.; Zhu, J.Z.; Ren, L.L.; Cai, M.J.; Wang, T.; Xu, Y.B.; Li, Z.J. High density bromide-based nanocomposite gel for temporary plugging in fractured reservoirs with multi-pressure systems. *J. Pet. Sci. Eng.* **2021**, *205*, 108778. [CrossRef]

8. Xu, T.H.; Wang, J.; Bai, Z.W.; Wang, J.H.; Zhao, P. Plugging performance of a new sulfonated tannin gel system applied in tight oil reservoir. *Geofluids* **2022**, *2022*, 3602242. [CrossRef]
9. Gao, H.; Xu, R.Z.; Xie, Y.G. Quantitative evaluation of the plugging effect of the gel particle system flooding agent using NMR technique. *Energy Fuels* **2020**, *34*, 4329–4337. [CrossRef]
10. Zhao, G.; Dai, C.; Chen, A.; Yan, Z.; Zhao, M. Experimental study and application of gels formed by nonionic polyacrylamide and phenolic resin for in-depth profile control. *J. Pet. Sci. Eng.* **2015**, *135*, 552–560. [CrossRef]
11. Jia, H.; Niu, C.C.; Yang, X.Y. Improved understanding nanocomposite gel working mechanisms: From laboratory investigation to wellbore plugging application. *J. Pet. Sci. Eng.* **2020**, *191*, 107214. [CrossRef]
12. Huang, B.; Zhang, W.S.; Zhou, Q.; Fu, C.; He, S.B. Preparation and experimental study of a low-initial-viscosity gel plugging agent. *ACS Omega* **2020**, *5*, 15715–15727. [CrossRef]
13. Liu, J.B.; Zhong, L.G.; Wang, C.; Li, S.H.; Yuan, X.N.; Liu, Y.G.; Meng, X.H.; Zou, J.; Wang, Q.X. Investigation of a high temperature gel system for application in saline oil and gas reservoirs for profile modification. *J. Pet. Sci. Eng.* **2021**, *202*, 108416. [CrossRef]
14. Sheng, Y.J.; Peng, Y.C.; Zhang, S.W.; Guo, Y.; Ma, L.; Wang, Q.H.; Zhang, H.L. Study on thermal stability of gel foam co-stabilized by hydrophilic silica nanoparticles and surfactants. *Gels* **2022**, *8*, 123. [CrossRef]
15. Li, X.; Pu, C.S.; Chen, X. A novel foam system stabilized by hydroxylated multiwalled carbon nanotubes for enhanced oil recovery: Preparation, characterization and evaluation. *Colloids Surfaces A Phys. Eng. Asp.* **2022**, *632*, 127804. [CrossRef]
16. Lai, N.J.; Zhao, J.; Zhu, Y.Q.; Wen, Y.P.; Huang, Y.J.; Han, J.H. Influence of different oil types on the stability and oil displacement performance of gel foams. *Colloids Surfaces A Phys. Eng. Asp.* **2021**, *630*, 127674. [CrossRef]
17. Zhang, F.; Schechter, D.S. Gas and foam injection with CO_2 and enriched NGL's for enhanced oil recovery in unconventional liquid reservoirs. *J. Pet. Sci. Eng.* **2021**, *202*, 108472. [CrossRef]
18. Li, D.Q.; Li, F.; Deng, S.; Liu, J.H.; Huang, Y.H.; Yang, S. Preparation of oil-based foam gel with nano-SiO_2 as foam stabilizer and evaluation of its performance as a plugging agent for fractured shale. *Geofluids* **2022**, *2022*, 9539999. [CrossRef]
19. Jin, F.Y.; Jiang, T.T.; Varfolomeev, M.A.; Yuan, C.D.; Zhao, H.Y.; He, L.; Jiao, B.L.; Du, D.J.; Xie, Q. Novel preformed gel particles with controllable density and its implications for EOR in fractured-vuggy carbonated reservoirs. *J. Pet. Sci. Eng.* **2021**, *205*, 108903. [CrossRef]
20. Dijvejin, Z.A.; Ghaffarkhah, A.; Sadeghnejad, S.; Sefti, M.V. Effect of silica nanoparticle size on the mechanical strength and wellbore plugging performance of SPAM/chromium (III) acetate nanocomposite gels. *Polym. J.* **2019**, *51*, 693–707. [CrossRef]
21. Ren, G.W. Assess the potentials of CO2 soluble surfactant when applying supercritical CO_2 foam. Part I: Effects of dual phase partition. *Fuel* **2020**, *277*, 118086. [CrossRef]
22. Sun, H.Q.; Wang, Z.W.; Sun, Y.H.; Wu, G.H.; Sun, B.Q.; Sha, Y. Laboratory evaluation of an efficient low interfacial tension foaming agent for enhanced oil recovery in high temperature flue-gas foam flooding. *J. Pet. Sci. Eng.* **2020**, *195*, 107580. [CrossRef]
23. Li, Q.; Yu, X.R.; Wang, L.L.; Qu, S.S.; Wu, W.C.; Ji, R.J.; Luo, Y.; Yang, H. Nano-silica hybrid polyacrylamide/polyethylenimine gel for enhanced oil recovery at harsh conditions. *Colloids Surfaces A Phys. Eng. Asp.* **2022**, *633*, 127898. [CrossRef]
24. Lin, R.Y.; Luo, P.Y.; Sun, Y.; Pan, Y.; Sun, L. Experimental study on the optimization of multi-level nano–microsphere deep profile control in the process of gas injection in fracture-type buried-hill reservoirs. *ACS Omega* **2021**, *6*, 24185–24195. [CrossRef]
25. Liang, S.; Hu, S.Q.; Li, J.; Xu, G.R.; Zhang, B.; Zhao, Y.D.; Yan, H.; Li, J.Y. Study on EOR method in offshore oilfield: Combination of polymer microspheres flooding and nitrogen foam flooding. *J. Pet. Sci. Eng.* **2019**, *178*, 629–639. [CrossRef]
26. Wang, T.F.; Fan, H.M.; Yang, W.P.; Meng, Z. Stabilization mechanism of fly ash three-phase foam and its sealing capacity on fractured reservoirs. *Fuel* **2020**, *264*, 116832. [CrossRef]

Article

Soft Movable Polymer Gel for Controlling Water Coning of Horizontal Well in Offshore Heavy Oil Cold Production

Jie Qu [1], Pan Wang [1], Qing You [1,*], Guang Zhao [2], Yongpeng Sun [2] and Yifei Liu [2,*]

1 School of Energy Resources, China University of Geosciences, Beijing 100083, China; 2006210060@email.cugb.edu.cn (J.Q.); wangpanworkmail@gmail.com (P.W.)
2 School of Petroleum Engineering, China University of Petroleum (East China), Qingdao 266580, China; zhaoguang@upc.edu.cn (G.Z.); sunyongpeng@upc.edu.cn (Y.S.)
* Correspondence: youqing@cugb.edu.cn (Q.Y.); liuyifei@upc.edu.cn (Y.L.)

Abstract: Horizontal well water coning in offshore fields is one of the most common causes of rapid declines in crude oil production and, even more critical, can lead to oil well shut down. The offshore Y oil field with a water cut of 94.7% urgently needs horizontal well water control. However, it is a challenge for polymer gels to meet the requirements of low-temperature (55 °C) gelation and mobility to control water in a wider range. This paper introduced a novel polymer gel cross-linked by hydrolyzed polyacrylamide and chromium acetate and phenolic resin for water coning control of a horizontal well. The detailed gelant formula and treatment method of water coning control for a horizontal well in an offshore field was established. The optimized gelant formula was 0.30~0.45% HPAM + 0.30~0.5% phenolic resin + 0.10~0.15% chromium acetate, with corresponding gelation time of 26~34 h at 55 °C. The results showed that this gel has a compact network structure and excellent creep property, and it can play an efficient water control role in the microscopic model. The thus-optimized gelants were successively injected with injection volumes of 500.0 m^3. The displacement fluid was used to displace gelants into the lost zone away from the oil zone. Then, the formed gel can be worked as the chemical packer in the oil–water interface to control water coning after shutting in for 4 days of gelation. The oil-field monitoring data indicated that the oil rate increased from 9.2 m^3/d to 20.0 m^3/d, the average water cut decreased to 60~70% after treatment, and the cumulative oil production could obtain 1.035×10^4 t instead of 3.9×10^3 t.

Keywords: polymer gel; water coning; chemical packer; horizontal well; offshore oil-field application

Citation: Qu, J.; Wang, P.; You, Q.; Zhao, G.; Sun, Y.; Liu, Y. Soft Movable Polymer Gel for Controlling Water Coning of Horizontal Well in Offshore Heavy Oil Cold Production. *Gels* **2022**, *8*, 352. https://doi.org/10.3390/gels8060352

Academic Editor: Mario Grassi

Received: 8 May 2022
Accepted: 31 May 2022
Published: 5 June 2022

Publisher's Note: MDPI stays neutral with regard to jurisdictional claims in published maps and institutional affiliations.

1. Introduction

Waterflooding is the most economical and commonly used method for oil fields, especially for offshore oil fields with high equipment investment [1–3]. Due to the reservoir heterogeneity, it is inevitable that injected water will break into the oil well at the later stage of development, resulting in a high water production rate. Multi-branch, long-distance horizontal wells are the main way of drilling and production in offshore oil fields [4–8]. This is mainly due to the rational use of the operation space, saving resources, and increasing the oil drainage area of the reservoir. Once the formation water enters the horizontal well, the water cut increases sharply. A high level of water production leads to an increase in corrosion and scale, a high load of fluid-handling facilities, and significant environmental protection concerns, which eventually results in shut-in wells without economic benefit [9–12]. Therefore, efficient and low-cost water control methods are particularly important for the development of offshore oil fields.

Polymer gel treatment is a cost-effective method to improve sweep efficiency in reservoirs and reduce excess water produced during oil and gas production. Polymer gel treatment can also improve the rheology and composition of crude oil [13–16]. The gel formed by polymer and cross-linker can be injected into the target formation with high water cut and fully or partially seal the formation at reservoir temperature [17–20] Polymer gels prepared with chromium acetate or phenolic resin are commonly used and relatively stable systems, but their gelation temperatures are very different [21–24]. Polymer gels prepared with chromium acetate or phenolic resin are commonly used and relatively stable systems, but their gelation temperature is very different. The activation temperature of phenol/formaldehyde cross-linked by acrylamide-based copolymer is above 80 °C which is suitable for use in middle-/high-temperature oil reservoirs. When the reservoir temperature is low, the cross-linking condensation reaction of this system will slow down and better formation effects cannot be guaranteed [25,26]. Since the formation of brine will affect the gelation time and stability of the dynamic gel, the researchers tried to develop a salt-insensitive gel system using a special salt-insensitive modified polymer (PAN, PVA, PAMPS, PSSA, etc.) [27]. Metal-based cross-linking agents (Cr^{3+}, Al^{3+}) are widely used in low-temperature reservoirs due to their low price, but the gelation time will decrease rapidly with the increase in gel temperature, which will lead to the problem of an uneven density of cross-links and poor gel stability. Organometallic cross-linkers can increase gelation time by slow release of metal ions [28,29]. Although the gel reaction rate can be controlled to some extent, the gel formed by metal complexation has poor toughness. It is more susceptible to damage during formation migration [30]. Moreover, the polymer gel should have good deformability, in addition to meeting the conditions of gel strength and gelation time, which easily leads to deeply migrating into the water-yielding formation and plugging water channels for enhanced oil recovery. The low-temperature composite cross-linked soft movable polymer gel can be cross-linked multiple times after being destroyed during the migration process. This gel with excellent strength and creep properties is more suitable for controlling water coning of horizontal wells in offshore oil fields.

The Y oil field is an offshore heavy oil field that has been put into scale development It is a Neogene marine sandstone reservoir. HX well is situated in the formation $NgII_{11+2}$. For the formation $NgII_{11+2}$, the reservoir temperature is 55 °C, the reservoir pressure is about 13.5 MPa, the average porosity is 30.0%, the effective permeability is 2.1 μm^2, and the salinity of formation water is 34,178.2 mg/L. The effective thickness of the sandstone reservoir is about 8.4 m. There is a water layer with sufficient bottom water energy under the production layer ($NgII_{11+2}$), resulting in a sharp rise in the water content of the production well. Recently, the water cut of the HX well has been up to 100%. When the bottom water breaks through, the great adverse mobility ratio of water to oil also promotes a sharp increase in water production and a significant decrease in oil production. In order to control water production, most horizontal wells need to take measures to block water channels and increase oil production. In this study, a novel soft movable polymer gel was introduced for controlling water coning of horizontal well in offshore heavy oil cold production. The strength and creep properties of gel were investigated, and its rheology and micromorphology are evaluated. Additionally, the treatment strategy was simulated and verified through a microscopic visual model. The oil-field application was implemented using the optimized treatment, and finally, good application results were achieved.

2. Experimental

2.1. Experimental Materials

Partially hydrolyzed polyacrylamide (HPAM, M_w: 1.2 × 10^7 Da, degree of hydrolysis: ~23%), chromium acetate (Cr^{3+}) acetate cross-linker, and phenolic resin (PR, M_w: 2 × 10^4~3 × 10^4 Da, sulfonation degree: ~20%) cross-linker were purchased from Shida Oil Field Service Company, Dongying, China. All samples were industrial grade (~95%). Y oil field HX well of reservoir temperature is 55 °C, and the formation water composition is shown in Table 1.

Table 1. Composition of Y oil-field formation water.

Components	Na^+ and K^+	Ca^{2+}	Mg^{2+}	HCO_3^-	SO_4^{2-}	Cl^-
Concentration (mg/L)	10,516.2	423.5	1524.6	139.8	2821.6	18,752.5

2.2. Preparation of Brine

High-density brine is an indispensable working fluid, which is usually determined by the density of the formation water. High-density brine is formulated by dissolving soluble inorganic salts. High-density brine used in oil-field construction usually has the following characteristics: (1) the density of brine is higher than that of the gelants; (2) the brine and formation water are compatible, and the viscosity is higher than that of the formation water; (3) the inorganic salts for preparing brine are cheap and environmentally friendly. Therefore, sodium chloride (NaCl) is an ideal water-soluble inorganic salt to prepare a high-density brine.

Gel and brine were prepared as the basis of the material, and then the performance of the gel was systematically studied. The flowchart of methodology is shown in Figure 1.

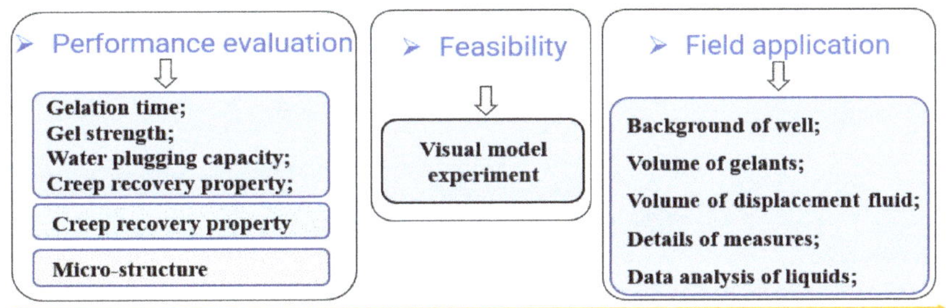

Figure 1. Flowchart of methodology.

2.3. Optimum Formula of Soft Movable Polymer Gels

(1) Gelation time

The polymer (0.3~0.45%) and cross-linker (0.3~0.6% PR + 0.05~0.20% Cr) were dissolved in tap water, and the pH value of the cross-linked glue solution is 6~6.5. Then, the mixed solution was placed in an oven at a temperature of 55 °C until the gel was completely gelled. Temperature triggered chromium ion release forms complex with a polymer carboxyl group, and phenolic resin is cross-linked with an amide group. Gelation time is the time when the gelants change from solution into code E according to the macroscopic visual inspection method (Sydansk's method) commonly used in the oil field [31,32], as shown in Table 2. The gel strength was characterized by the code from A to G.

(2) Water-plugging capacity

After the gelants enter the core pores, the polymer gels gelled by gelants can block the channeling channel. The water-plugging capacity was evaluated by comparing the permeability of the core after injection of gelants with the permeability of the original core. The schematic of the experimental setup is shown in Figure 2. The experimental procedures are as follows: (1) a sandbag was filled with dry sand as a reservoir unit, and its weight was measured; (2) the sandbag was evacuated overnight, saturated with water, and weighed. The volume of water (pore volume) was calculated from the mass of incoming water; (3) waterflooding was conducted, and the stable pressure was recorded; (4) 0.3 pore volume (PV) gelants were injected into a sand pack and placed into an oven, with reservoir temperature for a certain time; (5) waterflooding was implemented again, and the stable pressure was recorded; (6) the water-plugging capacity (C) and the residual resistance factor (F_{rr}) of polymer gel were calculated, using Equations (1) and (2) as follows:

$$C = (1 - k_b/k_a) \times 100\% \quad (1)$$

$$F_{rr} = k_b/k_a = \Delta P_a/\Delta P_b \quad (2)$$

where k_b—permeability of the sandbag before plugging; k_a—permeability of the sandbag after plugging.

Table 2. Gel strength ratings and descriptions [17].

Code	Detailed Gel Description
A	**No detectable gel formed:** gel has the same viscosity as the original polymer solution.
B	**Highly flowing gel:** gel is slightly more viscous than the original polymer solution.
C	**Flowing gel:** most of the gel flows to the vial/tube top upon inversion.
D	**Moderately flowing gel:** only a small portion (5~15%) of the gel does not readily flow to the vial/tube top upon inversion.
E	**Barely flowing gel:** gel barely flows to the vial/tube top; a significant portion (>15%) of the gel does not flow upon inversion.
F	**Highly deformable nonflowing gel:** gel does not quite reach the vial/tube top upon inversion.
G	**Moderately deformable nonflowing gel:** gel deforms about half the distance to the vial/tube top upon inversion.
H	**Slightly deformable nonflowing gel:** gel surface slightly deforms upon inversion.
I	**Rigid gel:** no gel surface deformation occurs upon inversion.
J	**Ringing rigid gel:** a tuning fork-like mechanical vibration occurs upon tapping.

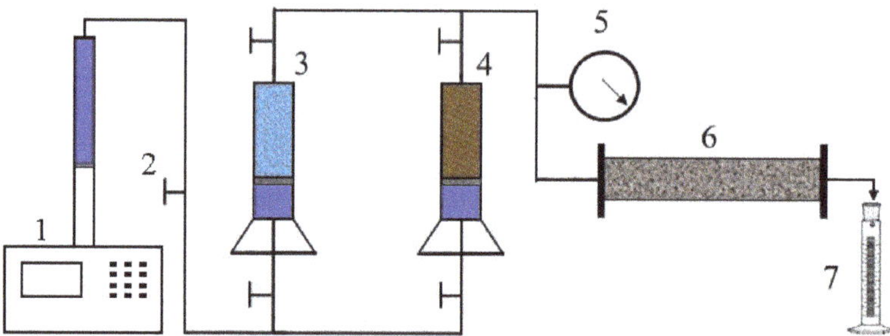

Figure 2. Schematic of the experimental set-up: 1—pump, 2—valve, 3—brine water container, 4—treatment fluid contianer, 5—pressure meter, 6—sand pack, and 7—graduated cylinder.

2.4. *Creep Recovery Property*

The creep and recovery properties of elastic polymer gels were measured using an MCR302 rheometer (Annton Parr Company, Germany), with a plate–plate at 55 °C. Constant shear stress of 2.4 Pa was applied on polymer gel, and the linear viscoelastic range was obtained. The strain of elastic polymer gel versus time was studied after eliminating the constant shear stress.

2.5. Microstructure of Gel

The gel microstructure was observed via environmental Quanta 200 FEG scanning electron microscopy (ESEM) (FEI Company, Hillsboro, OR, USA). During the experiments, a small piece of gel sample was directly placed on a covered ESEM grid at 25 °C.

2.6. Visual Physical Simulation Experiments

The visual physical model system was designed and built, as shown in Figure 3. The glass plate model consisted of an etched base glass plate combined with a cover glass plate, the boundaries of which were sealed with sealant. There were three ports at the bottom of the model—two water injection ports were located in the lower part, and one water outlet is located in the upper part. After the pumping device and the observation device were connected, the airtightness of the pipeline was checked, and the experiment of bottom-water coning and control in a horizontal well was simulated.

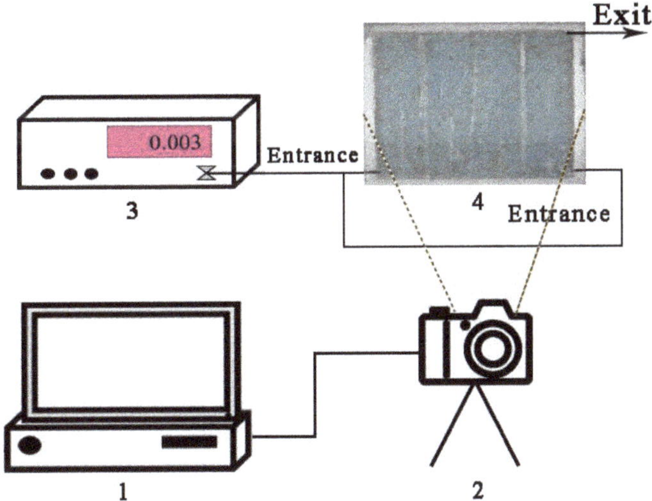

Figure 3. Schematic of visual physical simulation model: 1—computer, 2—video camera system, 3—syringe pump, and 4—model.

2.7. Flowchart of Soft Movable Polymer Gel for Controlling Water Coning in Horizontal Well

As shown in Figure 4, the horizontal well was located in the middle of the oil layer on a reservoir, and bottom water broke through and entered the wellbore from bottom to top (Figure 4a,b). Gravity separation is a core principle of controlling bottom-water coning technology, which relies on the density difference between fluids (oil and water) to cause gravity separation in the reservoir. In this study, first, high-density brine (0.99~1.02 g/cm^3) was injected into the reservoir through the wellbore, and it mainly migrated to the water–oil transition zone to protect the polymer gel from entering the bottom of the water layer (Figure 4c). In the second step, the gelants with a density (0.97~0.98 g/cm^3) between oil and brine were injected into the penetration zone (Figure 4d). Then, displacement fluid (~1.00 g/cm^3) was injected to displace the gelants in the wellbore into the water coning zone and away from the oil layer of the reservoir (Figure 4e). Finally, after the displacement fluid was flushed away with fresh water from the mine, the horizontal well was shut in for gelation, completing a chemical gel packer (Figure 4f).

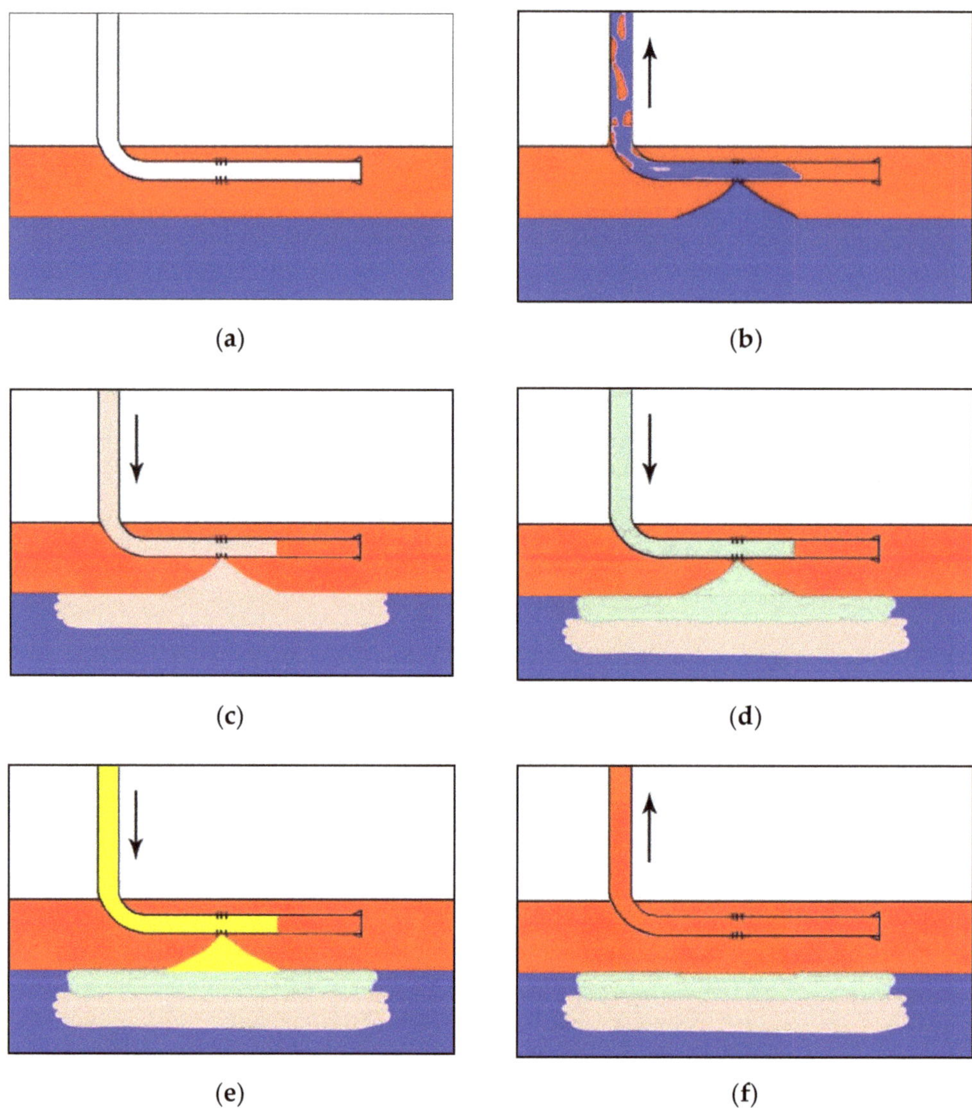

Figure 4. Schematic of polymer gel for controlling bottom-water coning in horizontal well: (**a**) reservoir with bottom water; (**b**) bottom water coning; (**c**) injecting high-density brine; (**d**) injecting the gelants; (**e**) injecting displacement fluid; (**f**) restarting well production: oil layer; aquifer; high-density brine; gelant; displacement fluid.

2.8. Construction Plan of HX Well

The water and oil production data of oil wells in recent years were first analyzed, after which the gel water control plan was determined. After calculating the usage of the three liquids, three working fluids had to be injected in sequence, mainly including high-density brine, gelants, and displacement fluid. The high-density brine was first injected into the bottom water layer to increase the density difference between the oil and water phases.

The gelants of polymer gel were injected into the bottom water layer through the water-cone water channel to act as a plug. The displacement fluid can drive the gelants of the wellbore and oil layer into the water layer to keep the oil flow channel unobstructed. After the injection of the three kinds of liquids, the HX well was shut in, which transformed the gelling agent form into a chemical gel packer near the interface between the oil and water layers, blocking the bottom-water coning. Finally, the data of water and oil production of oil wells after adopting gel water control were analyzed.

3. Results and Discussion

3.1. Density and Viscosity Properties of Brine

The density and viscosity of brine as a function of salt mass fraction were measured, as shown in Figure 5. The density of the brine should be higher than that of the formation water in Table 1, so the mass fraction of sodium chloride in the preparation of high-density brine should be higher than 8%. Furthermore, there are two factors to consider. On the one hand, high-density brine may have a negative impact on the gelation of the gelants; on the other hand, the density between the high-density saline and transition zone also decreases exponentially during high-density saline and gel injection. The calculation found that when the injected high-density saline was 8% NaCl solution, the transition-density saline was half of that in high-density saline.

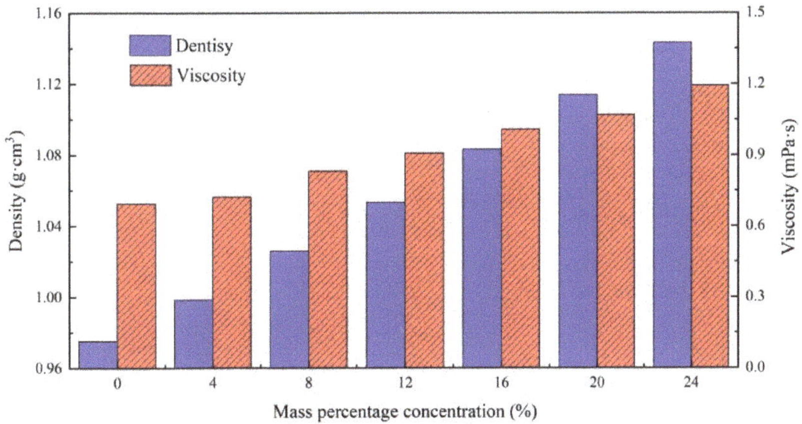

Figure 5. Density and viscosity of brine as a function of salt mass fraction at 55 °C.

3.2. Optimum Formula of Soft Movable Polymer Gels

Based on temperature (55 °C) and salinity conditions (34,178.2 mg/L) of the Y oil-field reservoir, a novel polymer gel prepared via HPAM + PR and Cr^{3+} cross-linker was investigated. The gelation time, gel strength, thermal stability, and water-plugging capacity of the soft movable polymer gels were tested, as shown in Figures 6 and 7, and Tables 3 and 4.

The experimental results showed that the gelation time of gel prepared using HPAM + PR and Cr^{3+} cross-linker was 15~50 h. Additionally, the gel strength of most gels was higher than code F. After aging for 30 days at 55 °C, the gel stability of these gels slightly decreased, and the thermal stability of gel strength was very good.

It is necessary to guarantee safe injectivity into the horizontal well by using gelants with a long gelation time. Both the near- and far-well formulas of gels must satisfy the gelation time of greater than 24 h for construction. The residual resistance factor of the gel used in the far well should not be too large (F_{rr} < 50), and the residual resistance factor of the gel used in the near well should be higher than that of the far well (F_{rr} < 200). According to practical experience and economic benefits, two kinds of formulas of polymer gels were selected: 0.30% HPAM + 0.3% PR and 0.1% Cr^{3+} cross-linker, as well as 0.45% HPAM + 0.5% PR and 0.15% Cr^{3+} cross-linker. Taking one formula of the polymer gel as an example, Figure 8 shows the gelation performance of the gel prepared via 0.30%

HPAM + 0.30% PR and 0.1% Cr^{3+} after aging 3 days and 30 days. It can be seen that the gel has no syneresis even after aging for 30 days.

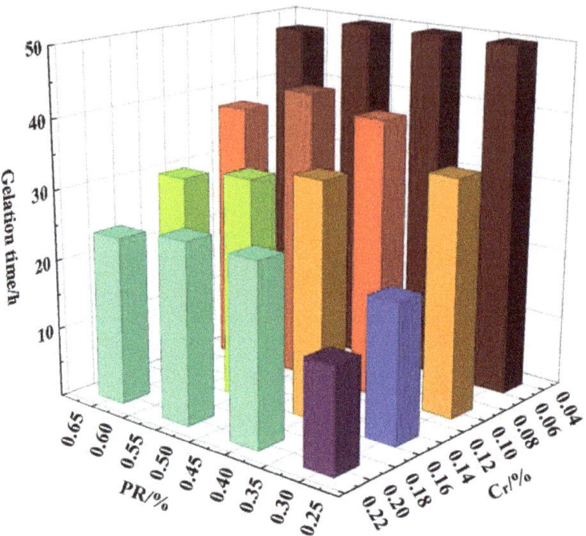

Figure 6. Histogram of gelation time (h) of 0.30% HPAM + PR and Cr^{3+} cross-linker (The color represents gelation time of gel).

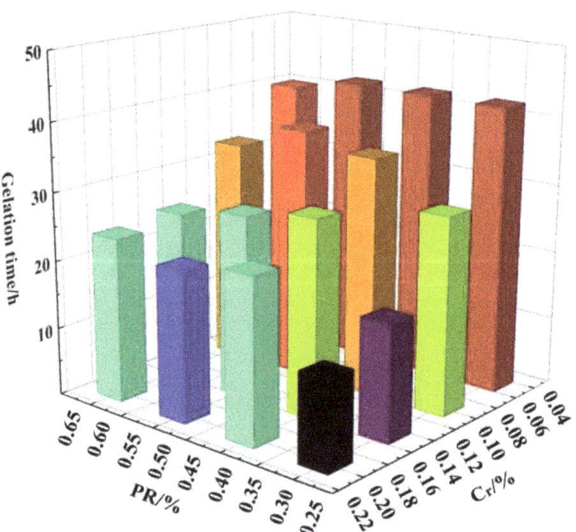

Figure 7. Histogram of gelation time (h) of 0.45% HPAM + PR and Cr^{3+} cross-linker. (The color represents gelation time of gel).

Table 3. Gel strength and long-term thermal stability of the gel system prepared via 0.30% HPAM + PR and Cr^{3+} cross-linker.

HPAM/%	PR/%	Cr^{3+}/%	Gel Strength/%	Aging 30 d at 55 °C
0.30	0.30	0.05	D~E	No syneresis
		0.10	F	No syneresis
		0.15	E	No syneresis
		0.2	E	No syneresis
	0.40	0.05	C~D	No syneresis
		0.10	F	No syneresis
		0.15	F	No syneresis
		0.20	E~F	No syneresis
	0.50	0.05	C	No syneresis
		0.10	F	No syneresis
		0.15	F~G	No syneresis
		0.20	F	No syneresis
	0.60	0.05	C	No syneresis
		0.10	E~F	No syneresis
		0.15	F~G	No syneresis
		0.20	F~G	No syneresis

Table 4. Gel strength and long-term thermal stability of the gel system prepared via 0.45% HPAM + PR and Cr^{3+} cross-linker.

HPAM/%	PR/%	Cr^{3+}/%	Gel Strength/%	Aging 30 d at 55 °C
0.45	0.30	0.05	F	No syneresis
		0.10	F	No syneresis
		0.15	F~G	No syneresis
		0.2	F~G	No syneresis
	0.40	0.05	D~E	No syneresis
		0.100	F~G	No syneresis
		0.15	F~G	No syneresis
		0.20	G	No syneresis
	0.50	0.05	C	No syneresis
		0.10	F	No syneresis
		0.15	G~H	No syneresis
		0.20	G~H	No syneresis
	0.60	0.05	C	No syneresis
		0.10	F	No syneresis
		0.15	G~H	No syneresis
		0.20	G~H	No syneresis

It can be seen from Table 5 that plugging efficiency was more than 98% for different formulas of gel systems. Moreover, with the increase in polymer and cross-linker concentration, both the plugging efficiency and residual resistance factor were improved. This indicates that the gel system has good plugging properties.

3.3. Creep Recovery Property

The creep recovery property of the gel is shown in Figure 9. The polymer gel good deformability, which easily leads to deep migration into water-yielding formation and water plugging for enhanced oil recovery. By analyzing the strain characteristics of the gel, the creep recovery process can be divided into two phases. In the first phase, the constant stress (2.4 Pa) acted on the gel samples for 120 s. The initial strain increased sharply and then tended to become stable over time. The connections between the main structural units in the elastic gels were stretched elastically, and the strain value was several times than that of initial state at the constant shear stress. In the second phase, the constant stress was removed from the gel samples, and the strain could be recovered. The metallic chromium cross-linked gel destroyed its viscoelasticity under shear stress. It can be concluded the deformation and recovery abilities of the gel prepared via HPAM + PR and Cr^{3+} cross-linker is strong, which means that the gel has strong deformability and good elasticity.

Figure 8. The gelation performance of gel prepared via PR and Cr^{3+} cross-linker after aging 3 d and 30 d, respectively.

Table 5. Plugging capacity of the gel system.

Gel Type	Gel Formula (wt)	Gelation Time/h	Initial Permeability/μm²	Plugging Efficiency/%	Residual Resistance Factor
HPAM + PR + Cr^{3+}	0.30% + 0.50% + 0.20%	26	3.89	99.85	769.2
	0.30% + 0.40% + 0.15%	34	3.09	99.16	109.1
	0.30% + 0.30% + 0.10%	32	4.24	96.94	32.2
	0.45% + 0.50% + 0.20%	22	3.87	99.65	356.8
	0.45% + 0.4% + 0.15%	24	3.08	98.75	80.0
	0.45% + 0.3% + 0.10%	24	2.83	95.75	23.5

3.4. Microstructure of the Gel Systems

ESEM is a good way to accurately investigate the microstructure of the gel system. This method can keep the gel system from damage and observe the gel samples in their natural state. The ESEM micrograph of the gel sample is shown in Figure 10. It can be noted that the gel sample had a uniform three-dimensional network. Additionally, the pore size ranged from 5.0 μm to 28.0 μm, and the border thickness changed from 10.0 μm to 30.0 μm for the gel sample. A single chromium metal cross-linking agent forms a large network hole and a thin omentum [25,26], which is easy to dehydrate under external force damage. It can be concluded that the composite cross-linked gel has a more compact structure than that of metal cross-linked gels, which contributes to the good stability of the gel system.

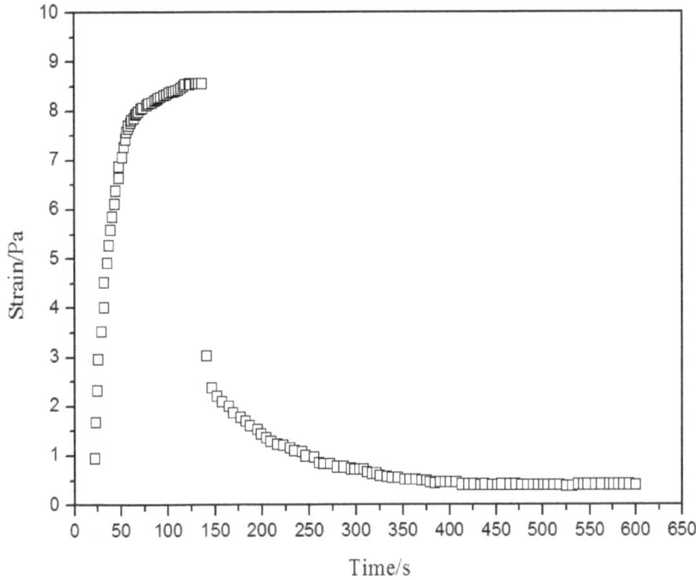

Figure 9. Elastic characteristics of the gel system at 55 °C.

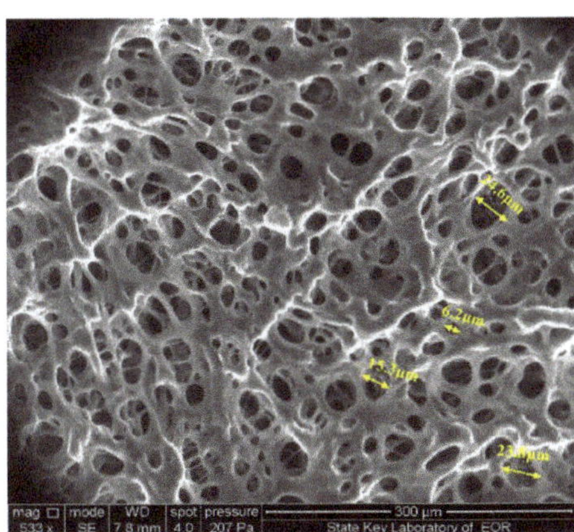

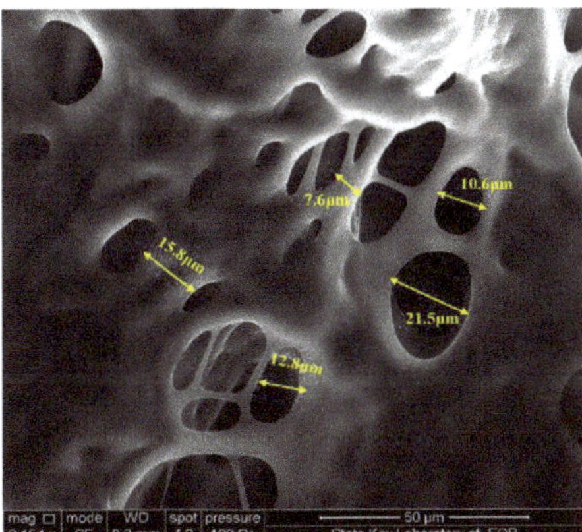

Figure 10. ESEM micrographs of the gel sample.

3.5. Visual Model Experiment of Water-Coning Control in the Horizontal Well

As shown in Figure 11, the reservoir model with bottom water was placed vertically, and the middle point at top of the model was used to simulate the water coning point of the horizontal well. The vertical stripe was the water-channeling channel in the matrix model. The simulated oil (0.89 g/cm^3, ~180 mPa·s) was injected from the water injection port at the bottom of the model to saturate the model, and then the water was injected after the crude oil aged and stabilized (Figure 11a). The simulated oil was produced from the top outlet in the model to mimic oil production from the horizontal well with bottom water (Figure 11b). As oil was produced from the top of the model, bottom-water coning would occur (Figure 11b). High-density brine has a higher density than plugging agent and higher viscosity than formation water, so it entered the bottom aquifer along the water coning channels. The brine had little effect on the performance of the following gel. The transitional density saline was injected into the well when the high-density saline was completed, and it separated the high-density saline and the polymer gel and weakened the effect of salt on gelant properties. The gelants entered the bottom aquifer along the water-coning channels and spread in a horizontal direction during their injection stage (Figure 11c). The displacement fluid displaced gelants in the vicinity of the oil–water interface (Figure 11d). Then, the top production port was shut in order to wait for the transformation of gelants into gels. The chemical gel packer can effectively solve horizontal bottom-water coning and increase oil production (Figure 11e).

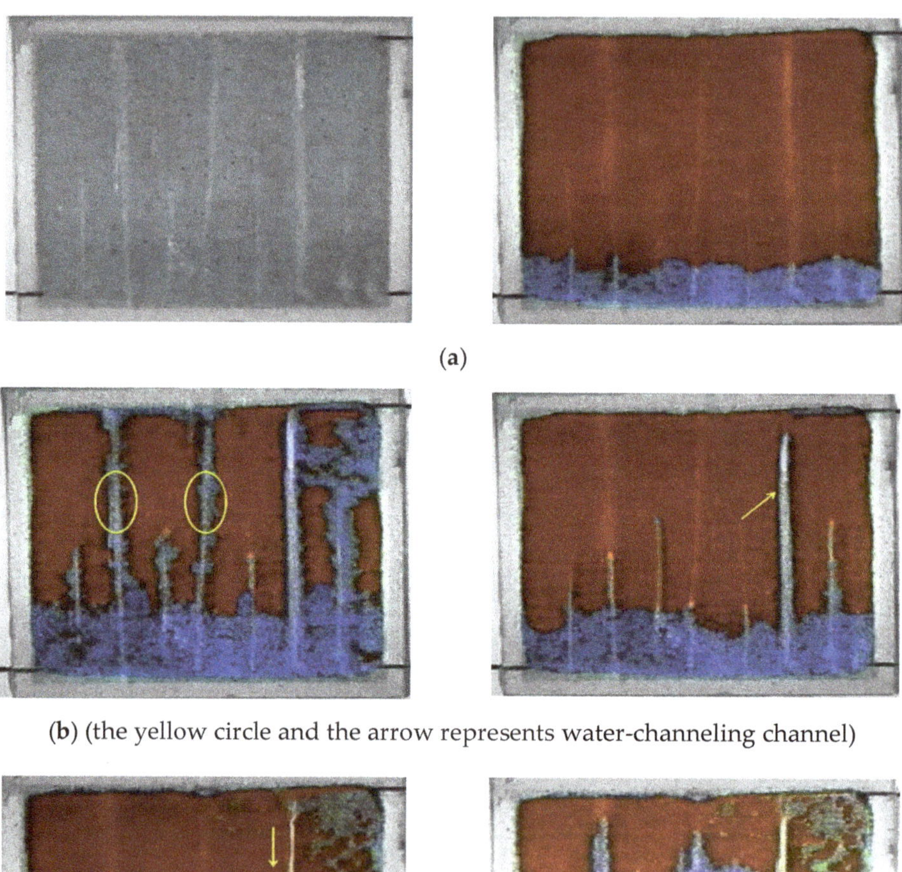

(**a**)

(**b**) (the yellow circle and the arrow represents water-channeling channel)

(**c**) (the arrow represents injection channel of gelant)

Figure 11. *Cont.*

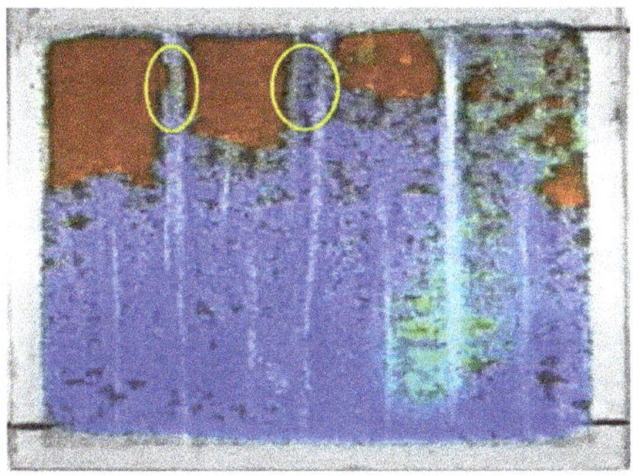
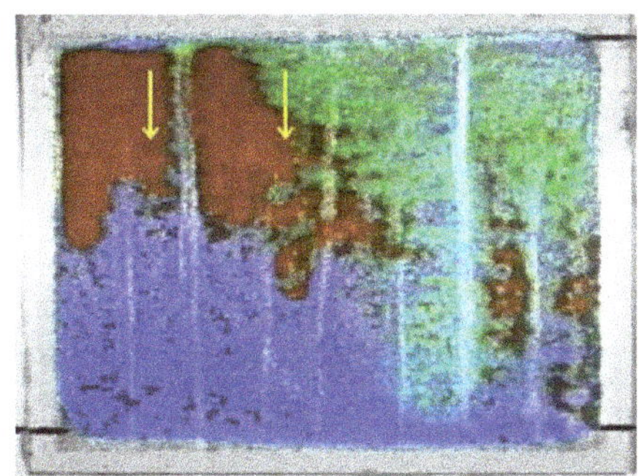

(**d**) (the yellow circle and the arrow represents injection channel of gelant)

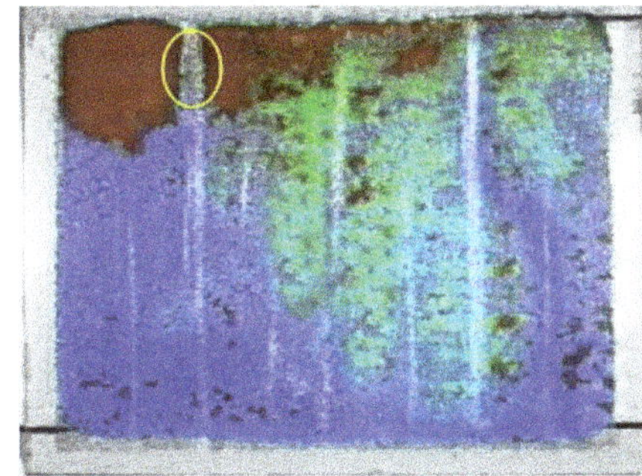

(**e**) (the yellow circle represents new water-channeling channel)

Figure 11. Visual physical simulation experiments: (**a**) horizontal well with bottom water; (**b**) bottom water coning; (**c**) injecting high-density brine and transition-density brine and injecting the gelants; (**d**) injecting displacement fluid; (**e**) resumption of production.

From the experimental results in the above figure, it can be seen that the injection of high-density brine increased the density difference between the bottom water and oil. The gelants' density ranged from 0.97 g/cm^3 to 0.98 g/cm^3, which is between high-density brine (ρ8%NaCl = 1.0255 g/cm^3) and oil (ρ_o = 0.886 g/cm^3). Thus, this chemical gel packer formed near the oil–water transition layer. When the chemical gel packer was completed to control bottom-water coning, it was difficult for the bottom water to flow into the horizontal well. The longer the gel packer stability was, the better the effect of controlling bottom-water coning was.

3.6. Field Application

3.6.1. Background of HX Well

The pay zone of the Y oil field is NgII$_{11+2}$, with a formation thickness of 8.4 m and a reservoir temperature of 55 °C. The depth of the HX well is 2045.0 m, and the horizontal section length is 382.0 m. The data of crude oil and liquids production in the HX well over time are presented in Figure 12. Due to a high water cut (>90%) in Mar 2015, measures were taken in the HX well to control excess water production using polymer gel in March 2016.

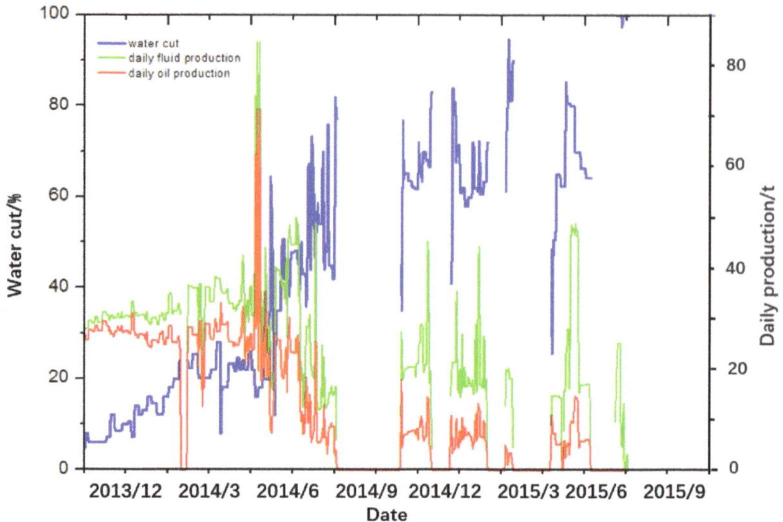

Figure 12. Data of crude oil and liquids production over time in HX well before treatment.

The main reasons for choosing a chemical gel packer are as follows: (1) The oil layer is thin (<10.0 m), and the horizontal well section is located lower and close to the oil–water interface; (2) there is no tight layer at the bottom of the oil layer to block bottom water; (3) the serious vertical heterogeneity of the reservoir is conducive to the gravity separation of injected fluid.

3.6.2. Gelants Formula (A + B) of Polymer Gel

(1) Gelant A formula for far from wellbore

Gelant A were made up of 0.30% HPAM + 0.30% PR and 0.10% Cr^{3+} cross-linker, and the gelation time was 36 h at 55 °C;

(2) Gelant B formula for near-wellbore zone

Gelant B were made up of 0.45% HPAM + 0.50% PR and 0.15% Cr^{3+} cross-linker, and the gelation time was 24 h at 55 °C.

3.6.3. Volume of Gelants

Equation (3) for calculating the volume of high-density brine is shown as follows:

$$V_1 = 2 \times (R_1 - R_2) \times L \times \varphi \times \beta \quad (3)$$

where V_1—the volume of high-density brine, m^3; R_1—radius of the brine layer, m; R_2—radius of gelants layer, m; L—length of horizontal segment, m; φ—porosity, %; β—the ratio of the width of vertical high permeability to the length of horizontal segment [18].

For HX well, L = 382.0 m, φ = 30.4%. Assuming R_1 = 12.0 m, R_2 = 11.0 m and β = 20%, then volume of high-density brine as follows:

$$V_1 = 2 \times (12 - 11) \times 382 \times 30.4\% \times 20\% = 47.0 \text{ m}^3$$

The designed V_1 was 50 m^3. The volume ratio of the high-density brine (8% NaCl) to transition density brine (4% NaCl) was 1:1, so each volume of them was 25.0 m^3.

Equation (4) for calculating the volume of polymer gel is shown as follows:

$$V_2 = 2 \times (R_2 - R_3) \times L \times H \times \beta \times \varphi \tag{4}$$

where V_2—the volume of polymer gel, m^3; R_2—radius of gelants, m; R_3—radius of displacement fluid, m; L—length of horizontal section, m; φ—porosity, %; β_1—the proportion of fluid channeling ($\leq$5%).

If L = 382.0 m, φ = 30.4%, R_2 = 11.0 m, R_3 = 3.0 m and β_1 = 3.1%, then volume of polymer gel as follows:

$$V_2 = 2 \times (11 - 3) \times 382 \times 8.4 \times 30.4\% \times 3.1\% = 500.0\ m^3$$

The designed V_2 was 500.0 m^3. The two gelants were divided into two parts according to the volume ratio of 3:2 (A:B), and the volume of gelant A for far from wellbore was 500.0 × 3/5 = 300.0 m^3, while the volume of gelant B for the transition zone was 500 × 2/5 = 200.0 m^3.

Equation (5) for calculating the volume of displacement fluid is shown as follows:

$$V_3 = 2 \times R_3 \times L \times H \times \varphi \times K \tag{5}$$

where V_3—volume of displacement fluid, m^3. Then, volume of displacement fluid as follows:

$$V_3 = 2 \times 3 \times 382 \times 8.4 \times 30.4\% \times 3.1\% = 181.0\ m^3$$

The gelants' density of the above gel formula was between 0.97 g/cm^3 and 0.98 g/cm^3, which is between high-density brine ($\rho_{8\%NaCl}$ = 1.0255 g/cm^3) and oil (ρ_o = 0.886 g/cm^3). Therefore, the chemical packer formed by gel was easily built up near the oil–water interface.

3.6.4. Detail of Measurements

(1) The HX well maintained production for 10 days and restored the existing water cone channel; (2) briefly, 50.0 m^3 high-density brine was injected into the HX well; (3) then, 300.0 m^3 far-wellbore gelant was injected, after which 200.0 m^3 near-wellbore gelant was injected into HX well; (4) then, 181.0 m^3 displacement fluid was injected into HX well; (5) seawater was used as a displacement fluid to drive gelants into percolation water layer; (6) the HX well was shut in for 4 days, and the gelants became a chemical gel packer; (7) the HX well was reopened, and production resumed.

3.6.5. Field-Test Results

The data of crude oil and liquids production over time in HX well after treatment is shown in Figure 13. It can be seen that the highest total oil production per day of the HX well was 29.6 t, and the water cut decreased to 8% and remained at a lower level, less than 50% for a long time after the treatment in the HX well. Lastly, the accumulated oil production was 1.035×10^4 t instead of 3.9×10^3 t. Oil-field construction results confirmed that this novel method is quite suitable for water coning control and has a broad application prospect.

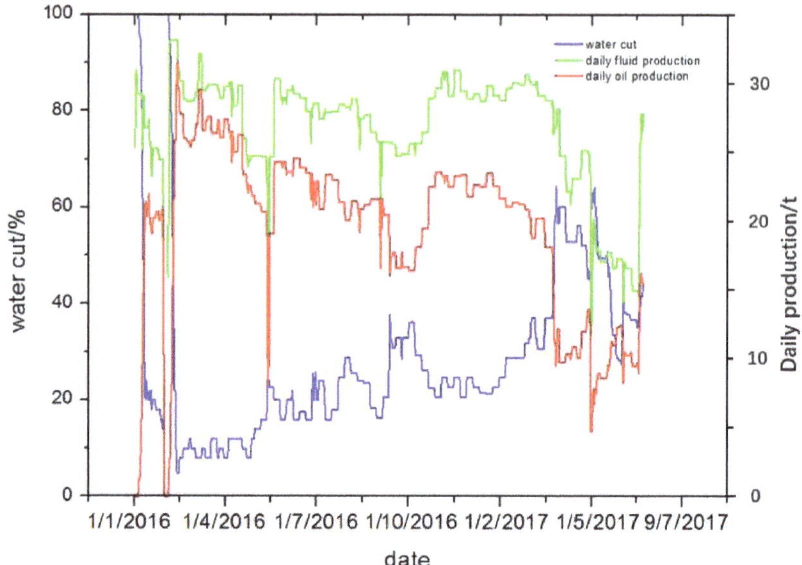

Figure 13. Data of crude oil and liquids production over time in HX well after treatment.

4. Conclusions

(1) The designed gelant formula of soft movable polymer gel was 0.30~0.45% HPAM + 0.30~0.5% phenolic resin + 0.10~0.15% chromium acetate, with a corresponding gelation time of 26~34 h at 55 °C. The formula meets the technological requirements of controlling bottom-water coning in horizontal wells in offshore oil fields.

(2) The visual glass plate simulation device can intuitively present the phenomenon of bottom-water coning and the effect of high-density brine injection and chemical packer control, which confirmed the feasibility of the process.

(3) Combined with the production status and reservoir conditions of the HX well in the Y oil field, the technology of soft movable polymer gel for controlling water coning was successfully applied, which provides a technical basis for cost reduction and efficiency improvement in offshore oil fields.

Author Contributions: Data curation, Y.S.; funding acquisition, Q.Y.; investigation, G.Z.; methodology, Q.Y. and Y.L.; visualization, P.W.; writing—original draft preparation, J.Q.; writing—review and editing, Q.Y. and Y.L. All authors have read and agreed to the published version of the manuscript.

Funding: The authors sincerely appreciate the financial support from the National Natural Science Foundation of China (No. 52074249, 51874261) and Fundamental Research Funds for the Central Universities (2-9-2019-103).

Institutional Review Board Statement: Not applicable.

Informed Consent Statement: Not applicable.

Data Availability Statement: Data are contained within the article.

Conflicts of Interest: The authors declare no conflict of interest.

References

1. Machale, J.; Majumder, S.K.; Ghosh, P.; Sen, T.K. Development of a novel biosurfactant for enhanced oil recovery and its influence on the rheological properties of polymer. *Fuel* **2019**, *257*, 116067. [CrossRef]
2. Liu, Y.; Zou, C.; Zhou, D.; Dai, C. Novel chemical flooding system based on dispersed particle gel coupling in-depth profile control and high efficient oil displacement. *Energy Fuels* **2019**, *33*, 3123–3132. [CrossRef]
3. Dehaghani, A.; Badizad, M.H. Impact of ionic composition on modulating wetting preference of calcite surface: Implication for chemically tuned water flooding. *Colloids Surfaces A* **2019**, *568*, 470–480. [CrossRef]
4. Wei, P.; Zheng, L.; Yang, M. Fuzzy-ball fluid self-selective profile control for enhanced oil recovery in heterogeneous reservoirs: The techniques and the mechanisms. *Fuel* **2020**, *275*, 117959. [CrossRef]
5. Bai, Y.; Lian, Y.; Ban, C.; Wang, Z.; Zhao, J.; Zhang, H. Facile synthesis of temperature-resistant hydroxylated carbon black/polyacrylamide nanocomposite gel based on chemical crosslinking and its application in oilfield. *J. Mol. Liq.* **2021**, *329*, 115578. [CrossRef]
6. Song, Z.; Bai, B.; Challa, R. Using Screen models to evaluate the injection characteristics of particle gels for water control. *Energy Fuels* **2018**, *32*, 352–359. [CrossRef]
7. You, Q.; Zhao, F. Research and application of deep plugging water in oil well. *Drill Prod. Technol.* **2007**, *30*, 85–87.
8. Bartosek, M.; Mennella, A.; Lockhart, T. Polymer gels for conformance treatments: Propagation on Cr (III) crosslinked complexes in porous media. In Proceedings of the SPE/DOE Symposium on Enhanced Oil Recovery, Tulsa, OK, USA, 17–20 April 1994.
9. Seright, R.S.; Lane, R.H.; Sydansk, R.D. A strategy for attacking excess water production. *SPE Prod. Facil.* **2003**, *18*, 158–169. [CrossRef]
10. Albonico, P.; Bartosek, M.; Malandrino, A. Studies on phenol-formaldehyde crosslinked polymer gels in bulk and in porous media. In Proceedings of the SPE International Symposium on Oilfield Chemistry, San Antonio, TX, USA, 14–17 February 1995.
11. Reddy, B.R.; Eoff, L.; Dalrymple, E.D.; Black, K.; Brown, D.; Rietjens, M. A natural polymer-based cross-linker system for conformance gel systems. *SPE J.* **2003**, *8*, 99–106. [CrossRef]
12. Zhao, F. *Chemicals for Oil Recovery*; China University of Petroleum Press: Dongying, China, 2001.
13. Zhao, F. *Principle of Enhanced Oil Recovery*; China University of Petroleum Press: Dongying, China, 2006.
14. Strelets, L.A.; Ilyin, S.O. Effect of enhanced oil recovery on the composition and rheological properties of heavy crude oil. *J. Pet. Sci. Eng.* **2021**, *203*, 108641. [CrossRef]
15. Ilyin, S.O.; Strelets, L.A. Basic Fundamentals of Petroleum Rheology and Their Application for the Investigation of Crude Oils of Different Natures. *Energy Fuels* **2018**, *32*, 268–278. [CrossRef]
16. Gorbacheva, S.N.; Ilyin, S.O. Morphology and rheology of heavy crude oil/water emulsions stabilized by microfibrillated cellulose. *Energy Fuels* **2021**, *35*, 6527–6540. [CrossRef]
17. Amir, Z.; Saaid, I.; Junaidi, M.; Baker, W. Weakened PAM/PEI polymer gel for oilfield water control: Remedy with silica nanoparticles. *Gels* **2022**, *8*, 265. [CrossRef] [PubMed]
18. Whitney, D.; Montgomery, D.; Hutchins, R. Water shutoff in the North sea: Testing a new polymer gel system in the heather field, UKCS Block 2/5. *SPE Prod. Facil.* **1996**, *11*, 108–112. [CrossRef]
19. Sydansk, R.D. Acrylamide-polymer/chromium (III)-carboxylate gels for near wellbore matrix treatments. *SPE Adv. Technol. Ser.* **1993**, *1*, 146–152. [CrossRef]
20. Southwell, G.P.; Posey, S.M. Applications and results of acrylamide-polymer/chrnmium (III) Carboxylate Gels. In Proceedings of the SPE 27779, SPE/DOE Improved Oil Recovery Symposium, Tulsa, OK, USA, 17–20 April 1994.
21. Larry, E.; Eldon, D.; Don, M.; Julio, V. Worldwide field applications of a polymeric gel system for conformance applications. *SPE Prod. Oper.* **2007**, *22*, 231–235.
22. Gino, D.; Phil, R. New insights into water control—A review of the state of the art. In Proceedings of the SPE77963, SPE Asia Pacific Oil and Gas Conference and Exhibition, Melbourne, Australia, 8–10 October 2002.
23. Liu, J.; Li, L.; Sun, Y. Biomimetic functional hydrogel particles with enhanced adhesion characteristics for applications in fracture conformance control. *J. Ind. Eng. Chem.* **2022**, *106*, 482–491. [CrossRef]
24. Zhao, G.; Dai, C.; You, Q. Characteristics and displacement mechanisms of the dispersed particle gel soft heterogeneous compound flooding system. *Pet. Explor. Dev.* **2018**, *45*, 464–473. [CrossRef]
25. Jia, H.; Pu, W.; Zhao, J.; Liao, R. Experimental investigation of the novel phenol-formaldehyde cross-linking HPAM gel system: Based on the secondary cross-linking method of organic cross-linkers and its gelation performance study after flowing through porous media. *Energy Fuels* **2011**, *25*, 727–736. [CrossRef]
26. Yang, H.; Iqbal, M.; Lashari, Z.; Kang, W. Experimental research on amphiphilic polymer/organic chromium gel for high salinity reservoirs. *Colloids Surf. A: Physicochem. Eng. Asp.* **2019**, *582*, 123900. [CrossRef]
27. Zhu, D.; Bai, B.; Hou, J. Polymer gel systems for water management in high-temperature petroleum reservoirs: A chemical review. *Energy Fuels* **2017**, *31*, 13063–13087. [CrossRef]
28. Pu, J.; Geng, J.; Han, P.; Bai, B. Preparation and salt-insensitive behavior study of swellable, Cr^{3+}-embedded microgels for water management. *J. Mol. Liq.* **2019**, *273*, 551–558. [CrossRef]
29. Zhang, W.; Yang, H.; Sarsenbekuly, B.; Zhang, M.; Aidarova, S. The advances of organic chromium based polymer gels and their application in improved oil recovery. *Adv. Colloid Interface Sci.* **2020**, *282*, 102214. [CrossRef] [PubMed]

30. Sabhapondit, A.; Borthakur, A.; Haque, I. Water soluble acrylamidomethyl propane sulfonate (AMPS) copolymer as an enhanced oil recovery chemical. *Energy Fuels* **2003**, *17*, 683–688. [CrossRef]
31. Sydansk, R.D.; Marathon Oil Co. A new conformance improvement treatment chromium(III) gel technology. In Proceedings of the SPE 17329 Presented at SPE Enhanced Oil Recovery Symposium, Tulsa, OK, USA, 17–20 April 1988.
32. Liu, Y.; Dai, C.; Wang, K.; Zou, C.; Gao, M.; Fang, Y.; You, Q. Study on a novel cross-linked polymer gel strengthened with silica nanoparticles. *Energy Fuel* **2017**, *31*, 9152–9161. [CrossRef]

 gels

MDPI

Article

Novel Acrylamide/2-Acrylamide-2-3 Methylpropanesulfonic Acid/Styrene/Maleic Anhydride Polymer-Based $CaCO_3$ Nanoparticles to Improve the Filtration of Water-Based Drilling Fluids at High Temperature

Zhichuan Tang [1], Zhengsong Qiu [1,*], Hanyi Zhong [1], Hui Mao [2], Kai Shan [1] and Yujie Kang [1]

1 School of Petroleum Engineering, China University of Petroleum (East China), No. 66 Changjiang West Road, Economic & Technical Development Zone, Qingdao 266580, China; b17020063@s.upc.edu.cn (Z.T.); zhonghanyi@126.com (H.Z.); b20020005@s.upc.edu.cn (K.S.); 17854210608@163.com (Y.K.)

2 State Key Laboratory of Oil & Gas Reservoir, Chengdu University of Technology, 1 East 3 Road, Chengdu 610059, China; maohui17@cdut.edu.cn

* Correspondence: teamo_tzc@163.com

Abstract: Filtration loss control under high-temperature conditions is a worldwide issue among water-based drilling fluids (WBDFs). A core–shell high-temperature filter reducer (PAASM-$CaCO_3$) that combines organic macromolecules with inorganic nanomaterials was developed by combining acrylamide (AM), 2-acrylamide-2-methylpropane sulfonic acid (AMPS), styrene (St), and maleic anhydride (MA) as monomers and nano-calcium carbonate (NCC). The molecular structure of PAASM-$CaCO_3$ was characterized. The average molecular weight of the organic part was 6.98×10^5 and the thermal decomposition temperature was about 300 °C. PAASM-$CaCO_3$ had a better high-temperature resistance. The rheological properties and filtration performance of drilling fluids treated with PAASM-$CaCO_3$ were stable before and after aging at 200 °C/16 h, and the effect of filtration control was better than that of commonly used filter reducers. PAASM-$CaCO_3$ improved colloidal stability and mud cake quality at high temperatures.

Keywords: filtration reducer; high temperature; water-based drilling fluid; nanomaterials; calcium carbonate

Citation: Tang, Z.; Qiu, Z.; Zhong, H.; Mao, H.; Shan, K.; Kang, Y. Novel Acrylamide/2-Acrylamide-2-3 Methylpropanesulfonic Acid/Styrene/Maleic Anhydride Polymer-Based $CaCO_3$ Nanoparticles to Improve the Filtration of Water-Based Drilling Fluids at High Temperature. *Gels* **2022**, *8*, 322. https://doi.org/10.3390/gels8050322

Academic Editor: Georgios Bokias

Received: 22 April 2022
Accepted: 18 May 2022
Published: 20 May 2022

1. Introduction

With the deep exploration and development of oil and gas resources, high-temperature water-based drilling fluid (WBDF) technology has become one of the key technologies in drilling engineering [1–3]. As the most commonly used high-temperature drilling fluid additive for a long time, additive materials (such as sulfonated phenolic resin, sulfonated lignite, sulfonated tannin) have been widely used to control rheology or filtration performance [4–7]. However, sulfonated materials still decompose easily at high temperatures [8], and some need to be used together to achieve the best results [9], most of which are environmentally unfriendly [10,11]. In view of the above disadvantages, since the 1980s, researchers have started to develop the application of multicomponent copolymers in high-temperature drilling fluids and have achieved good results [12–14].

As early as the 1980s, Giddings et al. [15] developed a terpolymer filtrate reducer. This agent was copolymerized with acrylamide, 2-acrylamide-2-methylpropanesulfonic acid, and 2-mercaptobenzoic acid as monomers. The rigid side chain and the large number of sulfonic acid groups in its molecule improve its high-temperature effect. Dickert et al. [16] developed a pH-adaptive high-temperature filtrate reducer by aqueous solution polymerization with acrylamide, 2-acrylamide-2-methylpropanesulfonic acid, and n-vinyl-alkyl amide as monomers. On the basis of Giddings, American scholar Patel [17] successfully prepared a high-temperature and high-salinity filtration reducer with 2-acrylamide-2-methylpropionic acid as the monomer and 2-mercaptobenzoic acid as the cross-linking agent based on precise control of the molecular structure. The agent can resist Ca^{2+} and Mg^{2+} ion pollution and has a good effect against temperatures exceeding 200 °C.

As the research progressed, the shortcomings of the descending agents of the polymers formed by copolymerization of AM monomers with alkene monomers and sulfonic acid monomers were gradually exposed [18–20]. Therefore, researchers began to develop environmentally friendly high-temperature and -salinity WBDF loss reducers on the basis of ethylene sulfonic acid monomer/acrylamide (or its derivatives). Thaemlitz et al. [21] developed a fluid loss reducer with high-temperature resistance and high salinity with N-vinyl carbazole (NVC), polystyrene sulfonic acid (PSS), and AMPS as monomers. Drilling fluids using it as a key additive can maintain good rheology and filtration after aging at high temperatures, and also have good anti-pollution and some shale inhibition effects.

In 2019, Soric and Heier [22] developed a high-temperature and high-salinity fluid loss reducer by aqueous solution polymerization. Its relative molecular weight was about 1 million and its temperature resistance exceeded 180 °C. The drilling fluid system constructed with it as a key treatment agent has been successfully applied to shale gas blocks in the Republic of Herwazka and exhibits a reservoir protection effect. The Exxon company has prepared an environment-friendly, high-temperature, and high-salinity WBDF system with synthetic polymer as the key treatment agent, which can resist the high temperature of 210 °C, and the waste drilling fluid can be directly discharged into the sea after being tested by the U.S. Environmental Protection Agency (EPA); Schlumberger also developed drilling fluid systems with a density of 2.20 g/cm^3 and temperature resistance exceeding 220 °C. It has been used in some sensitive areas such as marine blocks in the United States and is environmentally friendly. Bagum et al. used aloe additive to form four representative drilling fluid formulations along the base bentonite. A complete rheological test and filtration test of mud additives with different concentrations were carried out to study the feasibility of this new additive.

The team of Prof. Zhengsong Qiu at the University of Petroleum, China, thoroughly studied the mechanism of action of high-temperature, high-density, and high-salinity drilling fluids and developed a HTP-1 filtration reducer using amps, NVP, DAAC as monomers [23]. It worked well at 240 °C with a NaCl content of 20%wt and a density of 2.0 g/cm^3. Based on this, a novel filtrate reducer, FLR-1, was developed by introducing nanotechnology [24]. Its filtration loss effect is significant, and the HTHP filtration at 200 °C/16 h is only 20.5 mL. In addition, FLR-1 can also significantly improve the rheological properties of the drilling fluid system with excellent salt tolerance and meet the environmental protection standards. Researchers [25] have shown that nanomaterials can significantly improve the performance of drilling fluids and broaden their service conditions. Scientists have [26] applied zinc oxide nanoparticles prepared in the laboratory to WBDF. The results show that nanoparticles improve the rheological properties of WBDF. Adding a single nanomaterial to the WBDF will not significantly affect the API filtration volume. However, the mud cake thickness decreases with the concentration of nanoparticles. The results show that nanoparticles can improve the rheological properties. The application of waste nanomaterials [27,28] was summarized in rheological and lubricity testing, adequate rheological and filtration checks were performed on water-based drilling fluids, and the effect of waste as an additive was evaluated on drilling fluid performance. Minakov et al. [11] found that the yield stress and consistency index of nanoparticle drilling fluids increase with temperature. As particle size increases, their influence on the temperature dependence of drilling fluid viscosity increases. The addition of nanoparticles stabilizes the viscosity of drilling fluids relative to temperature.

In summary, great progress has been made in WBDF filtration reducer technology at high temperature in recent years, but the problems still exist [29,30]. At present, some vinyl copolymer drilling fluid reducing agents are nontoxic and environmentally friendly [31,32], but are easily (partially) degraded under high-temperature conditions [33,34]. Some of the degraded products may be toxic/lowly toxic, which will affect their eco-friendly performance [35–37].

Recent studies have shown [38–40] that nanoparticles have the advantages of high surface energy, thermal stability, and rigidity. By assembling block copolymers into the grid holes formed by polymer frameworks, the rigidity and thermal stability of nanomaterials can be combined with the advantages of salt resistance and toughness of polymers [41], thus further improving the high-temperature stability of modifiers, which also provides a new idea for the development of high-temperature WBDF agents.

Therefore, this paper summarizes the design concept of the molecular structure of the high-temperature WBDF filtration reducer (PAASM-$CaCO_3$), which has excellent properties. The rigidity and thermal stability of inorganic nanomaterials were combined, the nanoparticles were embedded into the grid pores formed by the polymer framework, and the organic–inorganic nanocomposites with excellent properties were developed.

2. Materials and Methods

2.1. Materials

The main reagents for the reaction are detailed in Table 1.

Table 1. Major Materials for Synthesis of PAASM-$CaCO_3$.

Materials	Purity	Suppliers
Acrylamide (AM)	CP	Shanghai Sinopharm Chemical Reagent Co., Ltd., Shanghai, China
2-Acrylamido-2-Methyl Propanesulfonic Acid (AMPS)	CP	Aladdin Reagent Co., Ltd., Shanghai, China
Styrene (St)	CP	Aladdin Reagent Co., Ltd., Shanghai, China
Maleic anhydride (Ma)	AR	Shanghai Sinopharm Chemical Reagent Co., Ltd.
$K_2S_2O_8$	AR	Shanghai Sinopharm Chemical Reagent Co., Ltd.
$NaHSO_3$	AR	Shanghai Sinopharm Chemical Reagent Co., Ltd.
NaOH	AR	Shanghai Sinopharm Chemical Reagent Co., Ltd.
Span 80	AR	Shanghai Sinopharm Chemical Reagent Co., Ltd.
Tween 60	AR	Shanghai Sinopharm Chemical Reagent Co., Ltd.
N-amyl alcohol	AR	Shanghai Sinopharm Chemical Reagent Co., Ltd.
Dimethyl sulfoxide	GC	Aladdin Reagent Co., Ltd., Shanghai, China
Nano$CaCO_3$ (NCC, particle size: 15 nm)	Ind	GreenSource Biotech Co., Ltd., Jinan, China

In addition, common filter reducer products were purchased to compare PAASM-$CaCO_3$ with PAASM-$CaCO_3$ from Shida Innovative Technology Co., Ltd., Dongying, China: Driscal D and D-4; high-temperature polymer filtration reducer, 80A51; high-temperature calcium-resistant fluid loss reducer, jt888; high-temperature salt-resistant filtration reducer, PJA-2; bitumen filtration reducer, FT-A; sulfonated phenolic resin, SMP-1.

2.2. Methods

2.2.1. Synthesis of Poly (AM-AMPS-St-MA)-$CaCO_3$

According to the designed molecular structure, monomers such as acrylamide (AM), maleic anhydride (MA), 2-acrylamide-2-methylpropanesulfonic acid (AMPS), and styrene (St) were used to prepare the high-temperature-resistant molecular framework, and then the pre-dispersed inorganic nanoparticles (NCC) were embedded into the prepared framework to ensure good temperature resistance. The reaction schematic of poly (AM-AMPS-St-MA)-$CaCO_3$ (PAASM-$CaCO_3$) is shown in Figures 1 and 2.

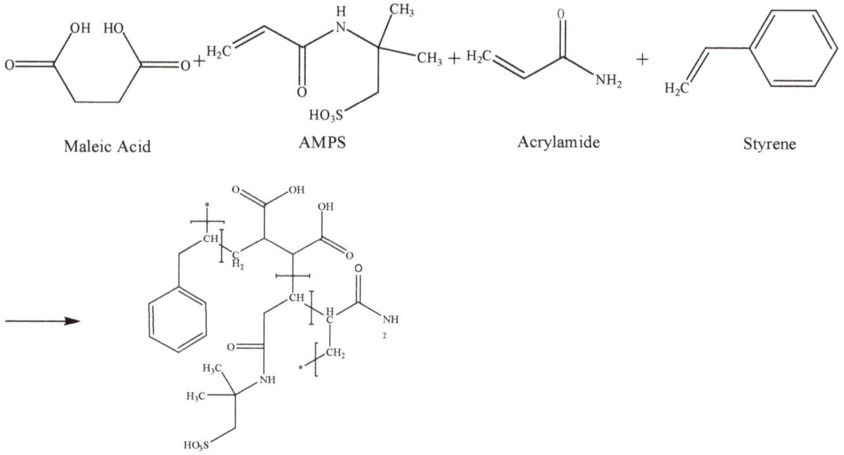

Maleic Acid AMPS Acrylamide Styrene

Figure 1. Schematic of Acrylamide/2-Acrylamide-2-methylpropanesulfonic Acid/Styrene/Maleic anhydride polymer (PAASM) synthesis.

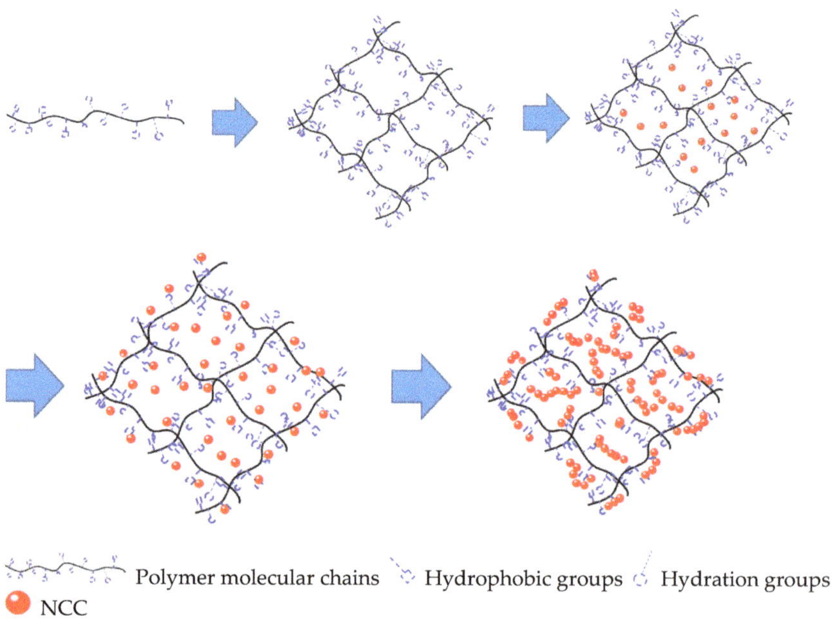

Figure 2. Schematic diagram of PAASM-$CaCO_3$ reaction.

An amount of 30 mL of sodium tetraborate buffer solution was prepared in beaker 1, and then 17.5 g of NCC was added. After high-speed stirring at 12,000 rpm for 15 min, 5 mL of dimethyl sulfoxide was added dropwise and placed in an ultrasonic cell disintegrator for dispersion. An amount of 50 mL of deionized water was added in beaker 2, and then 43.5 g of AMPS was added to dissolve with sufficient stirring. In beaker 3, 20 mL of deionized water was added and 29.8 g of AM was dissolved with stirring under heating. White oil, 1.25 g of tween 60, and 3.75 g of span 80 were added to beaker 4 and stirred well to homogeneity with a glass rod. The solutions in beaker 2 and beaker 3 were poured into a four-port flask, the pH values were adjusted to 6.0 with 15 mol/L of NaOH solution,

10.3 g of MA and 21.8 g of St were added, and the reaction system was stirred under the protection of N_2. Beaker 2 and beaker 3 were cleaned with deionized water and added to a four-port flask. The liquid in beaker 4 was added into the flask and stirred for 15 min. Then, 0.336 g of $K_2S_2O_8$ and $NaHSO_3$ was added in beaker 5, dissolved by deionized water, and then added dropwise to the four-port flask. The reaction system began to heat up. The suspension in beaker 1 was dripped into the reaction system after 2 h and then continued to react for 5 h. Then, it was rapidly cooled to room temperature. Ethanol and acetone were added to filter the resulting sediment and quickly washed out with dilute hydrochloric acid. A novel polymer-based NCC with a core–shell structure as a high-temperature filtrate reducer for drilling fluid (PAASM-$CaCO_3$) was obtained.

2.2.2. Characterization of PAASM-$CaCO_3$

(1) FTIR: 1 mg of dry PAASM-$CaCO_3$ powder and 20 mg of KBr were mixed fully. The mixture was loaded into the mold and compacted with 50 MPa of pressure. FTIR spectra of the compacted tableting were obtained on a NEXUS FTIR spectrometer.

(2) TGA: The thermogravimetric analysis of PAASM-$CaCO_3$ was carried out by a Mettler Teledo thermogravimetric analyzer in Switzerland. The temperature range was from room temperature to 1000 °C, the heating rate was 10 K/min, the atmosphere was nitrogen, and the gas flow rate was 50 mL/min.

(3) GPC: The relative molecular mass of PAASM (without NCC) was determined by the German SFD gel permeation chromatograph (GPC). The mobile phase was phosphate-buffered solution. The column was a SHODEX (K-806 M chloroform system) and the filler was styrene and two vinyl benzene copolymers.

(4) Surface hydroxyl number test: 18 mL of sodium dihydrogen phosphate-buffered solution was prepared, and then 12 mL of NaCl solution with a mass fraction of 0.2% was added.

After full mixing, the pH value of 0.5 mol/L of dilute hydrochloric acid sodium dihydrogen phosphate-buffered solution was adjusted to 5.5. The 0.6 g sample was fully stirred and dispersed, and the pH value of the liquid was measured with a precision pH meter. The suspension was titrated dropwise by a 1.5 mol/L NaOH solution to pH 9.0 for 20 s, which was the end point of the titration. The amount of NaOH consumed during this period was recorded

The hydroxyl number N on the surface of NCC can be calculated according to formula (1):

$$N = CVN_A 10^{-3}/Sm \tag{1}$$

where C is the concentration of NaOH solution, mol/L; V is the amount of NaOH used from the start of the titration to the end point, mL; N_A is the Avogadro constant, 6.02×10^{23}; S is the specific surface area of the particles, nm^2/g; m is the mass of sample involved in the titration, g.

2.2.3. Performance Evaluation of PAASM-$CaCO_3$

(1) Drilling fluid preparation and aging

An amount of 16 g of sodium-based bentonite was added to 400 mL of clear water and stirred at 8000 rpm on a high-speed blender for 30 min, and then 0.8 g of Na_2CO_3 was added, stirred for 20 min, and then pre-hydrated for 24 h. A certain amount of polymer was then added to the fluid, which was stirred at 8000 rpm for 30 min on a high-speed blender. The composites were aged at a set temperature for 16 h by aging and were cooled to room temperature before stirring at high speed for 20 min. Rheological and filtration properties of drilling fluid before and after rolling at a specific temperature/16 h were tested according to the drilling fluid performance evaluation standard SY/t5621-1993.16 g.

(2) API Static Filtration Test

The static API filterability of drilling fluid was tested with a ZNZ-D3 API medium pressure filter (Qingdao Haitong Instrument Co., Ltd.). A certain amount of drilling fluid was loaded into the filter kettle, the top was covered with API filter paper, and it was placed under 100 psi. The filtered volume (FL or FL_{API}) of the drilling fluid was recorded for 30 min, which is recommended by the API.

(3) High-temperature high-pressure static filtration loss test

The static HTHP filterability of drilling fluid was tested with a GGS424A high-temperature and high-pressure static filter instrument (Qingdao Haitongda Instrument Co., Ltd., Qingdao, China). A certain amount of drilling fluid was loaded into the filter kettle, the top cover was covered with HTHP filter paper, the top cover was tightened, and the difference between the upper and lower pressure was 3.5 MPa. The test temperature was the hot roll temperature (the test temperature is 180 °C when the hot roll temperature is higher than 180 °C). The filtered volume (FL_{HTHP}) of the drilling fluid was recorded for 30 min, which is recommended by the API.

(4) Rheological property test

The rheological parameters of the drilling fluid were tested according to the drilling fluid performance evaluation standard SY/T5621-1993. The apparent viscosity, plastic viscosity, and yield point of drilling fluid were measured with the ZNP-M7 6-speed rotating viscometer (Qingdao Haitongda Instrument Co., Ltd.). They measured the apparent viscosity, plastic viscosity, and yield point of drilling fluid with φ600 and φ300. The value of 300 was calculated according to the test program recommended by the API.

$$AV = \varphi 600/2 \quad (2)$$

$$PV = \varphi 600 - \varphi 300 \quad (3)$$

$$YP = \varphi 300 - \varphi 600/2 \quad (4)$$

where:

AV is the apparent viscosity (mPa·s);
PV is the plastic viscosity (mPa·s);
YP is the yield point (Pa);
φ600 is the dial reading of the 6-speed rotational viscometer at 600 r/min (dia);
φ300 is the dial reading of the 6-speed rotational viscometer at 300 r/min (dia).

2.2.4. Study of Filtrate Control Mechanism

(1) Zeta potential test

A Brookhaven zeta potential tester (Brookhaven instruments Ltd., New York, NY, USA) was used to test the zeta potential of drilling fluid before and after aging. An amount of 8 g of sodium montmorillonite was added to 400 mL of deionized water and placed on a magnetic stirrer for 24 h. Then, a certain amount of PAASM-$CaCO_3$ was added and stirred for 24 h to ensure that the various components of the mixed material were mixed sufficiently. Then, the drilling fluid was placed at a certain temperature and rolled for 16 h. When tested, drilling fluids were equipped with a microprocessor unit that automatically calculates the electron mobility of particles and converts them into ζ Potential. The average of the three tests was taken as the zeta potential of the drilling fluid.

(2) Particle size distribution test

An amount of 8 g of Na montmorillonite in 400 mL of deionized water was added and placed on a magnetic stirrer to stir for 24 h; then, a certain amount of PAASM-$CaCO_3$ was added, stirring was continued for 24 h to ensure the various components in the mixed material were fully mixed, and the particle size distribution of drilling fluid before and after aging was tested by a bettersize2000 laser particle size distribution instrument (Dandong baited Instrument Co., Ltd., Dandong, China).

3. Results and Discussion

3.1. Structural Characterizations

3.1.1. Fourier Transfer Infrared (FTIR) Analysis

The results of the infrared spectroscopic analysis of PAASM-$CaCO_3$ and NCC are shown in Figure 3. It can be seen from (a) that 3380 cm^{-1} is the O-H vibration absorption peak, 3195 cm^{-1} is the N-H stretching vibration absorption peak of the amide group, and 3004 cm^{-1} is the stretching vibration absorption peak. = C-H on the benzene ring, 2923 cm^{-1}, is the C-H stretching vibration absorption peak on the saturated carbon atoms, 1796 cm^{-1} is the stretching vibration absorption peak of C=O in the carboxyl group; 1664 cm^{-1} can be attributed to the stretching vibration of C=O in the amide group; 1604 cm^{-1} is for the stretching vibration of the C=C skeleton of the benzene ring; 1429 cm^{-1} can be assigned to C-O antisymmetric stretching vibration; S=O symmetric contraction vibration in sulfonic acid groups appears at 1120 cm^{-1}. The absorption peak at 875 cm^{-1} is assigned to the in-plane bending deformation vibration peak of $CaCO_3$ C-O, and the absorption peak at 632 cm^{-1} is assigned to the in-plane deformation vibration peak of O-C-O, which indicates that the product contains NCC. There are no vibrational peaks of olefin double bonds at 1000 cm^{-1}~900 cm^{-1}, indicating no residual monomers in the products. The FTIR analysis result shows that the chains of the synthesized products have chains bearing all comonomers.

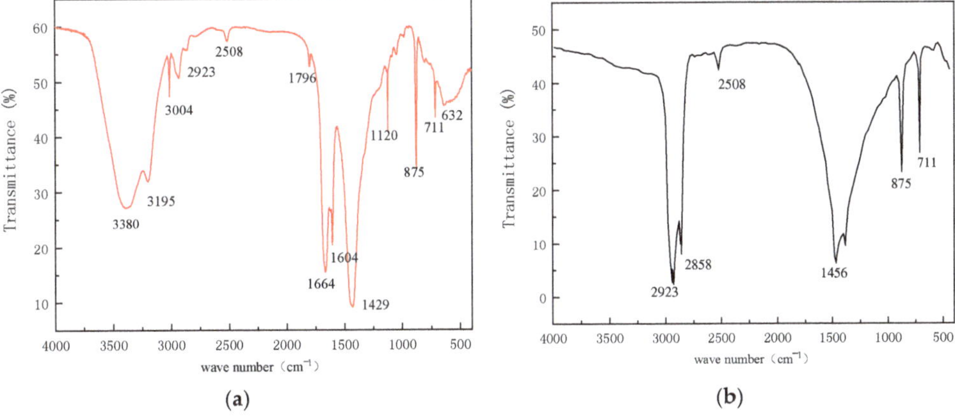

Figure 3. The FTIR analysis results. (**a**) PAASM-$CaCO_3$; (**b**) NCC.

3.1.2. Surface Hydroxyl Number Test

The results of the surface hydroxyl number test before and after NCC inlay copolymerization are shown in Table 2. From the test results, the number of surface hydroxyl groups of NCC is 0.1506/nm^2 before modification, and the number of surface hydroxyl groups of PAASM-$CaCO_3$ is drastically reduced to 0.0448/nm^2 after modification, indicating that a large number of hydroxyl groups on the surface of NCC participate in the reaction and the size of the nanoparticles matches well with the size of the network structure.

Table 2. Surface hydroxyl number test of NCC before and after modification.

	NCC	PAASM-$CaCO_3$
V/volume (mL)	0.411	0.129
N/Hydroxyl Number	0.1506	0.0448

3.1.3. Gel Permeation Chromatography (GPC) Test

The GPC experimental results of PAASM (without NCC) are shown in Table 3 and Figure 4. The results show that the weight-average molecular weight (Mw) of the polymer main chain is 6.98×10^5, and the number-average molecular weight (Mn) is 2.84×10^5 which gives PAASM a suitable relative molecular weight. It will be detrimental to the rheological regulation control of drilling fluid if the relative molecular weight of additive agents is too large. If the relative molecular weight is too small, it will be difficult to increase the viscosity, which will affect the effect of colloidal protection and filtration reduction and it will be difficult to guarantee high-temperature stability. The relative molecular mass of PAASM is moderate and has the potential to overcome the above drawbacks. At the same time, it can be found that the relative molecular weight distribution of PAASM is narrow and the polydispersity coefficient is 2.45, which indicates that the molecular mass distribution of PAASM polymer is relatively uniform.

Table 3. Relative molecular weight test results of PAASM.

Weight-Average Molecular Weight (M_W)	Number-Average Molecular Weight (Mn)	Polydispersity Coefficient (D)
697,500	284,600	2.45

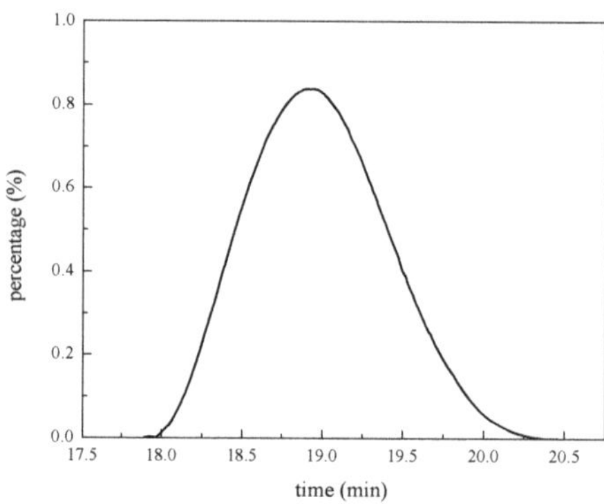

Figure 4. Results of GPC of PAASM.

3.1.4. Thermogravimetric Analysis

The thermogravimetric analysis results of PAASM-$CaCO_3$ are shown in Figure 5. As can be seen from Figure 5, there is no decomposition of PAASM-$CaCO_3$ from room temperature to 300 °C, and the weight loss is mainly due to the re-removal of adsorbed water. The side chain begins to decompose from 330 °C to 420 °C. When the temperature is higher than 420 °C, the molecular skeleton is completely destroyed and the final residual mass is about 45%, which is mainly composed of NCC and the carbonized main chain structure. The results of thermogravimetric analysis show that the total thermal weight loss of PAASM-$CaCO_3$ is about 55% in the range from room temperature to 500 °C. The thermal decomposition temperature of PAASM-$CaCO_3$ is much higher than those of traditional agents.

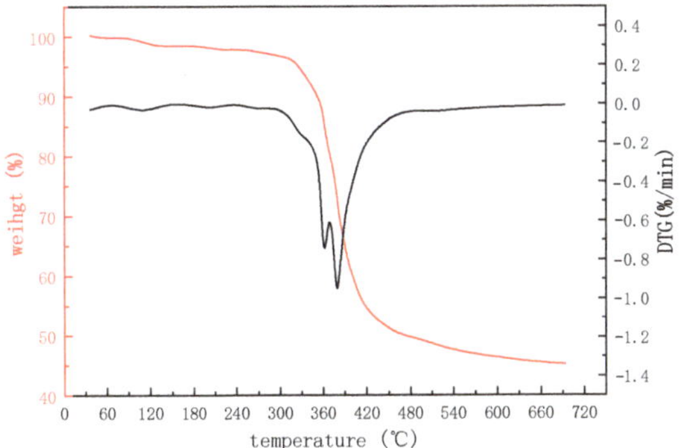

Figure 5. Thermogravimetric analysis results of PAASM-$CaCO_3$.

3.1.5. Micro-Morphology Test

TEM images of NCC and PAASM-$CaCO_3$ in an aquatic environment are shown in Figure 6. It can be seen that the NCC before modification is cubic with an average particle size of about 15 nm (Figure 6a). The modified particles are spherical and the particle size changes to about 200 nm (Figure 6b). The reason for this change is that the modified polymer is coated on the surface of NCC, and the polymer swells in water and partially dissolves in water, resulting in adhesion, resulting in the increase in NCC particle size in TEM, which also indicates that NCC has been successfully modified.

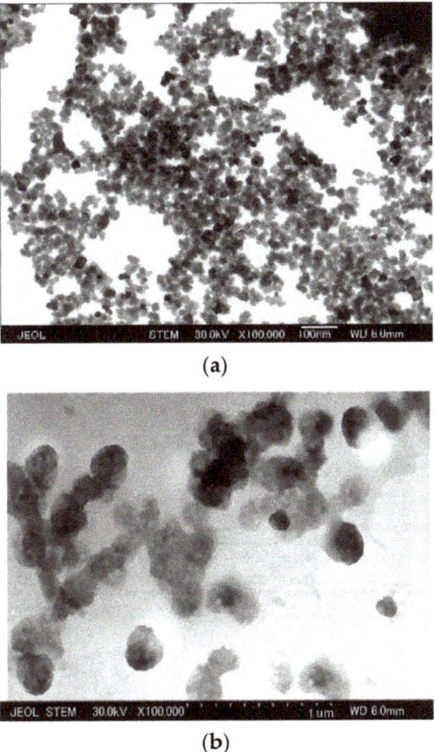

Figure 6. TEM test result. (**a**) NCC suspension; (**b**) PAASM-$CaCO_3$ suspension.

The SEM image of PAASM-$CaCO_3$ is shown in Figure 7. It can be seen that PAASM-$CaCO_3$ is mainly spherical with a large specific surface area, which can exert its surface energy advantage. The particles are closely packed, adhere to each other, and are basically connected by polymers. The particle size of the polymers is basically the same as that measured by TEM.

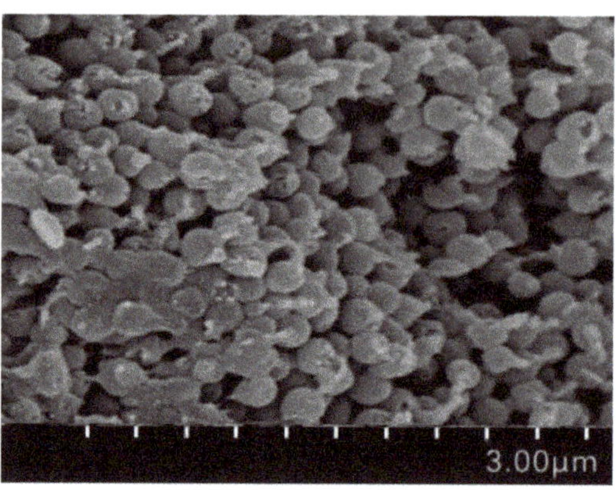

Figure 7. SEM photo of PAASM-$CaCO_3$.

3.1.6. Particle Size Distribution Test

The results of the particle size distribution of 0.1% PAASM-$CaCO_3$ in water are shown in Figure 8. As can be seen from the test results, the PAASM-$CaCO_3$ particle size distribution is narrow, with a D50 of approximately 259 nm, D10 of 1.88 nm, and D90 of 877 nm. The particle size test results are relatively close to the TEM results. Compared with the SEM results, the reason for the increase in particle size is mainly due to the water absorption of the polymer and the hydration in the solution.

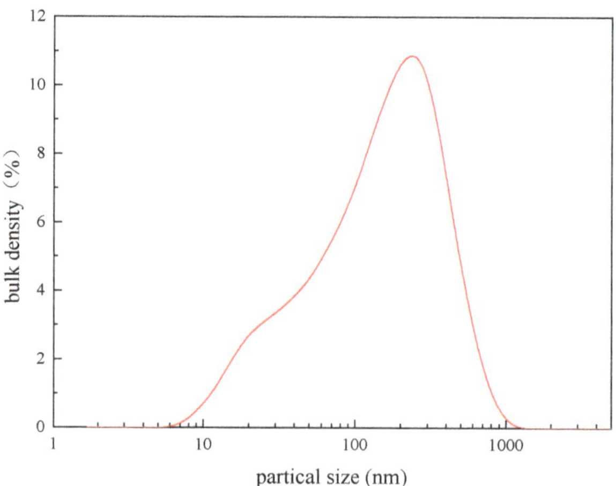

Figure 8. PAASM-$CaCO_3$ particle size distribution.

3.2. Performance Evaluation

3.2.1. Filtration Reduction Effect in 4% Bentonite Mud

The filtration volume under medium pressure and HTHP (200 °C/3.5 MPa) of PAASM-$CaCO_3$ and commonly used high-temperature-resistant polymer filter reducers (coded

Driscal-D and D-4) after aging at 200 °C/16 h was compared to further evaluate the filtration reduction capacity of PAASM-$CaCO_3$. Table 1 shows the formulations of different drilling fluids. The main components of drilling fluids are shown in Table 4.

Table 4. Preparation of drilling fluids.

Components	Amount (Concentration)			
	Base Fluid	1	2	3
Distilled water (mL)	400	400	400	400
sodium montmorillonite (g)	16	16	16	16
Na_2CO_3(g)	0.8	0.8	0.8	0.8
PAASM-$CaCO_3$	0	4	0	0
Driscal-D	0	0	4	0
D-4	0	0	0	4

The filtration effects of 1% PAASM-$CaCO_3$, 1% D-4, and Driscal-D in the base mud are shown in Figure 9. It can be seen that the filtration of base fluid decreases to some extent after adding different filtration reducers. After aging at 200 °C, D-4 has the best filtration reduction performance. The medium pressure filtration loss is 4.8 mL, and the HTHP filtration loss is 11 mL. The filtration reduction result of PAASM-$CaCO_3$ is close to that of D-4. After aging at 200 °C, the medium pressure filtration loss is 5.2 mL and the high-temperature and high-pressure filtration is 14.4 mL. After aging at 200 °C/16 h, the API filtration volume of Driscal-D is reduced by 19 mL. The test shows that PAASM-$CaCO_3$ has a good filtration control effect in 4% bentonite drilling fluid.

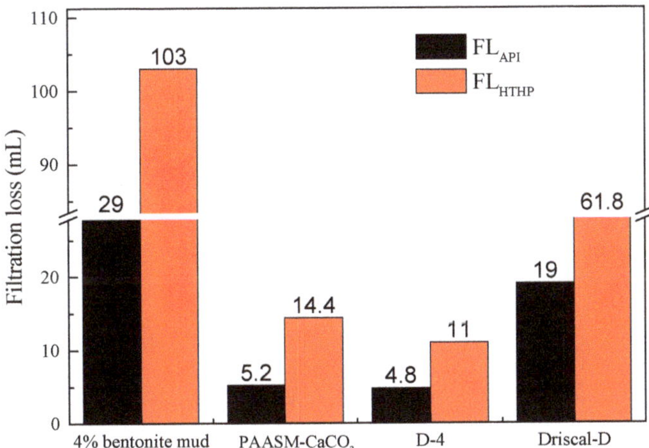

Figure 9. Filtration loss in based muds after aging at 200 °C.

3.2.2. Effect on Rheology of WBDF

In order to further study the influence of PAASM-$CaCO_3$ on the rheological properties of drilling fluids, the rheological properties and filtration properties of fluids were tested with different PAASM-$CaCO_3$ additions before and after aging at 200 °C/16 h. The base fluid is 400 mL deionized water + 16 g sodium montmorillonite + 0.8 g Na_2SO_3. The experimental results are shown in Figure 10.

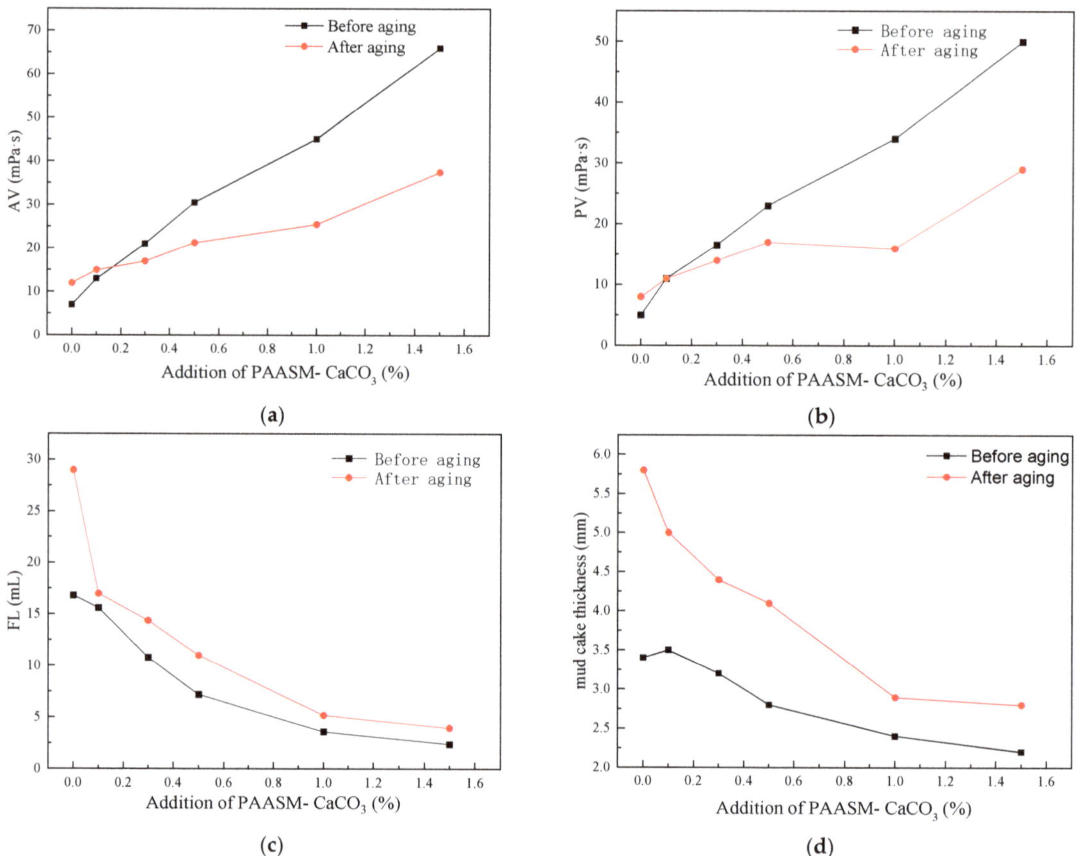

Figure 10. Effect of PAASM-$CaCO_3$ on rheological and filtration properties of 4% bentonite mud. (**a**) Apparent viscosity; (**b**) plastic viscosity; (**c**) filtration Volume; (**d**) mud cake thickness.

It can be seen from the results that with the increase in PAASM-$CaCO_3$ content in 4% bentonite mud, AV, PV, and YP gradually increase. When 1.5% PAASM-$CaCO_3$ is added, the apparent viscosity of the slurry before and after aging is 66 MPa·s and 37.5 MPa·s, respectively, which indicates that PAASM-$CaCO_3$ still has a good tackifying effect after aging. In addition, the yield point of the pulp before and after aging is stable under different PAASM-$CaCO_3$ additions. It is also found that with the continuous addition of PAASM-$CaCO_3$, the filtration loss of aging fluids decreases continuously. After aging at 200 °C/16 h, the filtration loss is 29 mL. Filtration loss is only 4 mL when 1.5% PAASM-$CaCO_3$ is added. At the same time, it is not difficult to see that with the increase in PAASM-$CaCO_3$, the mud cake thickness of fluid becomes thinner gradually after aging, which indicates that the mud cake quality gradually improves and starts to become thinner and tougher, which indicates that the addition of PAASM-$CaCO_3$ can improve the colloidal high-temperature stability of drilling fluids.

3.2.3. Comparison with Other Commonly Used Filtration Reducers

In order to further evaluate PAASM-$CaCO_3$'s filtration reduction ability, the filtration control effects of PAASM-$CaCO_3$ and other commonly used high-temperature-resistant filter agents before and after aging at 200 °C/16 h were compared. The results are shown in Table 5. It can be seen from the test results that the API filtration loss volume of D-4 and PAASM-$CaCO_3$ are smallest after hot-rolling at 200 °C/16 h, which are 4.8 mL and 5.2 mL, respectively. The mud cake thickness of PAASM-$CaCO_3$ drilling fluid after aging is similar to that of D-4 drilling fluid, which is 2.9 mm and 2.4 mm, respectively, which shows that PAASM-$CaCO_3$ prevents mud cake from becoming too thick and improves the quality, indicating that the newly developed PAASM-$CaCO_3$ has good filtration control and good heat resistance.

Table 5. Comparison of filtration control effects after aging (200 °C/16 h) of common filtration reducers.

No.	Components	Filtration Volume (mL)		MCT (mm)
		Before Aging	After Aging	
Base Fluid	4% bentonite mud	16.8	29.0	5.8
1	1#+1%PAASM-$CaCO_3$	3.6	5.2	2.9
2	1#+1%Driscal-D	8.8	19.0	4.7
3	1#+1%D-4	3.2	4.8	2.4
4	1#+1%80A51	7.2	20.0	5.0
5	1#+1%JT888	4.4	18.0	4.9
6	1#+4%PJA-2	6.2	17.0	5.0
7	1#+4%FT-A	6.6	19.0	4.9
8	1#+4%SMP-I	5.2	18.0	5.0

3.2.4. Evaluation of Temperature Resistance

The results of API filtration and HTHP filtration after aging at different temperatures are shown in Figure 11. It can be seen that with the increase in aging temperature, the API loss of drilling fluid gradually increases. As the temperature increases, the high-temperature and -pressure loss of drilling fluid increases rapidly at first, and then slowly and then rapidly. When the temperature is lower than 180 °C, the change in temperature has little effect on drilling fluid filtration. When the temperature exceeds 180 °C, the molecular structure of PAASM-$CaCO_3$ starts to become damaged under high temperature, the movement of water molecules becomes more serious, and the hydration group of PAASM-$CaCO_3$ cannot play its full role, which leads to the weakening of gel protection and the increase in filtration loss.

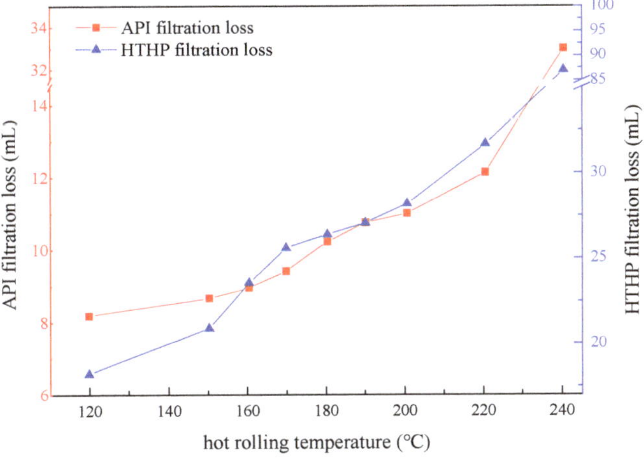

Figure 11. Filtration loss after aging at different temperatures.

3.3. Study of Filtration Control Mechanism

3.3.1. Particle Size Distribution Test

Large and small particles often coexist in drilling fluids and maintain a certain proportion. During filtration, large particles in drilling fluid can act as a bridge to support the main frame of the mud cake. Small particles play a filling role. The cage structure formed by large particles can be filled with a suitable proportion of small solid particles, thus improving the density of the filter cake and keeping the filtration performance of drilling fluids in a good state.

Figure 12 and Table 6 show the test results of the particle size distribution of drilling fluids after aging in the presence of PAASM-$CaCO_3$ at different concentrations. From the test results, it can be seen that the particle size distribution of basic mud is wide and multi-peaked. There are submicron, micron, and millimeter particles in the drilling fluid, and the median particle size is 28.80 μm. With the addition of PAASM-$CaCO_3$, the particle size distribution of mud gradually changes from multi-peaked to single-peaked, and the particle size distribution curve moves to the left. This shows that with the increase in the PAASM-$CaCO_3$, the number of small particles of micron and sub-micron size increases, and their particle size distribution becomes narrower, which makes the particles in mud more uniform. PAASM-$CaCO_3$ has a strong adsorptive functional group and rigid nanostructure, which has a high adsorptive energy and surface energy. It can appear on the surface of clay particles after adding in. In addition, it improves the thickness and diffusion of hydration film and the double layer on the surface of clay particles, enhancing the water and static repulsion between particles, and at the same time, its hydrophobic structure can form a film around clay particles. Therefore, the aggregation of clay particles is restrained and clay particles are decomposed into fine particles, which is beneficial to reducing mud filtration.

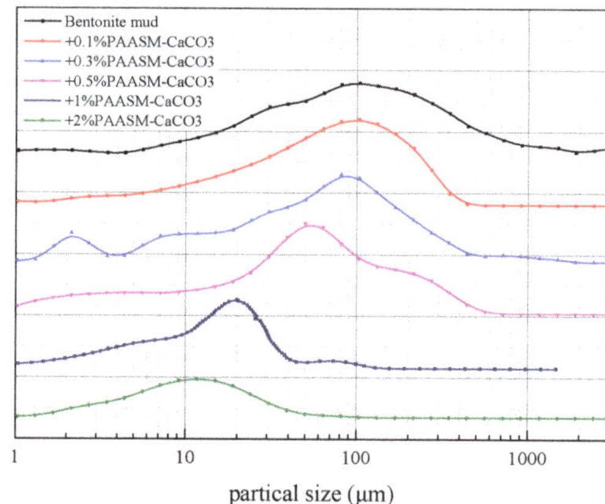

Figure 12. Particle size distribution after aging (in °C/16 hg).

Table 6. Partial size value of muds.

Concentration of PAASM-$CaCO_3$/%	Particle Size/μm		
	D10	D50	D90
0	5.119	28.8	139
0.1	1.772	19.01	58.75
0.3	2.016	18.58	66.99
0.5	1.171	14.46	60.71
1	3.171	14.48	31.27
2	2.772	12.65	23.65

The test results of the particle size distribution of drilling fluids before and after aging at different temperatures are shown in Figure 13 and Table 7. It can be seen that the particle size distribution of the drilling fluid is relatively wide before aging, which indicates that under this condition, the particles are conducive to the bridging action of large particles and filling action of small particles. After aging at different temperatures, the size distribution of the muds becomes narrower and the size becomes larger. It can be seen from the test results that within a certain temperature range (<210 °C), with the increase in aging temperature, the grain size distribution curve of the mud gradually moves to the left, indicating that the small particles in drilling fluid begin to increase and the large particles gradually decrease. In conclusion, PAASM-$CaCO_3$ gradually plays a role in this temperature range as the temperature increases. The clay particles affected by high-temperature dehydration and aggregation are dispersed to form a relatively stable colloidal suspension system. When the temperature increases further (>220 °C), the median particle size of the mud increases, the particle size distribution curve starts to show a bi-peak distribution, and the main peak starts to move to the right, indicating that the sub-micron particles begin to increase gradually and the clay particles in the muds begin to aggregate. The reason for this may be that the ionization balance of water is promoted with increasing temperature. At this time, the active H+ in water gradually increases and the attack probability of the PAASM-$CaCO_3$ molecular main chain increases greatly, which leads to the oxidation, deformation, and even decomposition of the PAASM-$CaCO_3$ molecular main chain, releasing rigid nanoparticles initially encapsulated in the polymer framework and combining them. Therefore, clay particles without PAASM-$CaCO_3$ protection also begin to dehydrate and aggregate at high temperatures, forming small submicron particle clusters and thinning the hydration film. The intermolecular hydration repulsion is weakened, resulting in an increase in particle size.

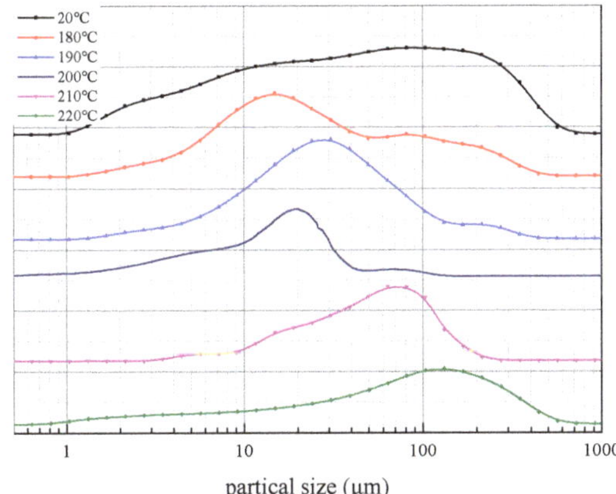

Figure 13. Particle size distribution curve of drilling fluids after aging.

Table 7. Particle size value of muds after aging.

Aging Temperature/°C	Particle Size/μm		
	D10	D50	D90
20	1.322	12.31	72.03
180	3.824	13.7	97.59
190	4.93	16.7	57.64
200	3.171	14.48	31.27
210	4.521	16.67	40.7
220	1.796	24.17	77.94

3.3.2. Zeta Potential Test

The zeta potential values of PAASM-$CaCO_3$ mud with different additions are shown in Figure 14. It can be seen from the test results that the zeta potential of the muds is negative throughout the experiment, which indicates that the charge properties of the surface of bentonite particles have not changed with the addition of PAASM-$CaCO_3$. Before aging, with the increase in PAASM-$CaCO_3$ concentration, the absolute zeta potential of the drilling fluid begins to increase, which indicates that more and more bentonite particles are adsorbed on the surface of PAASM-$CaCO_3$. When the concentration of PAASM-$CaCO_3$ increases to 1.5%, the absolute zeta potential of the mud does not increase significantly indicating that under this condition, PAASM-$CaCO_3$ forms saturated adsorption, continues to increase its dosage, and cannot adsorb more bentonite colloidal particles. The absolute zeta potential of the mud increases significantly with the increase in PAASM-$CaCO_3$ concentration after aging at 200 °C/16 h, which indicates that the hydrate groups in PAASM-$CaCO_3$ adsorb on the surface of bentonite particles at high temperature and improve the double layer thickness of diffusion electricity. The protective effect of PAASM-$CaCO_3$ on colloidal particles is more obvious with the increase in PAASM-$CaCO_3$ dosage. High-concentration PAASM-$CaCO_3$ can be adsorbed excessively on the surface of clay particles, inhibiting the adverse effects of high-temperature dehydration. In addition, according to the zeta potential of before and after aging, it can be seen that the zeta potential of bentonite after high-temperature treatment decreases slightly compared with the absolute value before aging, but the decrease is not significant, indicating that the zeta potential source of clay particles is less affected by temperature. In the presence of different concentrations of PAASM-$CaCO_3$, the absolute zeta potential of drilling fluids after aging also decreases, which indicates that some hydration and adsorption groups of PAASM-$CaCO_3$ decompose at high temperature, resulting in a decrease in the amount of adsorption on the surface of clay particles.

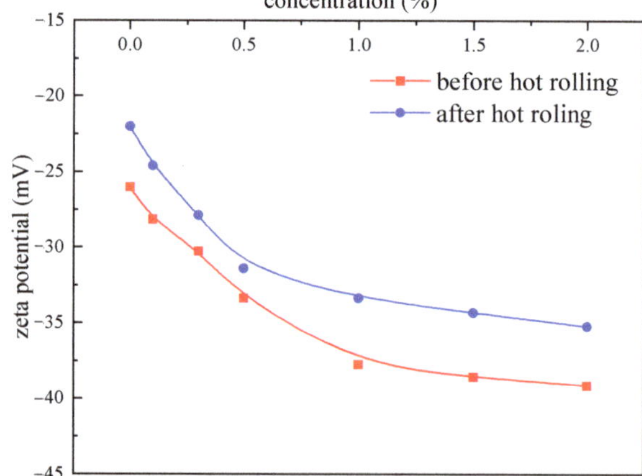

Figure 14. Effects of different concentrations of PAASM-$CaCO_3$ on zeta potential.

4. Conclusions

A high-temperature filtration reducer with a core–shell structure (PAASM-$CaCO_3$) was developed by mosaic copolymerization using AMPS, AM, St, and MA as monomers in combination with NCC. The monomers were successfully polymerized, and the weight-average relative molecular weight of the organic macromolecular backbone in the PAASM-$CaCO_3$ structure was 6.98×10^5. The PAASM-$CaCO_3$ initial thermal decomposition temperature was high, which was about 300 °C. The newly developed PAASM-$CaCO_3$ was in a spatially globular structure. PAASM-$CaCO_3$ had a better resistance to high temperature. The filtration loss of drilling fluid was basically stable before and after aging at 200 °C/16 h, whose effects were near or surpassed that of Driscal-d and was on top of commonly used filtration reducers. The rigid nanoparticles were properly introduced into the PAASM-$CaCO_3$, which enhanced the steric hindrance and stability of the molecular structure, and was beneficial in improving its high-temperature resistance.

Author Contributions: Data curation, K.S. and Y.K.; Funding acquisition, H.M.; Investigation, Z.T.; Project administration, Z.Q. and H.Z.; Writing—original draft, Z.T.; Writing—review & editing, Z.T. All authors have read and agreed to the published version of the manuscript.

Funding: This research received no external funding.

Institutional Review Board Statement: Not applicable.

Informed Consent Statement: Not applicable.

Data Availability Statement: Not applicable.

Conflicts of Interest: The authors declare no conflict of interest.

References

1. Saleh, T.A. Advanced trends of shale inhibitors for enhanced properties of water-based drilling fluid. *Upstream Oil Gas Technol.* **2022**, *8*, 100069. [CrossRef]
2. Akpan, E.U.; Enyi, G.C.; Nasr, G.; Yahaya, A.A.; Ahmadu, A.A.; Saidu, B. Water-based drilling fluids for high-temperature applications and water-sensitive and dispersible shale formations. *J. Pet. Sci. Eng.* **2019**, *175*, 1028–1038. [CrossRef]
3. Aftab, A.; Ali, M.; Sahito, M.F.; Mohanty, U.S.; Jha, N.K.; Akhondzadeh, H.; Azhar, M.R.; Ismail, A.R.; Keshavarz, A.; Iglauer, S. Environmental friendliness and high performance of multifunctional tween 80/ZnO-nanoparticles-added water-based drilling fluid: An experimental approach. *ACS Sustain. Chem. Eng.* **2020**, *8*, 11224–11243. [CrossRef]
4. Huang, Y.; Zheng, W.; Zhang, D.; Xi, Y. A modified Herschel–Bulkley model for rheological properties with temperature response characteristics of poly-sulfonated drilling fluid. *Energy Sources Part A Recovery Util. Environ. Eff.* **2020**, *42*, 1464–1475. [CrossRef]
5. LI, Y.; CHEN, Y.; Wang, H.; WU, X.-l. Application of sulfonated asphalt drilling fluid in geothermal exploration well of Guizhou. *Drill. Eng.* **2015**, *2*, 27–30.
6. Wang, Z.; Wu, Y.; Luo, P.; Tian, Y.; Lin, Y.; Guo, Q. Poly (sodium p-styrene sulfonate) modified Fe_3O_4 nanoparticles as effective additives in water-based drilling fluids. *J. Pet. Sci. Eng.* **2018**, *165*, 786–797. [CrossRef]
7. Lei, M.; Huang, W.; Sun, J.; Shao, Z.; Chen, Z.; Chen, W. Synthesis and characterization of high-temperature self-crosslinking polymer latexes and their application in water-based drilling fluid. *Powder Technol.* **2021**, *389*, 392–405. [CrossRef]
8. Sharghi, H.; Shiri, P.; Aberi, M. An overview on recent advances in the synthesis of sulfonated organic materials, sulfonated silica materials, and sulfonated carbon materials and their catalytic applications in chemical processes. *Beilstein J. Org. Chem.* **2018**, *14*, 2745–2770. [CrossRef]
9. Lin, L.; Luo, P. Amphoteric hydrolyzed poly(acrylamide/dimethyl diallyl ammonium chloride) as a filtration reducer under high temperatures and high salinities. *J. Appl. Polym. Sci.* **2014**, *132*. [CrossRef]
10 Shi, J.; Wu, Z.; Deng, Q.; Liu, L.; Zhang, X.; Wu, X.; Wang, Y. Synthesis of hydrophobically associating polymer: Temperature resistance and salt tolerance properties. *Polym. Bull.* **2021**, 1–11. [CrossRef]
11. Mm, A.; Hma, B.; Xz, A.; As, C.; Mm, C.; Msk, A. Okra mucilage as environment friendly and non-toxic shale swelling inhibitor in water based drilling fluids. *Fuel* **2022**, *320*, 123868.
12. Vecchi, S.; Demarchi, C.; Vigano, L.; Bassi, G.; Chiavacci, D.; Giuseppe, L.B. Filtrate Reducer for Drilling Muds. U.S. Patent 11/615,475, 12 July 2007.
13. Ma, X.; Yang, M.; Zhang, M. Synthesis and properties of a betaine type copolymer filtrate reducer. *Chem. Eng. Process.* **2020**, *153*, 107953. [CrossRef]
14. Liu, F.; Zhang, Z.; Wang, Z.; Dai, X.; Chen, M.; Zhang, J. Novel lignosulfonate/N,N-dimethylacrylamide/γ-methacryloxypropyl trimethoxy silane graft copolymer as a filtration reducer for water-based drilling fluids. *J. Appl. Polym. Sci.* **2020**, *137*, 48274. [CrossRef]
15. Giddings, D.M.; Williamson, C.D. Terpolymer Compositions for Aqueous Drilling Fluids. U.S. Patent 4,678,591, 7 July 1987.
16. Dickert, J.J.; Heilweil, I.J. Additive Systems for Control of Fluid Loss in Aqueous Drilling Fluids at High Temperatures. U.S. Patent 4,626,362, 2 December 1986.
17. Patel, D.A. Thermally Stable Drilling Fluid Additive Comprised of a Copolymer of Catechol-Based Monomer. U.S. Patent 4,595,736, 17 June 1986.
18. Liu, L.; Pu, X.; Rong, K.; Yang, Y. Comb-shaped copolymer as filtrate loss reducer for water-based drilling fluid. *J. Appl. Polym. Sci.* **2017**, *135*, 45989. [CrossRef]
19. Zhan, W.; Vaidya, R.N.; Suryanarayana, P.V. Simulation of Dynamic Filtrate Loss During the Drilling of a Horizontal Well with High-Permeability Contrasts and Its Impact on Well Performance. *Spe Reserv. Eval. Eng.* **2009**, *12*, 886–897.

20. Zamani, A.; Bataee, M.; Hamdi, Z.; Khazforoush, F. Application of smart nano-WBM material for filtrate loss recovery in wellbores with tight spots problem: An empirical study. *J. Pet. Explor. Prod. Technol.* **2019**, *9*, 669–674. [CrossRef]
21. Thaemlitz, C.; Patel, A.; Coffin, G.; Conn, L. A New Environmentally Safe High-Temperature, Water-Base Drilling Fluid System. In Proceedings of the Spe/iadc Drilling Conference, Amsterdam, The Netherlands, 4–6 March 1997.
22. Tomislav, S.; Robert, H.; Pavel, M. Uniquely Engineered Water-Base High-Temperature Drill-In Fluid Increases Production, Cuts Costs in Croatia Campaign. In Proceedings of the Spe/iadc Drilling Conference, Amsterdam, The Netherlands, 19–21 February 2003.
23. Zhu, A.C.; Qiu, Z.S.; Yuan, X.J.; Hu, H.F. Nanometer-modified CMC:Preparation, Structure Characterization, and Performance Evaluation. *Drill. Fluid Completion Fluid* **2007**, *5*, 45–53.
24. Mao, H.; Wang, W.; Ma, Y.; Huang, Y. Synthesis, characterization and properties of an anionic polymer for water-based drilling fluid as an anti-high temperature and anti-salt contamination fluid loss control additive. *Polym. Bull.* **2021**, *78*, 2483–2503. [CrossRef]
25. Ikram, R.; Jan, B.M.; Sidek, A.; Kenanakis, G. Utilization of Eco-Friendly Waste Generated Nanomaterials in Water-Based Drilling Fluids; State of the Art Review. *Materials* **2021**, *14*, 4171. [CrossRef]
26. Ahasan, M.H.; Alvi, M.F.A.; Ahmed, N.; Alam, M.S. An investigation of the effects of synthesized zinc oxide nanoparticles on the properties of water-based drilling fluid. *Pet. Res.* **2022**, *7*, 131–137. [CrossRef]
27. Ikram, R.; Jan, B.M.; Vejpravova, J. Towards recent tendencies in drilling fluids: Application of carbon-based nanomaterials. *J. Mater. Res. Technol.* **2021**, *15*, 3733–3758. [CrossRef]
28. Ikram, R.; Jan, B.M.; Vejpravova, J.; Choudhary, M.; Chowdhury, Z.Z. Recent Advances of Graphene-Derived Nanocomposites in Water-Based Drilling Fluids. *Nanomaterials* **2020**, *10*, 2004. [CrossRef] [PubMed]
29. Tingji, D.; Ruihe, W.; Jiafang, X.; Jiayue, M.; Xiaohui, W.; Jiawen, X.; Xiaolong, Y. Synthesis and application of a temperature sensitive poly (N-vinylcaprolactam-co-N, N-diethyl acrylamide) for low-temperature rheology control of water-based drilling fluid. *Colloids Surf. A Physicochem. Eng. Asp.* **2022**, *644*, 128855. [CrossRef]
30. Sun, J.; Zhang, X.; Lv, K.; Liu, J.; Xiu, Z.; Wang, Z.; Huang, X.; Bai, Y.; Wang, J.; Jin, J. Synthesis of hydrophobic associative polymers to improve the rheological and filtration performance of drilling fluids under high temperature and high salinity conditions. *J. Pet. Sci. Eng.* **2022**, *209*, 109808. [CrossRef]
31. Yang, J.; Sun, J.; Bai, Y.; Lv, K.; Zhang, G.; Li, Y. Status and Prospect of Drilling Fluid Loss and Lost Circulation Control Technology in Fractured Formation. *Gels* **2022**, *8*, 260. [CrossRef]
32. Wei, Z.; Zhou, F.; Chen, S.; Long, W. Synthesis and Weak Hydrogelling Properties of a Salt Resistance Copolymer Based on Fumaric Acid Sludge and Its Application in Oil Well Drilling Fluids. *Gels* **2022**, *8*, 251. [CrossRef]
33. Ricky, E.; Mpelwa, M.; Wang, C.; Hamad, B.; Xu, X. Modified Corn Starch as an Environmentally Friendly Rheology Enhancer and Fluid Loss Reducer for Water-Based Drilling Mud. *SPE J.* **2022**, *27*, 1064–1080. [CrossRef]
34. Yunxiang, L.; Ling, L.; Wenke, Y.; Xin, L.; Han, G. Synthesis and Evaluation of Betaine Copolymer Filtrate Reducer for Drilling Mud. *Clays Clay Miner.* **2022**, 1–18. [CrossRef]
35. Long, W.; Zhu, X.; Zhou, F.; Yan, Z.; Evelina, A.; Liu, J.; Wei, Z.; Ma, L. Preparation and Hydrogelling Performances of a New Drilling Fluid Filtrate Reducer from Plant Press Slag. *Gels* **2022**, *8*, 201. [CrossRef]
36. da Silva, I.G.; Lucas, E.F.; Advincula, R. On the use of an agro waste, Miscanthus x. Giganteus, as filtrate reducer for water-based drilling fluids. *J. Dispers. Sci. Technol.* **2022**, *43*, 776–785. [CrossRef]
37. Li, J.; Sun, J.; Lv, K.; Ji, Y.; Liu, J.; Huang, X.; Bai, Y.; Wang, J.; Jin, J.; Shi, S. Temperature-and Salt-Resistant Micro-Crosslinked Polyampholyte Gel as Fluid-Loss Additive for Water-Based Drilling Fluids. *Gels* **2022**, *8*, 289. [CrossRef]
38. Liu, Y.; Chen, L.; Tang, Y.; Zhang, X.; Qiu, Z. Synthesis and characterization of nano-SiO_2@ octadecylbisimidazoline quaternary ammonium salt used as acidizing corrosion inhibitor. *Rev. Adv. Mater. Sci.* **2022**, *61*, 186–194. [CrossRef]
39. Mba, B.; Jmac, E.; Th, D.; Mehf, G. An experimental study to develop an environmental friendly mud additive of drilling fluid using Aloe Vera. *J. Pet. Sci. Eng.* **2022**, *211*, 110135.
40. Huang, W.A.; Wang, J.W.; Lei, M.; Li, G.R.; Duan, Z.F.; Li, Z.J.; Yu, S.F. Investigation of regulating rheological properties of water-based drilling fluids by ultrasound. *Pet. Sci.* **2021**, *18*, 11. [CrossRef]
41. Liu, F.; Yao, H.; Liu, Q.; Wang, X.; Dai, X.; Zhou, M.; Wang, Y.; Zhang, C.; Wang, D.; Deng, Y. Nano-silica/polymer composite as filtrate reducer in water-based drilling fluids. *Colloids Surf. A Physicochem. Eng. Asp.* **2021**, *627*, 127168. [CrossRef]

gels

Article

Experimental and Numerical Investigation on Oil Displacement Mechanism of Weak Gel in Waterflood Reservoirs

Hongjie Cheng [1], Xianbao Zheng [2], Yongbin Wu [3,*], Jipeng Zhang [4], Xin Zhao [2] and Chenglong Li [2]

1 Xinjiang Oilfield Company, Petrochina, Keramay 834000, China; chjie@petrochina.com.cn
2 Research Institute of Petroleum Exploration & Development, Petrochina Daqing Oilfield, Daqing 163000, China; zhengxianbao@petrochina.com.cn (X.Z.); zhaoxin530@petrochina.com.cn (X.Z.); lcl716@126.com (C.L.)
3 Research Institute of Petroleum Exploration & Development, Petrochina, Beijing 100083, China
4 School of Energy, China University of Geosciences (Beijing), Beijing 100083, China; 18152696912@163.com
* Correspondence: wuyongbin@petrochina.com.cn

Abstract: The production performance of waterflood reservoirs with years of production is severely challenged by high water cuts and extensive water channels. Among IOR/EOR methods, weak gel injection is particularly effective in improving the water displacement efficiency and oil recovery. The visualized microscopic oil displacement experiments were designed to comprehensively investigate the weak gel mechanisms in porous media and the numerical simulations coupling equations characterizing weak gel viscosity induced dynamics were implemented to understand its planar and vertical block and movement behaviors at the field scale. From experiments, the residual oil of initial water flooding mainly exists in the form of cluster, column, dead end, and membranous, and it mainly exists in the form of cluster and dead end in subsequent water flooding stage following weak gel injection. The porous flow mechanism of weak gel includes the preferential plugging of large channels, the integral and staged transport of weak gel, and the residual oil flow along pore walls in weak gel displacement. The profile-control mechanism of weak gel is as follows: weak gel selectively enters the large channels, weak gel blocks large channels and forces subsequent water flow to change direction, weak gel uses viscoelastic bulk motion to form negative pressure oil absorption, and the oil droplets converge to form an oil stream, respectively. The numerical simulation indicates that weak gel can effectively reduce the water-oil mobility ratio, preferentially block the high permeability layer and the large pore channels, divert the subsequent water to flood the low permeability layer, and improve the water injection swept efficiency. It is found numerically that a weak gel system is able to flow forward under high-pressure differences in the subsequent water flooding, which can further improve oil displacement efficiency. Unlike the conventional profile-control methods, weak gels make it possible to displace the bypassed oil in the deep inter-well regions with significant potential to enhance oil recovery.

Keywords: weak gel; oil recovery; numerical simulation; displacement efficiency

Citation: Cheng, H.; Zheng, X.; Wu, Y.; Zhang, J.; Zhao, X.; Li, C. Experimental and Numerical Investigation on Oil Displacement Mechanism of Weak Gel in Waterflood Reservoirs. *Gels* **2022**, *8*, 309. https://doi.org/10.3390/gels8050309

Academic Editors: Qing You, Guang Zhao and Xindi Sun

Received: 12 April 2022
Accepted: 13 May 2022
Published: 17 May 2022

1. Introduction

How to effectively shut in the high permeability channels and divert injected water to the bypassed low permeability regions, is a crucial question for the water flooding process in the late stage of water injection. After years of in-depth development in waterflood reservoirs, the residual oil is highly dispersed and mainly distributed in the deep area of the inter-well region [1], conventional modification of good pattern and injection and production parameters is less effective [2], and profile-control by traditional gels or polymers is no longer feasible as the high-concentration gels consolidate quickly and can only be effective in plugging the wellbore vicinity [3–6]. On the contrary, weak gels have a competitive edge over conventional gels in the following aspects: conventional gel has a relatively high concentration of polymer and crosslinker, high gel strength, short gel formation time, and high cost. Therefore, it can only be used for permeability adjustment in the near-well zone, but cannot effectively solve the problem of deep reservoir heterogeneity [6–8]. Weak gel however, uses delayed crosslinking technology to inject a certain concentration of polymer and crosslinking agent into the deep reservoir, forming a polymer gel of a certain strength in the high permeability zone far from the injector vicinity, forcing the subsequent fluid to shift into the low permeability zone with high oil saturation, expanding the swept volume and improving the oil displacement efficiency [9,10].

Extensive laboratory investigation and field applications of weak gels have been performed in past years [7,8]. Different gel systems including ultra-high molecular weight HPAM/phenolic weak gel system, particle gels, gel agents with a high salt resistance solvent–polymer weak gels, colloidal dispersion gels, et al., have been developed for oil reservoirs with various conditions [9–14], which effectively expand the applicability and potential of weak gels. Moreover, the hybrid process of weak gels with other recovery methods is also a new trend, such as the combination of weak gels and microorganisms and the CO_2-gel fracturing system in shale oil reservoirs [15,16].

Evaluation of gelation systems directly influences the type, operational parameters and application performance of weak gels [17–19], in which numerical simulations were performed to study the influence factors of gelling performance [20,21], and laboratory experiments including NMR and coreflooding were used to investigate the displacement mechanisms [22–24].

However, the flow characteristics of the weak gel system in the porous media and the micro-mechanisms of oil displacement by reducing relative permeability of water phase in high permeability channels and expanding the water flooding swept area are not fully understood. The numerical simulation coupling experiment phenomenon and the equations characterizing migration dynamics of weak gel system in porous media, which describes the process of gel deformation and migration under the influences of pressure gradient, velocity, retention, time, and other factors, was yet to be performed previously The visualized micro-scale experiments and the field scale numerical simulations were performed in this study to further investigate the oil displacement dynamics of weak gels and subsequent waterflooding, which is significant to guide the extensive application of weak gels.

2. Experiment

2.1. Experimental Equipment and Materials

The experimental oil is the simulated oil prepared using crude oil and kerosene in the laboratory, with a viscosity of 3.8 mpa·s at 45 °C. The experimental water uses the field-produced water with a salinity of 4800 mg/L. The weak gel used in the experiment is a polymer and chromium crosslinking agent (1600 mg/L × 2000 mg/L), in which the molecular weight of the polymer is 800 million. The experimental temperature is 45 °C to simulate the formation and reservoir conditions.

Experimental instruments include the microlithographic glass model, magnetic stirrer, Brookfield viscosimeter, HW-2B thermostat with temperature control accuracy ±0.5 °C, electronic balance with sensing accuracy ±0.01 g, analytical balance with sensing accuracy ±0.0001 g, one micro-pump (range 0.5–10 mL/h), one micro-pump control computer, one high-magnification microscope, and one camera. The experimental equipment and process are shown in Figure 1.

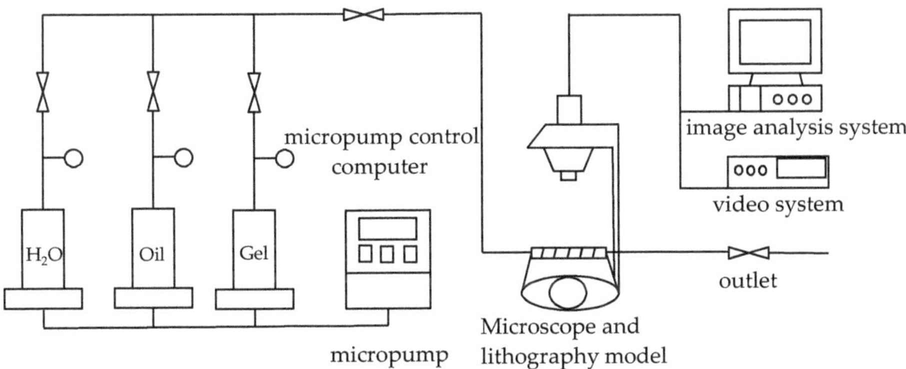

Figure 1. Visual connection diagram of micro-displacement experimental device.

2.2. Experimental Procedure

The experiments were carried out according to the following procedures:

1. Preparing the weak gel system required by the experiment: 0.4 g polymer was added into 200 mL produced formation water and stirred for 1 h. Then 0.9 g crosslinking agent and 0.2 g stabilizer were added to the solution and stirred for 1 h.
2. Filtering the crude oil to avoid blocking the model.
3. Saturating oil after the model is vacuumed.
4. Water flooding (constant speed 0.03 mL/h), observe the microscopic seepage process in the process of water flooding, and film the distribution and morphology of residual oil in the micro model until no oil is produced by water flooding.
5. Injecting 0.3 PV weak gel system into the lithography model while recording the microscopic oil displacement process by micro-camera.
6. Placing the lithographic model in a 45 °C incubator for 2 h, and wait for the weak gel system to gel.
7. The photolithography model was displaced by subsequent water until no oil was produced in the model. The microscopic oil flooding process was recorded by micro-camera.
8. Image analysis.
9. Cleaning the visualized model and preparing for the next experiment.

3. Analysis of Experimental Results

3.1. Phenomenon Description

3.1.1. Oil Displacement Dynamics and Residual Oil Types in Water Flooding Stage

The oil displacement photos at different stages were recorded by micro-camera to describe the experimental phenomena. It is shown in Figure 2 that the injection end is in the lower right corner, and the production end is in the upper left corner. In the stage of water flooding, under the action of injection pressure at both ends of the photolithography model, the injected water communicates a curved water flow channel along the direction of the injection end and production end. The injected water front surges forward in the large channel area with low flow resistance, forming an obvious dominant channel, which is manifested as a fingering phenomenon. Due to the large flow resistance of injected water in the small pore, the small pore area with large flow resistance becomes the main enrichment area of residual oil.

After the water flooding stage, the residual oil in the pores and throats is evenly distributed along both sides of the dominant channel. The residual oil mainly exists in the form of cluster, column, dead end, and membrane, in which the cluster residual oil remains in small throat pore clusters surrounded by unobstructed large pores in water flooding, the columnar residual oil remains in isolated plug shape or columnar shape at the throat of connected pores, the dead end residual oil is isolated in the form of droplets in the dead angle of the pores where the injection water cannot sweep, and it is usually the residual oil with one end closed or one end very difficult to flow, and the membranous residual oil, as shown in Figure 3, is an oil film attached to the pores or inner wall of throats during water flooding.

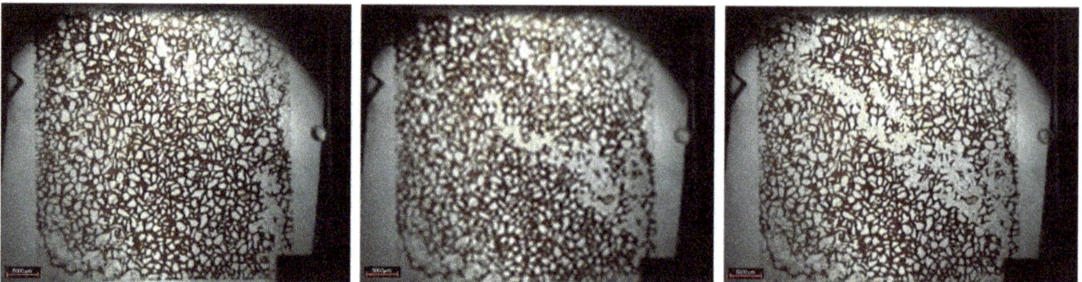

Figure 2. Water–oil displacement stage.

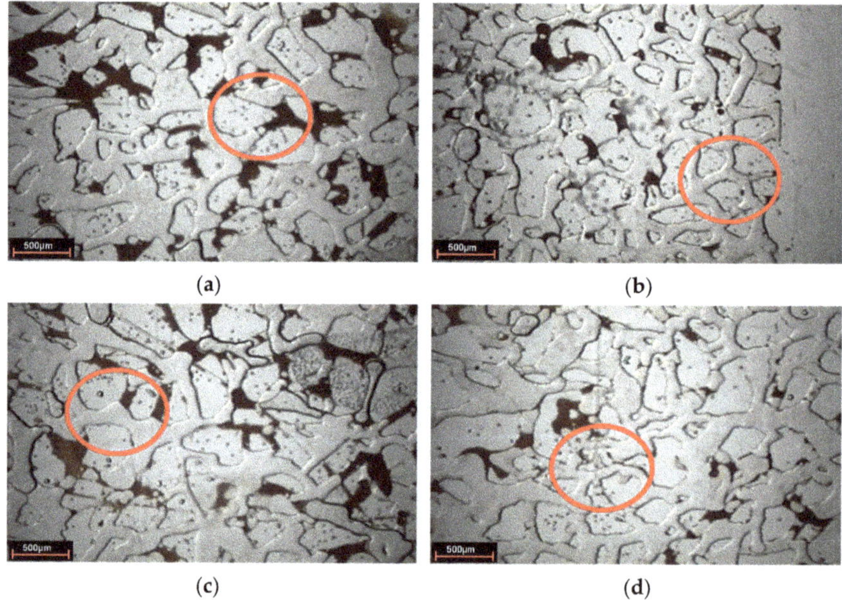

Figure 3. Distribution of residual oil after water flooding. (**a**) Cluster residual oil; (**b**) dead end residual oil; (**c**) columnar residual oil; (**d**) membranous residual oil.

3.1.2. Oil Displacement Dynamics and Residual Oil Characteristics in Weak Gel Injection Stage

At the stage of weak gel injection, under the action of injection pressure, the weak gel front advances forwards in a circular shape, and the advance speed of the main streamline is faster. Weak gel preferentially migrates forward in the preferential flow channel generated in water flooding (Figure 4). In addition, in the areas on both sides of the water flow channel, the injected volume increased significantly, and the residual oil in the smaller channels that were not previously driven by injected water was also driven forward for a certain distance by the weak gel. The displacement front was relatively more uniform than water flooding, indicating that the weak gel played a role in adjusting the displacement profile.

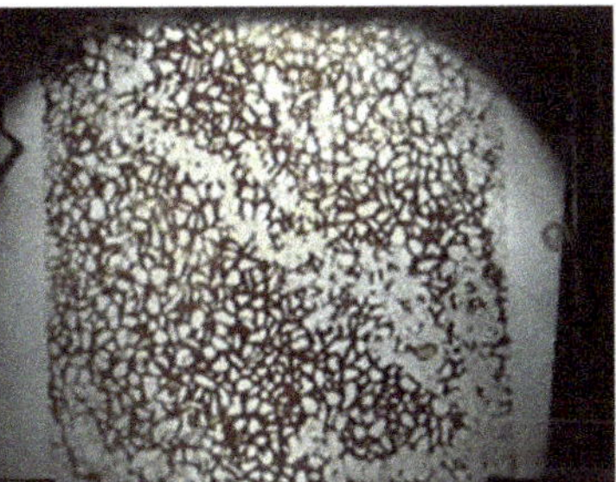
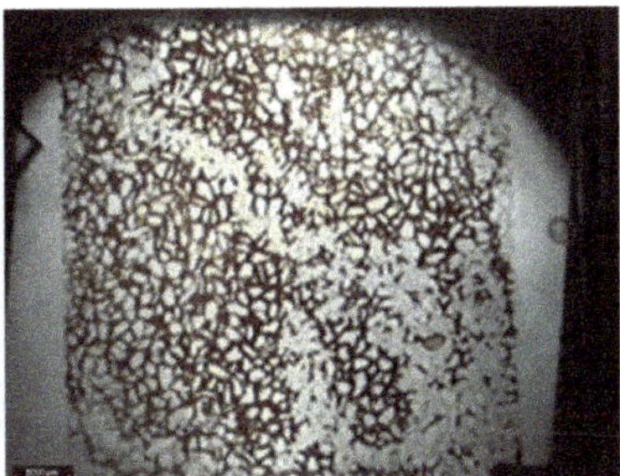
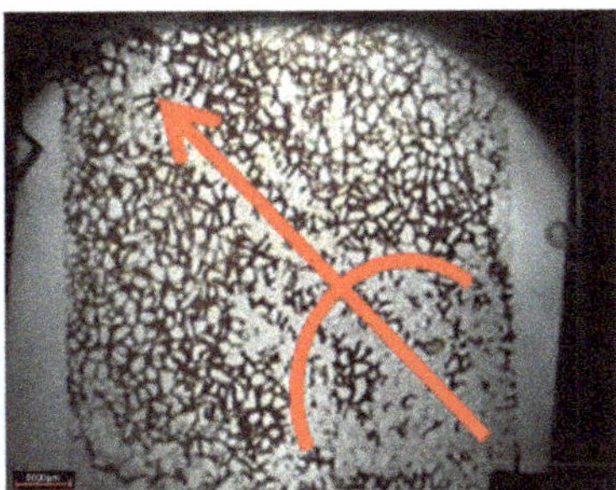

Figure 4. Weak gel injection stage (The arrow is the demonstration of aquauous weak gel displacement orientation, and the red arc is the swept area).

Due to the influence of profile-control of weak gel, the residual oil near the injection end with high flow resistance was displaced. By the end of the weak gel injection, the residual oil mainly exists in the pores and throats close to the production end with high flow resistance and on both sides of the weak gel injection channel, presenting a relatively uniform distribution on the whole and relatively scattered distribution in local areas.

Changing the microscope lens and using the camera to take photos to observe the present form of a weak gel in the pores after injection, as indicated in Figure 5. In the subsequent water flooding stage, the weak gel system after gelation exists in the form of gel groups in the large pore channels, thus playing the role of selective plugging.

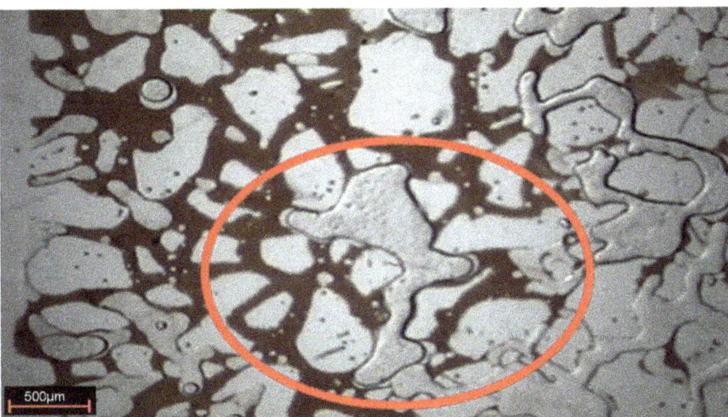

Figure 5. Weak gel group (The red circle is the weak gel group).

3.1.3. Oil Displacement Dynamics and Residual Oil Characteristics in Subsequent Water Flooding Stage

As shown in Figure 6, in the subsequent water flooding stage, due to the gelation of the weak gel system blocking the original water channel, the subsequent injection of water bypassed the original flow channel with the increase of water injection, and displaced the previously undeveloped oil in the small pores to the producing end, indicating that the weak gel plays a role in adjusting the displacement profile. Moreover, the injected water displaces the continuous weak gels along the pore extension direction, indicating that the weak gel also plays an oil displacement role. By comparing the pictures of the water flooding stage and subsequent water flooding stage, it can be seen that the sweep area of injected water has been significantly improved. A large number of residual oil in pores and throats is displaced by subsequent water, and the distribution of residual oil is very scattered. Microscopic observation shows that the residual oil mainly exists in the pore and throat in the form of clusters and dead ends after the subsequent water flooding stage.

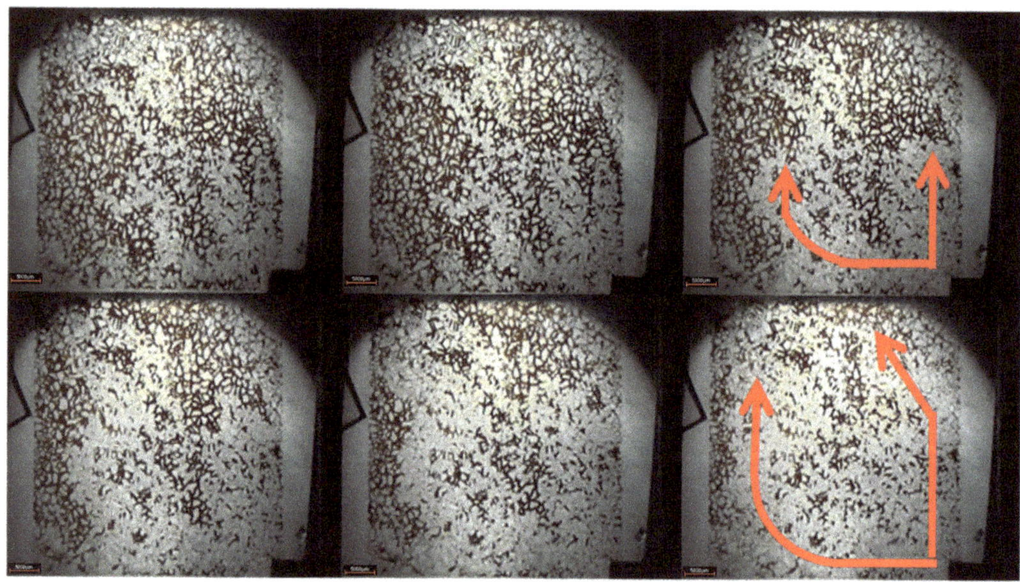

Figure 6. Subsequent water flooding stage (The arrow is the demonstration of subsequent water displacement orientation).

3.1.4. Description of Local Oil Displacement

In the experiment of weak gel microcosmic oil displacement, the phenomenon of oil displacement is different for parallel channels with different pore sizes in different oil displacement stages. In the stage of water flooding, for parallel channels with different sizes, the injected water preferentially replaces crude oil in larger channels, as shown in Figure 7. In the weak gel injection stage, for parallel channels with different sizes, the injected water preferentially replaces the crude oil in the smaller channels, as demonstrated in Figure 8. In the subsequent water flooding stage: for parallel channels with different sizes, the injected water preferentially replaces crude oil in larger channels, as indicated in Figure 9.

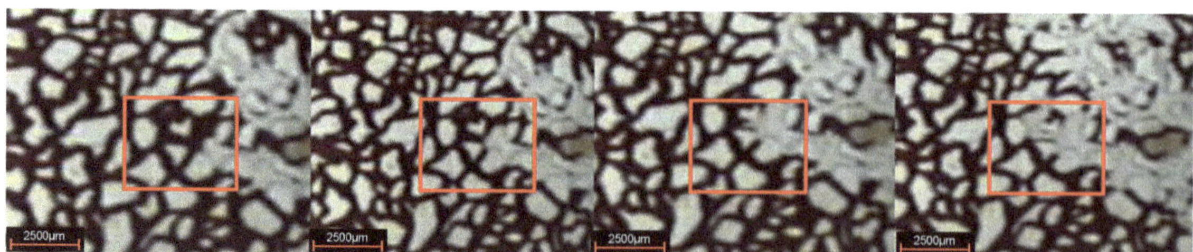

Figure 7. Oil displacement in water flooding stage.

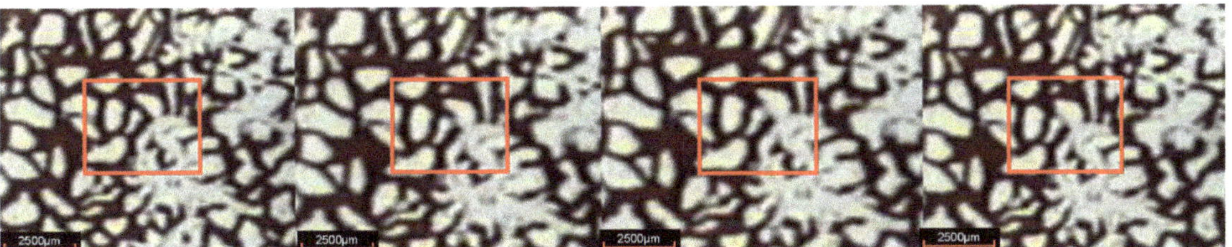

Figure 8. Oil displacement in weak gel injection stage.

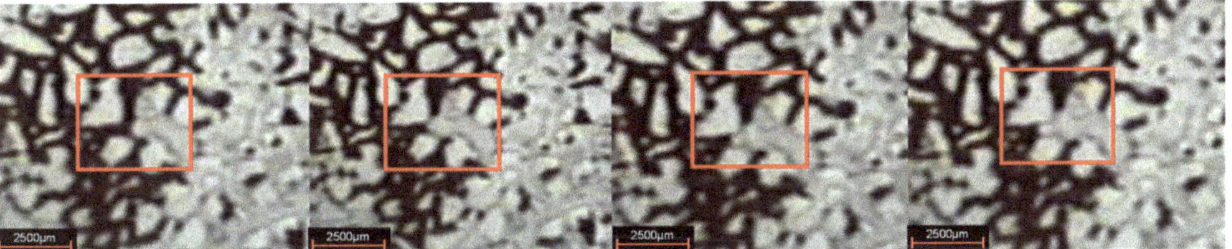

Figure 9. Oil displacement in subsequent water flooding stage.

3.2. The Porous Flow Mechanism

The visualized microscopic oil displacement experiment of weak gel acting on the photoengraving model shows that weak gel first flows through the large pore channels during injection, so the subsequent injected water cannot pass through the large pore channels occupied by weak gel, then it is forced to flow to the pore channel not swept by the early water flooding. With the increase of injection water, the injected water pushes part of the weak gel forward, stops again when it meets small pores, and changes the direction of water flow to drive out the residual oil that has not been used in the stage of water flooding, improving the swept volume of injected water.

Through the analysis of the experimental results, the main mechanism of weak gel profile-control is obtained as follows: the preferential plugging of large channels, the integral and staged transport of weak gel, and the residual oil flow along pore walls in weak gel displacement.

1. Preferential plugging of large channels

In the phase of weak gel injection, with the increase of injected volume, at the injection end, the weak gel is displaced to the producing end in a circular shape. As the flow resistance of the gel in the high permeability area is less than that in the low permeability area, the main line direction of the weak gel is the direction of the high permeability area with low flow resistance and high porosity. After 2 h of the coagulation stage, the large pore channel is preferentially blocked.

As demonstrated in Figure 10, weak gel preferentially flows into large channels with low seepage resistance. The residual oil in the circle is driven by weak gel, indicating that the weak gel has preferentially entered the large pore channel along the water flow channel generated by water flooding oil.

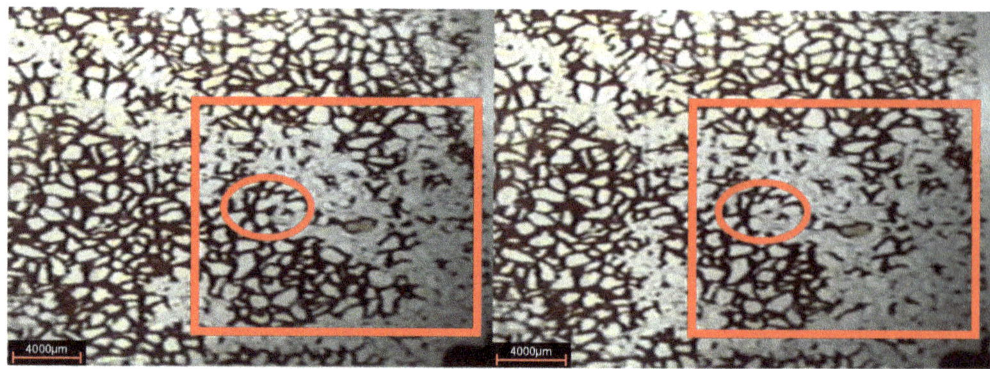

Figure 10. Preferential plugging of large channels.

2. The integral and staged transport of weak gel

As weak gel has a certain viscoelasticity, it mostly flows in the pore and throat structure in the form of gel groups in the seepage process, and the migration has strong integrity and coherence. Under the action of injection pressure, it is like a snake whose shape can change along the pore structure, advancing in the pore and throat. In addition, the migration of weak gel in the pore is phased. Before entering a certain pore, the weak gel accumulates liquid amount and pressure first. When the pressure reaches the threshold pressure to break through the pore, it immediately pushes into the pores.

As demonstrated in Figure 11, in the process of residual oil displacement, the weak gel pushes the residual oil forward as a whole, like a snake whose shape can change along the pore structure, and its migration has integrity and stages.

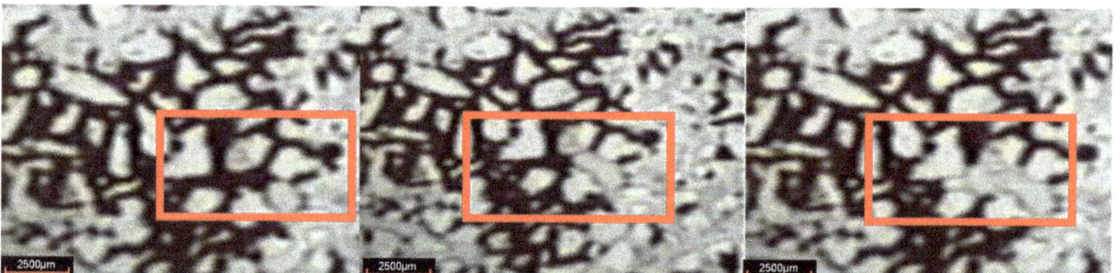

Figure 11. Weak gel transport.

3. The residual oil flow along pore walls in weak gel displacement

Due to the viscoelasticity of weak gel and the integrity of its migration, the weak gel can migrate forward close to the pore wall when it migrates in the pores, so the residual oil in membrane form attached to the pore wall can also be displaced by weak gel, as shown in Figure 11 above.

3.3. The Profile-Control Mechanism

Through further analysis of the experimental results, the main mechanism of weak gel profile-control is obtained as follows: weak gel selectively enters the large channels, weak gel blocks large channels and forces subsequent water flow to change direction, weak gel uses viscoelastic bulk motion to form negative pressure oil absorption, and the oil droplets converge to form an oil stream.

1. The weak gel selectively enters the large pore channels

The main flow path of weak gel injected into the model is along the direction of low flow resistance and high porosity, which is often consistent with the path of water flooding in the early stage. As the weak gel injected into the model has a certain viscosity, it has a certain integrity and continuity in the flow process. However, due to the large size of the large pores, the shear rate of the weak gel is relatively low, so the weak gel can flow more smoothly while maintaining its integrity and continuity. Therefore, under a certain injection pressure, the weak gel will overcome the resistance along the path and migrate in the large channels with low resistance, as exhibited in Figure 10 above.

2. The gelation blocks the large pore channels and diverts the subsequent water flow direction

In the subsequent water flooding stage, the viscosity of weak gel after gelation is relatively high, which can form a blockage in pores and produce an end face effect, and the critical pressure entering small pores is much higher than that of oil and water. Therefore, under a certain pressure, the migration of weak gel is difficult, and the probability of entering small pores is very small. According to the microscopic experimental results, in the subsequent water flooding process, due to the blocking phenomenon of weak gel in the channels under a certain pressure, the liquid flow of injected water is forced to change direction and flow to the small-channel area with relatively small resistance, so that the residual oil that is undeveloped in the stage of water flooding is displaced and the swept volume is increased.

As shown in Figure 12, in the subsequent water flooding stage, due to the weak gel after gelation blocked the pores, the subsequent injection of water is forced to flow into the small pores to displace the residual oil, which plays a role in adjusting the water imbibition profile.

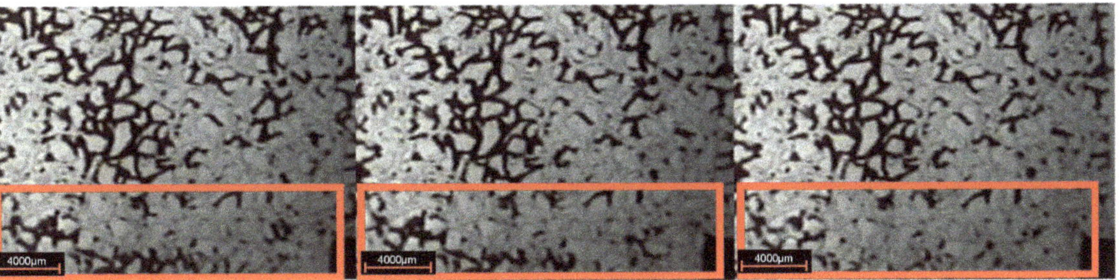

Figure 12. Subsequent water flow redirection.

3. The viscoelastic gelation moves integrally to absorb oil by negative pressure

For the weak gel, the chemical structure of the cross-linked polymer molecular coils is relatively stable, and the stability after gelation is better, which mainly reflects good viscoelasticity, integrity, and coherence during migration, and obvious water–glue interface with subsequent water. When the injection pressure reaches a certain level, the weak gel group flows through the hole like a snake through the grass, and flows rapidly in the direction of relatively little resistance. Due to the instantaneous speed of the integral migration of the weak gel group, and its good viscoelasticity, integrity, and coherence during migration, the subsequent fluid cannot be filled in time, resulting in instantaneous 'negative pressure' (the pressure of the main channel is lower than that of the surrounding channels). At this point, the oil or water in the surrounding porous channel overcomes the internal threshold pressure and extrudes out of the porous channel and is sucked into the oil stream.

Figure 13 indicates that in the weak gel injection stage, negative pressure is formed due to the viscoelastic bulk movement of the weak gel, which sucked out the residual oil in the subsequent water flooding stage.

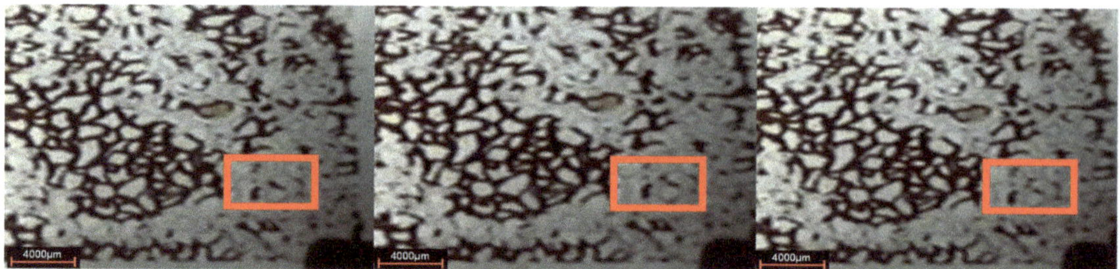

Figure 13. Negative pressure oil absorption.

4. Oil droplets converge to form oil stream

In the microscopic displacement experiment, it is also found that in the process of weak gel displacement, a large amount of oil in small pores is displaced, and gradually accumulates into oil droplets in large pores, which move forward and form continuous oil droplets. The thickness of oil droplets thickens with the displacement, and finally, they migrate to the outlet in the form of continuous oil flow. As shown in Figure 14, three small oil droplets converge to form a large oil droplet when passing through a narrow throat and migrate forwards in the form of an oil stream.

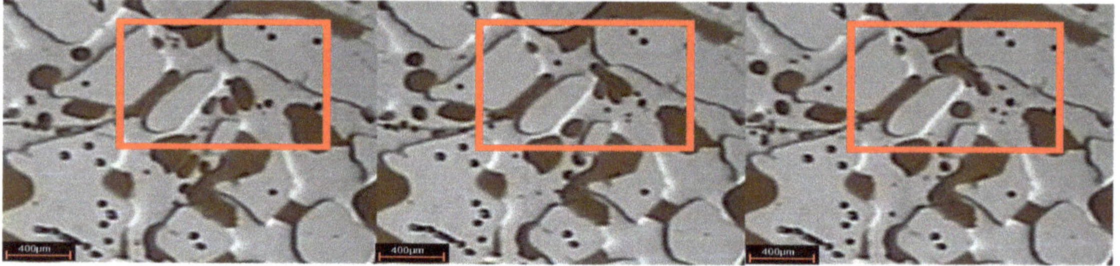

Figure 14. Accumulation of oil droplets into oil stream.

4. Numerical Simulation

4.1. Simulation Model Parameters

The numerical simulation is carried out using the CMG-STARS reservoir numerical simulator. The model is established using a Cartesian grid system, which is divided into 19 × 19 grids in X and Y directions with a grid size of 20 m, and 10 grids in the Z direction with each grid thickness of 10 m. The five-point well pattern was used which includes one injector in the center and four corner producers. The basic parameters and their values of the sector model are shown in Table 1.

Table 1. Basic data of the geological model.

Reservoir Parameters	Parameter Value
Reservoir top depth (m)	1400
Permeability (mD)	50, 55, 60, 80, 380, 400, 390, 180, 100, 50
Porosity (%)	0.23
Original formation pressure (MPa)	16

In this model, the gel injection rate is 0.097 PV/a, the maximum injection pressure is 20 MPa, the minimum bottomhole flow pressure is 3 MPa, and the cumulative gel injection volume is 0.29 PV. The basic data required for reservoir fluid modeling are shown in Table 2. The established 3D numerical model is demonstrated in Figure 15.

Table 2. Basic data of the fluid model.

Reservoir Parameters	Parameter Value
Underground oil viscosity (mPa·s)	8
Surface crude oil density (kg/m^3)	860
Water viscosity (mPa·s)	0.5
Oil volume coefficient (m^3/m^3)	1.09

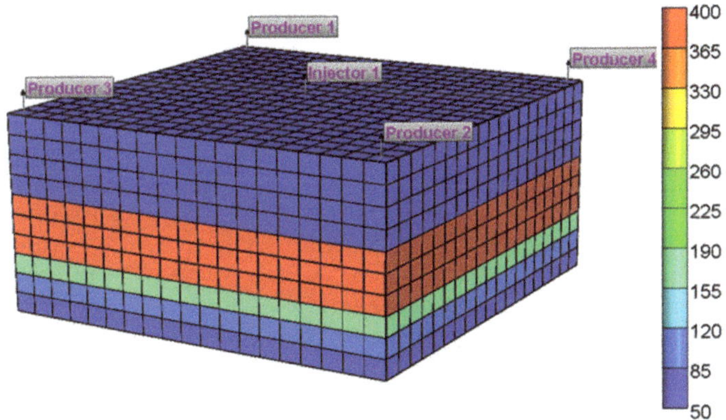

Figure 15. A 3D numerical model of reservoir.

4.2. Equations Influencing Weak Gel Displacement Dynamics

The efficiency of water flooding is largely related to the mobility ratio of displacing and displaced fluids. Definition of mobility ratio:

$$M = \frac{Displacing\ phase\ mobility}{Displaced\ phase\ mobility} = \frac{k_w/\mu_w}{k_o/\mu_o} = \frac{k_w}{k_o} \times \frac{\mu_o}{\mu_w} \tag{1}$$

where, k_w—Relative permeability of water; μ_w—Viscosity of water; k_o—Relative permeability of oil; μ_o—Viscosity of oil.

Due to the high mobility ratio, the injected water moves faster than the displaced oil, resulting in fingering. The injected water bypassed the displaced oil and migrated toward the producing well. Most of the oil is unswept by the water because of fingering, which is the water flow channel toward the producing well. Once the water channel is formed, water will bypass residual oil in the reservoir and flows directly from the injection well to the producing well.

After the injection of weak gel, the liquid viscosity can be calculated according to the nonlinear mixing Equation (1):

$$Ln(\mu_\alpha) = \sum_{i=1}^{n_c \in s} f(f_{\alpha i}) \times Ln(\mu_{\alpha i}) + N \times \sum_{i=1}^{n_c \notin s} f_{\alpha i} \times Ln(\mu_{\alpha i}) \tag{2}$$

where, μ_α—Mixed viscosity of aqueous phase ($\alpha = \text{w}$) or oil phase ($\alpha = \text{o}$); $\mu_{\alpha i}$—Viscosity of aqueous phase ($\alpha = \text{w}$) or oil phase ($\alpha = \text{o}$) component "i"; $f_{\alpha i}$—Weight factor of non-critical component "i" in nonlinear mixing calculation of aqueous phase ($\alpha = \text{w}$) or oil phase ($\alpha = \text{o}$); $f(f_{\alpha i})$—Weight factor of key component "i" in nonlinear mixed calculation of aqueous phase ($\alpha = w$) or oil phase ($\alpha = o$); $n_c \in s$—Key group fraction in liquid phase; $n_c \notin s$—Scores of other groups except the key group; N—Normalized factor.

By incorporating the equations above characterizing injected liquid viscosity and mobility ratio induced migration dynamics of weak gel system in porous media into the simulation model, it is able to simulate gel deformation and migration dynamics under the influences of pressure gradient, velocity, retention, time and other factors. The aqueous phase viscosity distribution field maps in different periods of weak gel injection are given through numerical simulation, as exhibited in Figure 16. The simulation results indicate that weak gel can increase the viscosity of the aqueous phase, thus reducing the mobility of water, reducing the fingering phenomenon, improving the plane heterogeneity, and enhancing the recovery factor.

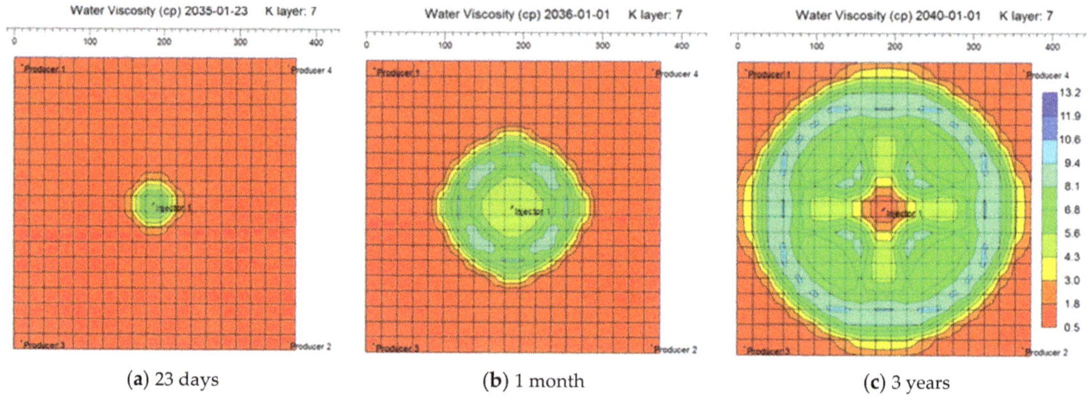

(**a**) 23 days (**b**) 1 month (**c**) 3 years

Figure 16. Aqueous phase viscosity distribution at different stages of weak gel injection.

4.3. *Dynamics of Weak Gel Injection*

As shown in Figure 17, numerical simulation results demonstrated that the higher the permeability, the lower the flow resistance, so the weak gel will preferentially enter the high permeability layer and increase its flow resistance. Consequently, the subsequent injected water will enter the low permeability layer or the low permeability area and expand the swept volume.

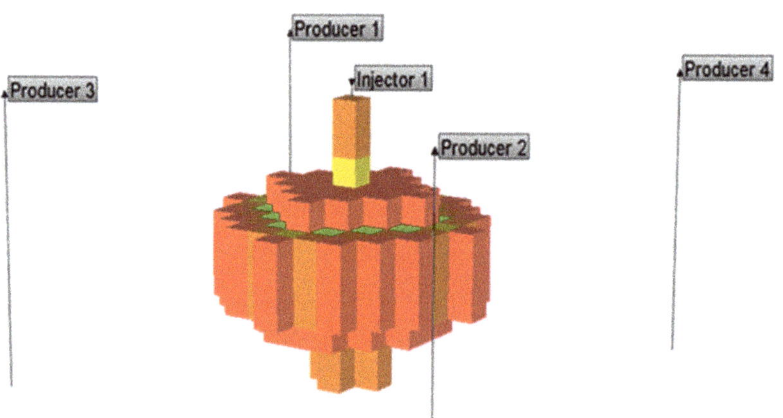

Figure 17. *Cont.*

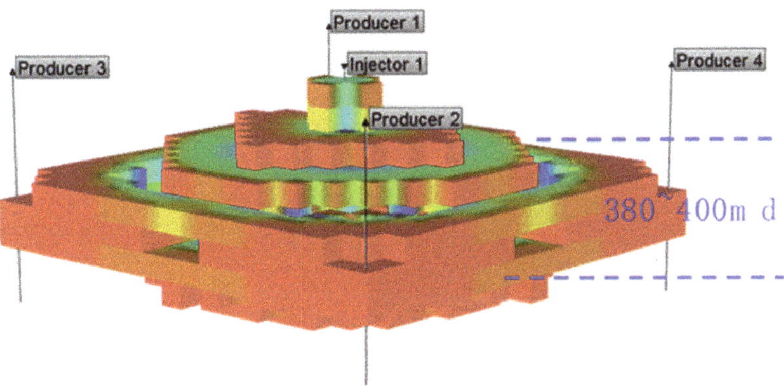

Figure 17. A 3D distribution of weak gel mole fraction.

According to the numerical simulation results, the plane distribution of weak gel mole fraction in Figure 18 shows that the main route of weak gel flow is along the direction of the high permeability flow channel, which is often the channel of water breakthrough in the early waterflood stage. After weak gel profile-control flooding, water pushes the weak gel to flow forward. Therefore, weak gels continuously expand the sweep region and displace the residual oil in this region to the producing well, so weak gels play two roles adjusting the heterogeneity and displacing oil.

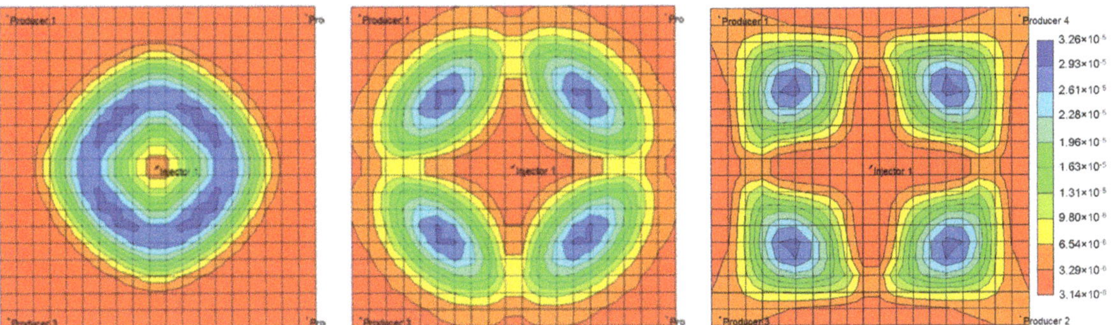

Figure 18. Planar distribution of weak gel mole fraction.

4.4. Diversion of Subsequent Water Flooding

Based on the established injection-production system of a five-point well pattern, the comparison of three-dimensional streamline distribution of initial water flooding and subsequent water flooding following weak gel was simulated numerically. It can be seen from Figure 19 that in the process of water flooding, injected water mainly flows along the highly permeable layer, and weak gel first enters the large pore originally occupied by water after injection. Under the effect of subsequent injection water, the weak gel continues to move forward along the large pore channels with low resistance. Meanwhile, the presence of weak gel increases the flow resistance of the large pore channels, forcing the injection water to change direction. Figure 20 indicates that the subsequent injected water can enter the low permeability zone unswept in the initial water injection, thus improving the sweep efficiency and ultimate recovery of the waterflood.

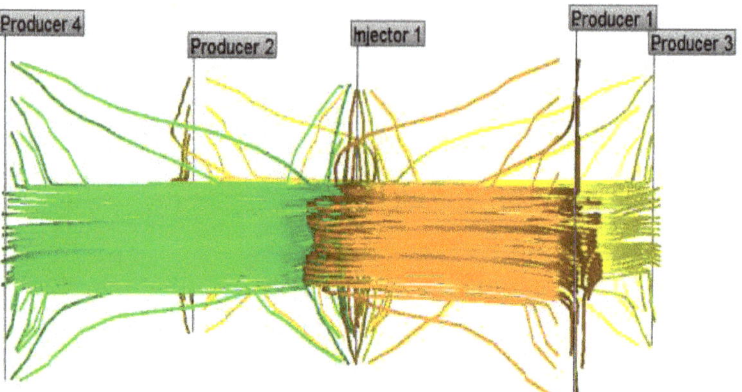

Figure 19. A 3D streamline distribution of initial water flooding.

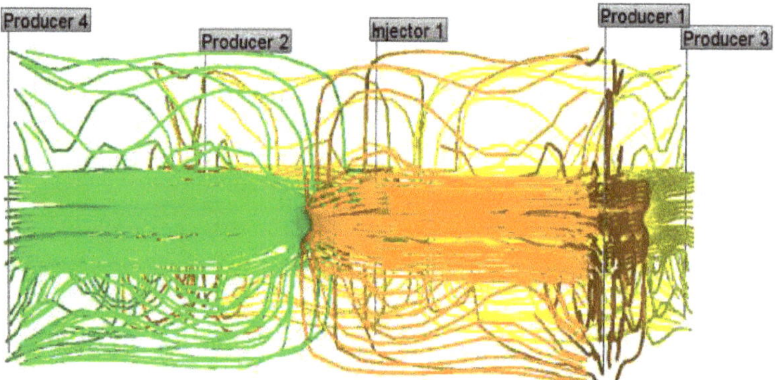

Figure 20. A 3D streamline distribution of subsequent water flooding.

The comparison of the planar streamline distribution in Figures 21 and 22 shows that weak gel will change the flow direction of other fluids when it flows in porous media. The reservoir itself has macroscopic and microscopic heterogeneity, and microscopic heterogeneity is mainly reflected in the difference in pore size, pore distribution, and pore surface properties in the rock pore structure. Therefore, in the process of water flooding, the injected water always tends to enter the large pore channels with low resistance, and the residual oil always remains in the small pores or in the form of oil droplets in the center of the large pores. According to channel flow theory, when the oil phase loses its continuity, it becomes residual oil and cannot be recovered. After the weak gel is injected into the porous media, the flow direction of the subsequent injected water can be diverted, as shown in Figure 22, forcing the injected water to enter the unswept area, thus improving the sweep efficiency of the subsequent water flooding.

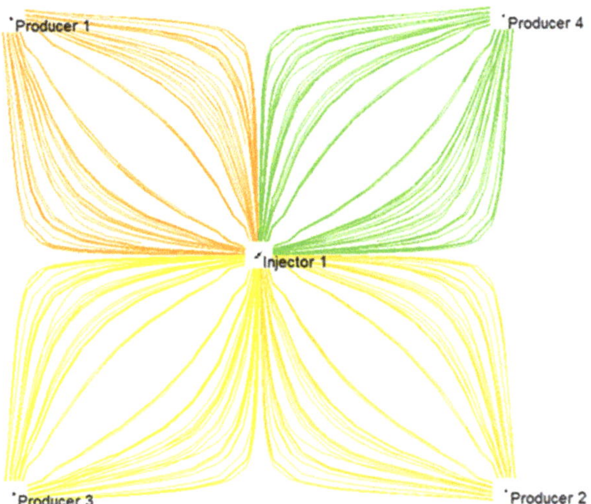

Figure 21. Planar streamline distribution of initial water flooding.

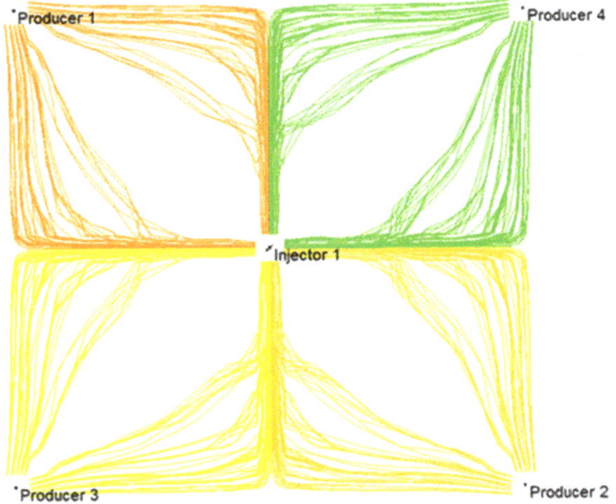

Figure 22. Planar streamline distribution of subsequent water flooding.

5. Conclusions

1. By analyzing the distribution patterns of residual oil in pores and throat through visualized microscopic oil displacement experiments at different stages, the residual oil of initial water flooding mainly exists in the form of cluster, column, dead end, and membranous, and it mainly exists in the form of cluster and dead end in subsequent water flooding stage following weak gel injection.
2. The porous flow mechanism of weak gel includes the preferential plugging of large channels, the integral and staged transport of weak gel, and the residual oil flow along pore walls in weak gel displacement.
3. The profile-control mechanism of weak gel is as follows: weak gel selectively enters the large channels, weak gel blocks large channels and forces subsequent water flow to change direction, weak gel uses viscoelastic bulk motion to form negative pressure oil absorption, and the oil droplets converge to form an oil stream, respectively.

4. Numerical simulation coupling equations characterizing weak gel viscosity induced dynamics indicate that weak gel can effectively reduce the water-oil mobility ratio, preferentially block the high permeability layer and the large pore channels, divert the subsequent water to flood the low permeability layer, and improve the water injection swept efficiency. A weak gel system is able to flow forward under high-pressure difference, which can further improve oil displacement efficiency besides flow diversion.

Author Contributions: Conceptualization and supervision, H.C.; methodology, X.Z. (Xianbao Zheng); investigation, Y.W. and X.Z. (Xin Zhao); data curation, J.Z.; writing—original draft preparation, Y.W.; writing—review and editing, C.L. All authors have read and agreed to the published version of the manuscript.

Funding: This work received the financial support of the China National Key Project (Waterflood performance improvement and EOR investigation for marginal oilfields: 2016E0209).

Institutional Review Board Statement: Not applicable.

Informed Consent Statement: Not applicable.

Data Availability Statement: The data of this article is available on request through e-mail.

Acknowledgments: The valuable comments made by the anonymous reviewers are sincerely appreciated.

Conflicts of Interest: The authors declare no conflict of interest.

References

1. Liu, Y.; Xiong, C.; Luo, J. Studies on indepth fluid diverting in oil reservoirs at high water cut stages. *Oilfield Chem.* **2006**, *23*, 248–251.
2. Yu, J. Study on reservoir property change in the late stage of waterflooding development in Daqing Oilfield. *Oil Gas Field Surf. Eng.* **2007**, *26*, 15–16.
3. Zhao, X.; Yang, M.; Zhang, L.; Liu, L. Research progress of deep profile control technology in oilfield. *Chem. Sci. Technol.* **2015**, *5*, 75–79.
4. Lu, X.; Ca, B.; Xie, K.; Cao, W.; Liu, Y.; Zhang, Y.; Wang, X.; Zhang, J. Enhanced oil recovery mechanisms of polymer flooding in a heterogeneous oil reservoir. *Pet. Explor. Dev.* **2021**, *48*, 169–178. [CrossRef]
5. El-Karsani, K.S.M.; Al-Muntasheri, G.A.; Hussein, I.A. Polymer Systems for Water Shutoff and Profile Modification: A Review over the Last Decade. *SPE J.* **2014**, *19*, 135–149. [CrossRef]
6. Quezada, G.R.; Toro, N.; Saavedra, J.H.; Robles, P.; Salazar, I.; Navarra, A.; Jeldres, R.I. Molecular Dynamics Study of the Conformation, Ion Adsorption, Diffusion and Water Structure of Soluble Polymers in Saline Solutions. *Polymers* **2021**, *13*, 3550. [CrossRef] [PubMed]
7. Zhang, H. Dynamic Weak-Gel Profile-Control in Conventional Heavy-Oil Reservoir. *Spec. Oil Gas Reserv.* **2018**, *25*, 125–128.
8. Jia, H. Research on deep profile control and water shutoff technology in Daqing Oilfield based on geology. *China Pet. Chem. Stand. Qual.* **2019**, *39*, 250–252.
9. Cui, C.; Zhou, Z.; He, Z. Enhance oil recovery in low permeability reservoirs: Optimization and evaluation of ultra-high molecular weight HPAM/phenolic weak gel system. *J. Pet. Sci. Eng.* **2020**, *195*, 107908. [CrossRef]
10. Akbar, I.; Hongtao, Z. The opportunities and challenges of preformed particle gel in enhanced oil recovery. *Recent Innov. Chem. Eng. (Former. Recent Pat. Chem. Eng.)* **2020**, *13*, 290–302. [CrossRef]
11. Al-Lbadi, A.; Civan, F. Experimental Investigation and Correlation of Treatment in Weak and High-Permeability Formations by Use of Gel Particles. *SPE Prod. Oper.* **2013**, *28*, 387–401.
12. Li, Z.; Zhao, G.; Xiang, C. Synthesis and Properties of a Gel Agent with a High Salt Resistance for Use in Weak-Gel-Type Water-Based Drilling Fluid. *Arab. J. Sci. Eng.* **2022**, 1–11. [CrossRef]
13. Williamson, N.H.; Dower, A.M.; Codd, S.L.; Broadbent, A.L.; Gross, D.; Seymour, J.D. Glass Dynamics and Domain Size in a Solvent-Polymer Weak Gel Measured by Multidimensional Magnetic Resonance Relaxometry and Diffusometry. *Phys. Rev. Lett.* **2019**, *122*, 068001. [CrossRef] [PubMed]
14. Diaz, D.; Somaruga, C.; Norman, C.; Romero, J.L. Colloidal Dispersion Gels Improve Oil Recovery in a Heterogeneous Argentina Waterflood. In Proceedings of the SPE Symposium on Improved Oil Recovery, Tulsa, OK, USA, 20–23 April 2008.
15. Qi, Y.B.; Zheng, C.G.; Lv, C.Y.; Lun, Z.M.; Ma, T. Compatibility between weak gel and microorganisms in weak gel-assisted microbial enhanced oil recovery. *J. Biosci. Bioeng.* **2018**, *126*, 235–240. [CrossRef] [PubMed]
16. Li, S.; Zhang, S.; Zou, Y.; Zhang, X.; Ma, X.; Wu, S.; Zhang, Z.; Sun, Z.; Liu, C. Experimental study on the feasibility of supercritical CO_2-gel fracturing for stimulating shale oil reservoirs. *Eng. Fract. Mech.* **2020**, *238*, 107276. [CrossRef]
17. Zhang, J.; Feng, J. Study on performance evaluation of weak gel control flooding system. *China Dly. Chem. Ind.* **2019**, *49*, 87–91.

8. Pi, Y. Performance evaluation and influence factors of weak gel. *Hangzhou Chem. Ind.* **2021**, *51*, 10–12.
9. Ji, G.; Li, H.; Wu, X.; Xu, C.; Li, X. Performance of weak gel plugging high permeability zone and conditions for starting low permeability layer. *Sci. Technol. Eng.* **2017**, *17*, 204–209.
10. Li, Y.; Su, Y.; Ma, K.; Wang, L. Numerical simulation of indoor displacement experiments in weak gel. *Daqing Pet. Geol. Dev.* **2014**, *33*, 117–120.
11. Zhang, Y. Study on influence factors of gelling performance of weak gel flooding system. *Oil Gas Reserv. Eval. Dev.* **2013**, 3, 50–53.
12. Zhao, G.; Dai, C.; You, Q. Characteristics and displacement mechanisms of the dispersed particle gel soft heterogeneous compound flooding system. *Pet. Explor. Dev.* **2018**, *45*, 481–490. [CrossRef]
13. Di, Q.; Zhang, J.; Hua, S.; Chen, H.; Gu, C. Visualization experiments on polymer-weak gel profile control and displacement by NMR technique. *Pet. Explor. Dev.* **2017**, *44*, 294–298. [CrossRef]
14. Hua, S.; Di, Q.; Wang, W.; Yang, P.; Zhang, J.; Ye, F. Study on weak gel's mobility in porous media using nuclear reasonable technique. *Spec. Top. Rev. Porous Media* **2018**, *9*, 13–20. [CrossRef]

 gels

MDPI

Article

Wellbore Stability through Novel Catechol-Chitosan Biopolymer Encapsulator-Based Drilling Mud

Zhichuan Tang [1], Zhengsong Qiu [1,*], Hanyi Zhong [1], Yujie Kang [1] and Baoyu Guo [2]

1 School of Petroleum Engineering, China University of Petroleum (East China), No. 66 Changjiang West Road, Economic & Technical Development Zone, Qingdao 266580, China; b17020063@s.upc.edu.cn (Z.T.); zhonghanyi@126.com (H.Z.); 17854210608@163.com (Y.K.)

2 Drilling Technology Research Institute, Shengli Petroleum Engineering Corporation Limited of SINOPEC, Donying 257064, China; guobaoyu719.slyt@sinopic.com

* Correspondence: teamo_tzc@163.com

Abstract: The problem of wellbore stability has a marked impact on oil and gas exploration and development in the process of drilling. Marine mussel proteins can adhere and encapsulate firmly on deep-water rocks, providing inspiration for solving borehole stability problem and this ability comes from catechol groups. In this paper, a novel biopolymer was synthesized with chitosan and catechol (named "SDGB") by Schiff base-reduction reaction, was developed as an encapsulator in water-based drilling fluids (WBDF). In addition, the chemical enhancing wellbore stability performance of different encapsulators were investigated and compared. The results showed that there were aromatic ring structure, amines, and catechol groups in catechol-chitosan biopolymer molecule. The high shale recovery rate demonstrated its strong shale inhibition performance. The rock treated by catechol-chitosan biopolymer had higher tension shear strength and uniaxial compression strength than others, which indicates that it can effectively strengthen the rock and bind loose minerals in micro-pore and micro-fracture of rock samples. The rheological and filtration property of the WBDF containing catechol-chitosan biopolymer is stable before and after 130 °C/16 h hot rolling, demonstrating its good compatibility with other WBDF agents. Moreover, SDGB could chelate with metal ions, forming a stable covalent bond, which plays an important role in adhesiveness, inhibition, and blockage.

Keywords: wellbore stability; biopolymer; encapsulator; water-based drilling fluids; chemical strengthening

Citation: Tang, Z.; Qiu, Z.; Zhong, H.; Kang, Y.; Guo, B. Wellbore Stability through Novel Catechol-Chitosan Biopolymer Encapsulator-Based Drilling Mud. *Gels* **2022**, *8*, 307. https://doi.org/10.3390/gels8050307

Academic Editors: Qiang Chen and Mario Grassi

Received: 22 April 2022
Accepted: 11 May 2022
Published: 16 May 2022

1. Introduction

The problem of wellbore stability in oil and gas drilling engineering has always been a worldwide technical problem. Especially in the complex deep-seated, deep-sea oil and gas, special structural wells, unconventional reservoirs, and other drilling projects, the problem of borehole wall instability is more prominent. According to incomplete statistics, the cost of dealing with borehole instability in drilling construction accounts for 50% of the cost of drilling fluid [1–3].

Hydration of shale is a complex process. Shale has low permeability and clay minerals such as montmorillonite are contained in the pores [4]. Na-montmorillonite consists of negatively charged clay sheets, which are similar to pyrophyllite layers, whose negative charge is compensated by interlayer sodium ions to maintain electrical neutrality [5]. The surface hydration first occurs after the montmorillonite meets water. At this stage, the expansion pressure generated by hydration is large and the volume expansion is small. With the hydration, the expansion pressure decreases rapidly [6].

The surface hydration causes crystallization expansion. Ionic hydration begins when the surface is hydrated [7]. Ion hydration brings hydration film to clay and hydration ions compete with water molecules for the connection position of crystal plane of clay. After surface hydration and ionic hydration of clay minerals such as sodium montmorillonite, the hydrated ions dissociate in the liquid away from the surface of clay minerals, forming a diffusion double layer between clay minerals [8]. The combined action of double layer repulsion and osmotic pressure produces hydration, that is, osmotic hydration. The volume of clay minerals expands under this action. At this stage, 1 g montmorillonite can absorb 10 cm of water. The volume increases by nearly 20 times, and the crystal layer spacing increases to 13 μm [9].

Therefore, under the huge drilling pressure difference, drilling fluid filtrate is very easy to invade the formation along the micro-porous and micro-fissures of shale [10]. The micro-voids and fractures of shale are rich in clay minerals such as montmorillonite. As sodium ions are very liable to occur in the above-mentioned hydration process in wet environment [11], the lattice of montmorillonite is very liable to cause interlayer crystal distance expansion after meeting water, which is manifested as macroscopic hydration expansion and dispersion. Once shale meets water, it will expand and generate huge expansion force, which will directly endanger wellbore integrity. At the same time, water intruding into shale crevices will also cause shale micro-cracks to expand and extend, resulting in wall instability [12,13].

In recent years, researchers have developed a series of additives which can improve the borehole pressure bearing capacity [14–17]. In terms of physical plugging, micro-nano particles are mainly used for direct plugging at present. There are many micro-nano pores in shale formation. The size of conventional plugging agent particles is too large to enter into the micro-nano pore in shale, which is easy to accumulate on the surface and cannot form effective plugging [18,19]. At present, nano-particles such as nano-silica, nano-magnesium oxide, iron oxide, zinc oxide, and graphene have been directly used or surface modified and used as plugging agent in the field of drilling fluids [20–23]. However, due to the high activity of nanoparticles, they are easy to agglomerate and lose their characteristics. At the same time, inorganic nanoparticles have a high rigidity and are easy to disengage and migrate after entering formation pore throat, which is not conducive to plugging [24,25].

In terms of chemical inhibition, different kinds of shale inhibitors have been developed, including inorganic salts, organic salts, macromolecular polymers, low molecular organic amines, polyalcohols, and so on [26–29]. These inhibitors mainly inhibit hydration expansion and dispersion of shale by restraining hydration of clay surface, encapsulating shale particles, changing wettability of shale surface, and controlling water activity of drilling fluids [28–31]. With the development of science and technology, the main applications in water-based drilling fluids are mainly silicates, organosilicons, polyalcohols, aluminum-based, and asphalts. With the changes in temperature, pH, salinity, or cloud point effect after they enter the formation, precipitation or insoluble substances are generated to plug the micro-pore and micro-fracture of shale, and then the purpose of sealing and consolidating the borehole wall is achieved. R. Schlemmer [32–34] of M-I Drilling Fluid Company has successfully developed an encapsulator to replace sugars and acrylates. At 4% addition, the 8-h expansion rate of shale can be reduced from 73.3% to 15.6%, which has been successfully applied in the Beihai Sea. 5high performance silicate-based muds were chosen by M. Al-booyeh [35] out of 81 different formulations of drilling fluids, which showed good rheology properties and all of them were thermally stable. The five performance silicate-based muds lost less than 10 mL before and after hot roll at 160 C/16 h, and the rolling recovery of shale cuttings also increased from less than 20% to nearly 80%. Qi Chu [36] developed a An organosilicon quadripolymer of acrylamide (AM), 2-acrylamido-2-methyl-1-propane sulfonic acid (AMPS), *N*-vinylpyrrolidone (NVP) and a kind of organosilicon monomer by solution free radical polymerization. The test result showed that the organosilicon quadripolymer drilling fluid performance was better than corresponding terpolymer without organosilicon group and shows favorable inhibitive property. EI-Monier [37] developed an environmentally friendly inorganic aluminum/aluminum shale inhibitor encapsulator-A through special size and structure design. The agent is a mixture of inorganic aluminum and aluminum in a certain proportion. It can effectively prevent the active clay from absorbing water, and the expansion rate can be reduced by about 90% after 8 h, while the water absorption of sodium montmorillonite can be reduced by 30%. It has low toxicity and can effectively seal formation pore under high temperature and strong acid environment.

However, there are some limitations of the above additives, such as limited bearing capacity of the well eye, poor compatibility of drilling fluid, weak effect of high temperature, and unfriendly to the environment. The results show that wellbore instability of shale formation is the result of both physical and chemical factors [38–40]. So, it is very important to develop an encapsulator which takes physical plugging and chemical cementing into account to solve the problem of wellbore instability in shale formation.

Mussel can firmly adsorb to rocks under strong winds and waves in the ocean. Its foot silk protein is a powerful adhesive and has been widely used in biological and medical fields. At the same time, the surface becomes denser and stronger after mussels adhere to rocks, hulls, etc., which also shows the potential of physical consolidation [41,42].

The active ingredient of this mussel adhesive protein is a catechol group. Inspired by this characteristic, catechol was grafted onto the main chain of chitosan by Schiff base-reduction reaction method. Finally, a novel environmentally friendly biopolymer encapsulators in WBDF, named SDGB, was developed, and the structural characterization and performance evaluation are conducted.

2. Results and Discussion

2.1. Orthogonal Experiment of SDGB

The main factors affecting polymer performance include the monomer ratio of catechol to chitosan, reaction time, initiator concentration, and reaction temperature. Orthogonal tests with three levels and four factor designs are shown in Table 1, and the results are shown in Table 2. The rolling recovery of shale rock chips was evaluated in SDGB aqueous solution. The temperature has the most significant effect on the rolling recovery of shale rock chips. The following optimal conditions were determined by rolling recovery of shale rock chips: catechol to chitosan molar ratio of 1:2, initiator concentration of 7 wt%, reaction time of 10 h, and reaction temperature of 10 °C.

Table 1. Orthogonal tests with three levels and four factor designs.

Factor Level	Mole Ratio of Catechol/Chitosan	Initiator Concentration * /%	Time /h	Temperature /°C
1	1:1	5	8	10
2	1:2	7	10	25
3	1:3	10	12	40

* Initiator amount is the weight percentage of initiator in total monomer.

Table 2. Optimized results of SDGB orthogonal experiments.

Num.	Monomer Ratio	Initiator Concentration /%	Time /h	Temperature /°C	Rolling Recovery /%
1	1:1	5	8	10	75.1
2	1:1	7	10	25	73.6
3	1:1	10	12	40	67.5
4	1:2	5	10	40	74.6
5	1:2	7	12	10	78.9
6	1:2	10	8	25	75.1
7	1:3	5	12	25	73.4
8	1:3	7	8	40	74.3
9	1:3	10	10	10	78.4

Table 2. *Cont.*

Num.		Monomer Ratio	Initiator Concentration /%	Time /h	Temperature /°C	Rolling Recovery /%
Level	k1	0.721	0.744	0.748	0.775	-
	k2	0.762	0.756	0.755	0.74	-
	k3	0.754	0.737	0.733	0.721	-
R		0.041	0.019	0.023	0.053	-

2.2. Characterization of SDGB

2.2.1. Fourier Transform Infrared Spectroscopy (FT-IR) Characterization Test

The result of FT-IR is shown in Figure 1. It can be seen that the amino group has a stretching vibration absorption peak at 1706 cm^{-1}, and the peak at 1380 cm^{-1} belongs to the methyl vibration absorption peak. The typical absorption peaks of O–C stretching vibration and the stretching vibration of aromatic C=C at 1090 cm^{-1} and 1530 cm^{-1} existed in chitosan, which proved that catechols had reacted with chitosan successfully. Meanwhile the IR absorption peak at 1300–1090 cm^{-1} was the symmetric vibration caused by the covalent bond between anthraquinone and Fe^{2+}, which indicated that the structure of catechol in SDGB had strong reactivity and could produce strong reactivity with metal ions [43,44].

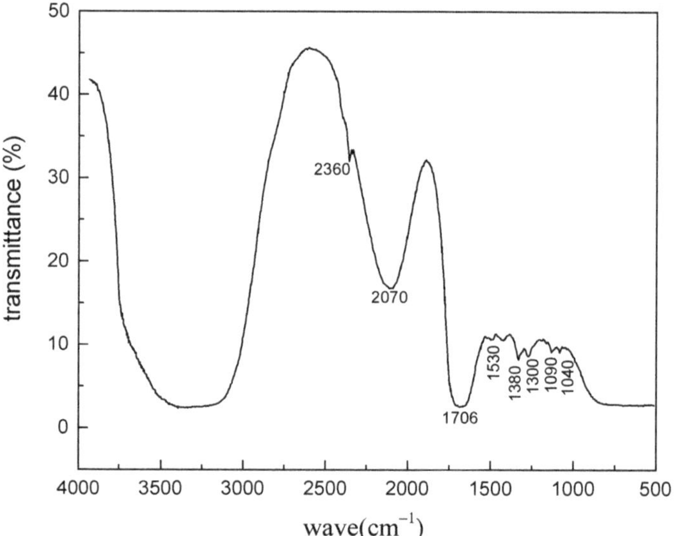

Figure 1. Fourier transform infrared spectrum of SDGB.

2.2.2. NMR Hydrogen Spectroscopy (HNMR) Analysis

The HNMR results of SDGB are shown in Figure 2. It can be seen that there is a peak of protons on the main chain of chitosan at chemical displacement δ of 3.6–4.0. The chemical shift δ at 3.1 corresponds to the C-2 proton peak of chitosan, and the chemical shift δ at 1.9 corresponds to the methyl proton peak of acetyl group in chitosan. The peak of chemical shift δ at 6.7 is the proton peak of benzene ring on catechol, and the peak of chemical shift δ at 4.1 is the proton peak of methylene linked to benzene ring, indicating that the polymer containing catechol structure has successfully reacted with chitosan.

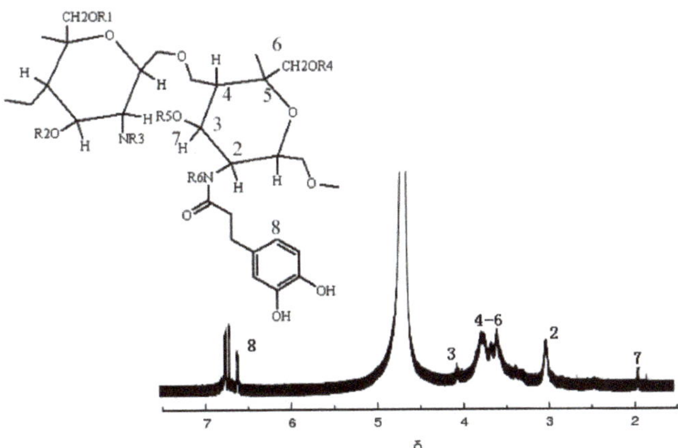

Figure 2. Spectrogram of HNMR.

2.2.3. Thermogravimetric Analysis

The thermogravimetric and differential curves of SDGB are shown in Figure 3. The results showed that the weight loss of SDGB could be divided into four stages. From room temperature to about 150 °C, this stage is mainly due to the loss of adsorbed water. From 180 °C to around 245 °C, there is an obvious weight loss at this stage, mainly due to the degradation of the main polymer chains, and the weight loss is about 30%. From 280 °C to 330 °C, this stage was mainly due to the decomposition of side chains and catechol structure. From 330 °C to 360 °C. The chitosan backbone was destroyed and the mass of SDGB remained unchanged after decomposition.

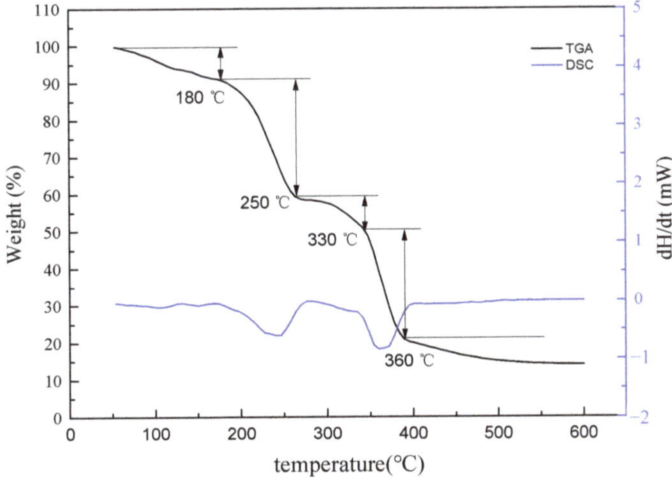

Figure 3. Thermogravimetric analysis results of SDGB.

2.2.4. Gel Permeation Chromatography (GPC) Test

The GPC experimental results of SDGB are shown in Table 3. Gel permeation chromatography showed that the number average molar weight of SDGB was 25,974 and the weight average molecular weight was 33,424. The relative molecular weight is moderate, which ensures that it not only has sufficient adsorption strength but also can enter shale microprobes or cracks. Meanwhile, its polydispersity index was low, indicating narrow molecular weight distribution and good homogeneity.

Table 3. Relative molecular weight test results of SDGB.

Mn (Dalton)	Mw (Dalton)	Mp (Dalton)	Mz (Dalton)	Mz_{+1} (Dalton)	Polydispersity Index
25,974	33,424	17,339	46,291	66,870	1.66731

2.3. *Evaluation of Improving Borehole Stability*

2.3.1. Tensile Shear Strength Test

The results of the tensile shear strength test of the lapped samples after treated by amphoteric polymer, SDGB, emulsified asphalt, aluminum humate, polyol, and Na_2SiO_3 are shown in Figure 4. It can be seen that the lap joint samples treated with SDGB, emulsified asphalt and aluminum humate have higher tensile shear strength in air. The tensile shear strength of 4% SDGB solution in air is about 0.501 MPa, which is the highest among the six samples. The tensile shear strength in water is significantly lower than that in air (0.21 MPa), while the other samples are zero. As can be seen from Figure 5 after treated by SDGB solution, the surface of the metal lap sample is discolored and completely damaged, indicating that SDGB has better adhesion performance. According to Deming TJ et al. [45], catechol group can be oxidized in water to form anthraquinone structure, which can chelate with metal ions or metal oxides on the rock surface to form stable covalent bonds, and then firmly adsorb on the rock surface to prevent rock samples from dispersing in water.

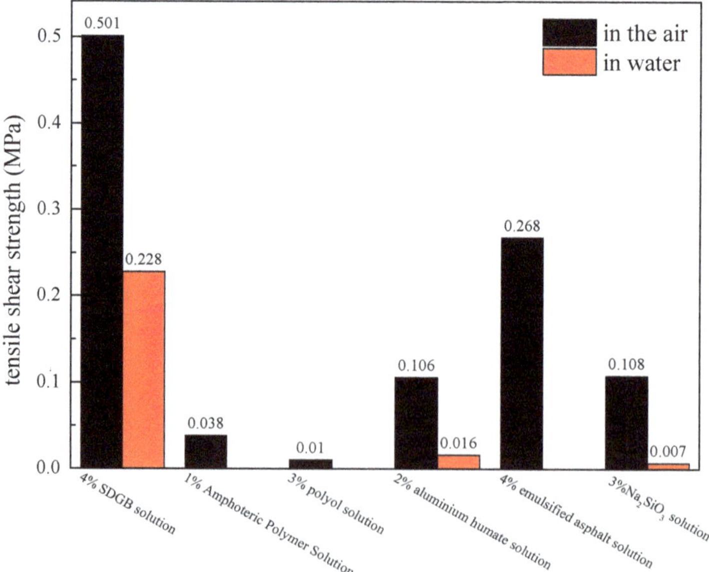

Figure 4. The result of tensile shear strength.

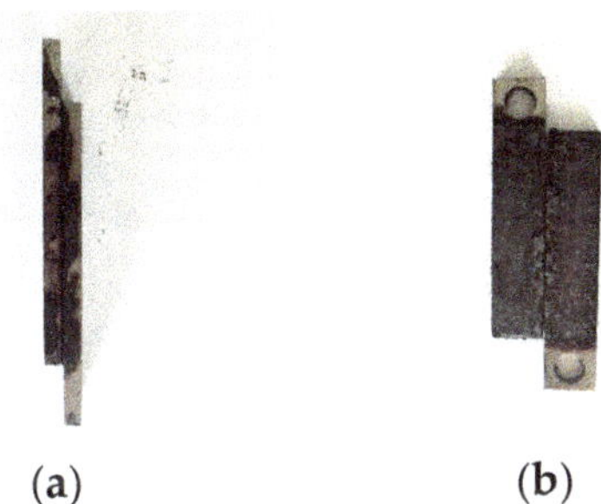

Figure 5. The metal sample treated by 4 wt%SDGB solution. (**a**) Before tested; (**b**) After tested.

2.3.2. Uniaxial Compressive Strength Test

The compressive strength test results of rock samples treated with different wellbore stabilizers are shown in Figure 6. The uniaxial compressive strength of dry rock samples is 12.4 MPa, reduced by about 50% after soaking in clean water. After being treated with different wellbore stabilizers, the compressive strength of rock samples is improved. Among them, 2% of SDGB has the best performance. The compressive strength reaches 8.1 MPa, which is about 25% higher than that of clean water. The results show that SDGB can effectively improve the compressive strength of rock samples.

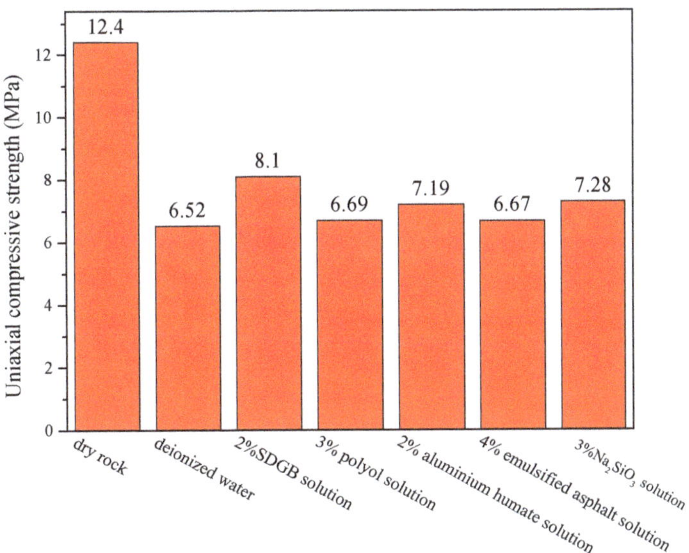

Figure 6. Compressive strength of rock samples treated with different solutions.

2.3.3. Inhibitory Shale Hydration Test

The main mineral composition of shale used for inhibitive experiments is shown in the Tables 4 and 5. It can be seen that the clay mineral content of the shale selected for the experiment is 53%. Illte/smectite minerals are the main clay minerals, indicating that the shale used for this experiment has a high content of active minerals and a high potential hydration ability. 2% SDGB solution can effectively inhibit the dispersion of shale cuttings, and the rolling recovery is 91.06% at 70 °C (Figure 7a). With the increase of temperature, the rolling recovery decreases slightly. However, the recovery after 150 °C/16 h hot rolling is still as high as 75.6% (Figure 7b), indicating that SDGB can still effectively inhibit hydration and dispersion at high temperature. The swelling rates of rock samples treated at 70 °C and 150 °C for 8 h are 42% and 57% respectively, which is lower than 74% of fresh water indicating that the higher the temperature, the worse the inhibition performance.

Table 4. The main mineral composition of shale used for inhibitive experiments.

Quartz	Plagioclase	Calcite	Hematite	Clay Mineral
32	8	2	2	53

Table 5. The main clay composition of shale used for inhibitive experiments.

Kaolinite	Chlorite	Illite	Illte/Smectite	Interlayer Ratio (%)
1	0	2	97	75

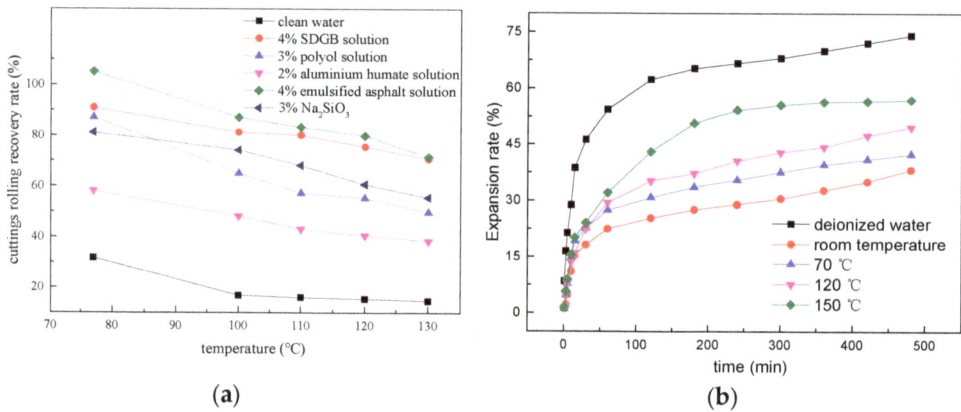

Figure 7. Results of inhibition tests. (**a**) Rolling recovery of different wellbore stabilizers; (**b**) variation of expansion rate at different temperature treated by SDGB.

2.3.4. Hydroscopicity Test

The hydroscopicity test results of shale core treated by different wellbore stabilizers are shown in Figure 8 and Table 6. The rock sample has strong water absorption capacity, which tends to be stable in about 2 h, and the total water absorption reaches 9.1%. After treatment with 2% SDGB solution, the total water absorption of rock samples decreased to 2.75%, nearly 70% lower than that of clean water. The results show that SDGB can firmly adsorb and cement rock samples, so as to inhibit the water absorption capacity of rocks and ensure the stability of wellbore.

Table 6. The hydroscopicity result of different wellbore stabilizers.

Wellbore Stabilizer	Water Absorption of Rock Core (%)	Water Absorption Reduction Rate (%)
Clean water	9.10	0
4% emulsified asphalt solution	8.05	11.53
3% polyol solution	5.38	40.88
3% Na_2SiO_3 solution	4.70	48.35
2% aluminium humate solution	3.24	64.40
2% SDGB solution	2.75	69.78
Clean water	9.10	0
4% emulsified asphalt solution	8.05	11.53

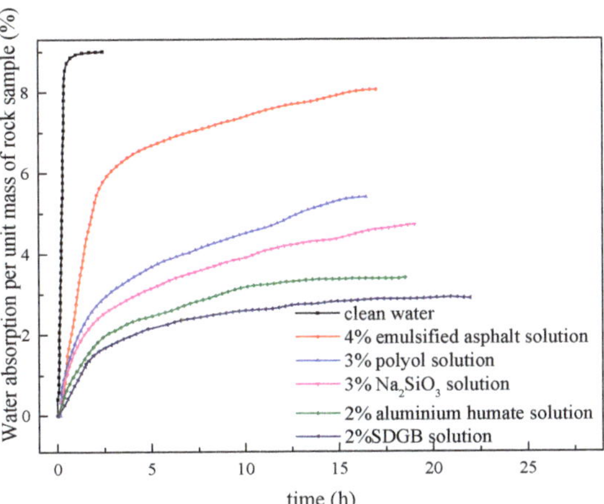

Figure 8. The hydroscopicity curve of different wellbore stabilizers.

2.3.5. Compatibility Evaluation in Drilling Fluid

Different amounts of SDGB were added to 4 wt% bentonite water-based drilling fluid, and its apparent viscosity, plastic viscosity and filtration were tested after hot rolling at 130 °C/16 h. The results are shown in Figures 9–11. It is not difficult to find that there is an optimal concentration of SDGB in the drilling fluid, and the addition of SDGB has no adverse effect on the water-based drilling fluid. After hot rolling at 130 °C, it has good rheological and filtration properties, and has the function of increasing viscosity and reducing filtration.

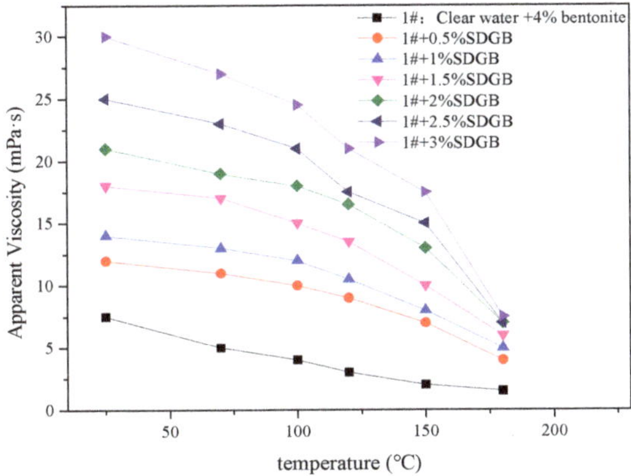

Figure 9. Variation of apparent viscosity of drilling fluid with temperature.

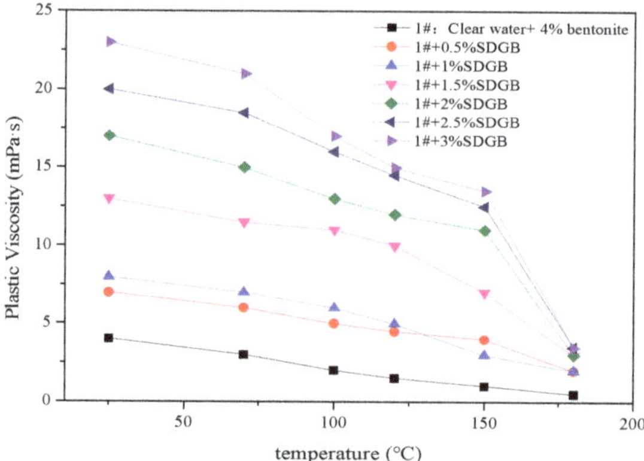

Figure 10. Variation of filtrate loss of drilling fluid with temperature.

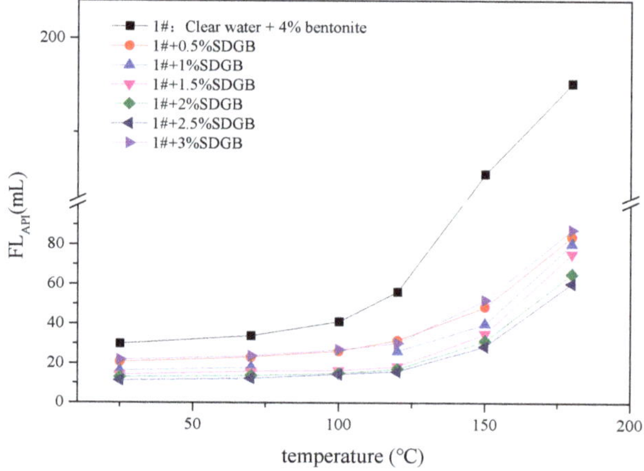

Figure 11. Variation of filtrate loss of drilling fluid with temperature.

2.4. Mechanism of SDGB

2.4.1. Adsorption Isotherm Test

Figure 12 shows the UV absorption spectrum and adsorption standard curve of SDGB, from which it can be seen that the absorption peak of SDGB is the most obvious at the wavelength of 277.8 nm, and the measured adsorption standard curve is roughly linear. Figure 13 is the curve of sodium adsorption amount of bentonite with different surface temperatures. It shows that the adsorption curves of SDGB at different temperatures are L-shaped, indicating that SDGB mainly adsorbs through a single layer on clay surface. The adsorption capacity of SDGB increases rapidly with increasing concentration. When a certain concentration is reached, the adsorption capacity tends to be flat. The saturated adsorption capacity at 25 °C is 117 mg/g, which shows a strong adsorption capacity. With the increase of temperature, the adsorption capacity of SDGB decreases, and the saturated adsorption capacity can still reach 94 mg/g at 90 °C. SDGB has strong high temperature desorption capacity. The analysis shows that the surface of clay particles is negatively charged and SDGB molecules contain more oxygen-containing functional groups which can be firmly adsorbed on the surface of clay particles by hydrogen bonds.

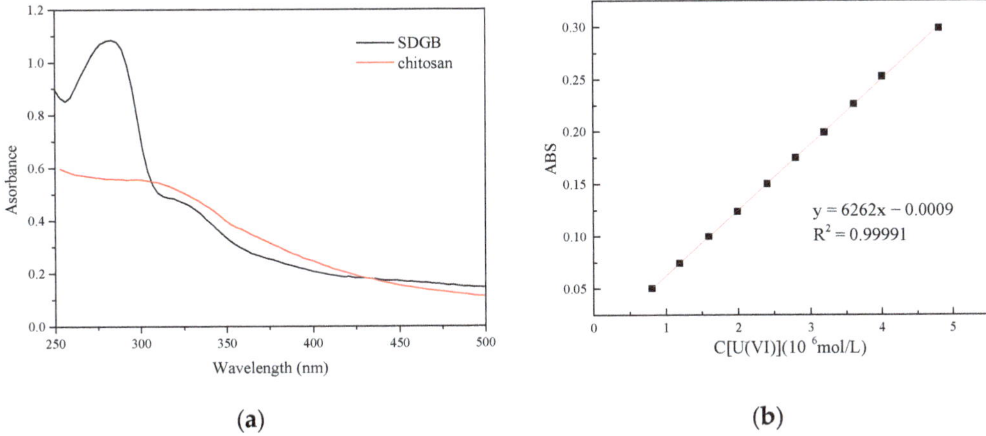

Figure 12. The UV absorption spectra and adsorption standard curve of SDGB. (**a**) UV absorption spectra of SDGB and chitosan; (**b**) SDGB adsorption standard curve.

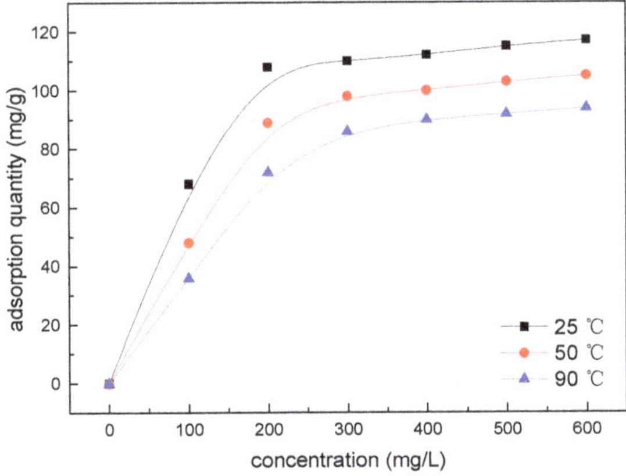

Figure 13. Adsorption isotherm of SDGB.

Shale surfaces are usually rich in metal ions and oxides. Catechol groups can chelate with metal substances to form stable covalent bonds and strongly adsorb on the surface of wellbore rock. In order to further study the adsorption characteristics of SDGB on rock surface, the effects of different concentrations of metal ions (taking ferrous ions as examples) on the adsorption of SDGB were tested. Figure 14 shows the adsorption capacity of SDGB in the presence of Fe^{2+} at different concentrations. The results showed that the adsorption capacity of catechol-chitosan biopolymer capsules on the surface of montmorillonite particles gradually increased with the increase of Fe^{2+} concentration, indicating that the presence of Fe^{2+} can effectively improve the adsorption capacity of catechol-chitosan biopolymer capsules on the surface of montmorillonite particles. According to Deming et al. [45], in the presence of Fe^{2+} ions or other metal ions, oxygen atoms in the catechol structure of SDGB molecules can interact with metal ions to form relatively stable chelation covalent structures, and some can even be oxidized to anthraquinone structures, thereby enhancing the adsorption strength of catechol structures on rock surfaces.

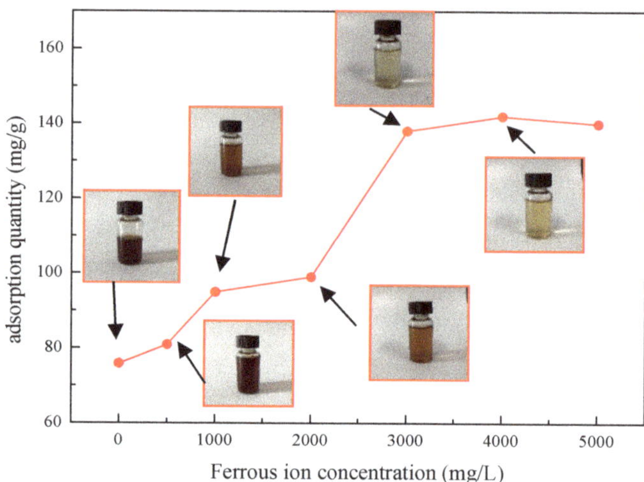

Figure 14. Effect of Fe^{2+} on the adsorption of SDGB.

2.4.2. Scanning Electron Microscope Test

Figure 15 shows the internal shape of cores before and after different treatments. It can be seen from (a) that micropore and microfissure are well developed in the core and clay minerals adhere to the grain surface. (b) It can be seen that after SDGB treatment, clay minerals originally attached to the grain surface are cemented and filled in the micropore of rock samples, which significantly reduces the micropore in rock samples and makes them more compact. It is found that SDGB not only acts as a seal, but also adsorbs, cements and solidifies clay minerals in micro-cracks of rock samples, thus improving the compressive strength of rock.

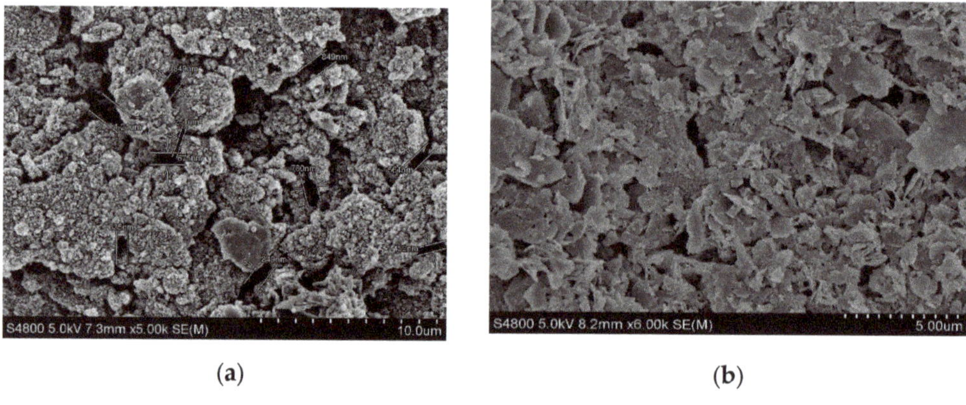

Figure 15. The shale sample treated by SDGB. (**a**) Before treatment (×5.0 k); (**b**) After treatment (×6.0 k).

2.4.3. FT-IR Test

In order to study the interaction mode and mechanism between SDGB and sodium montmorillonite, the infrared spectra of bentonite, SDGB and SDGB modified bentonite were tested, and the result is shown in Figure 16.

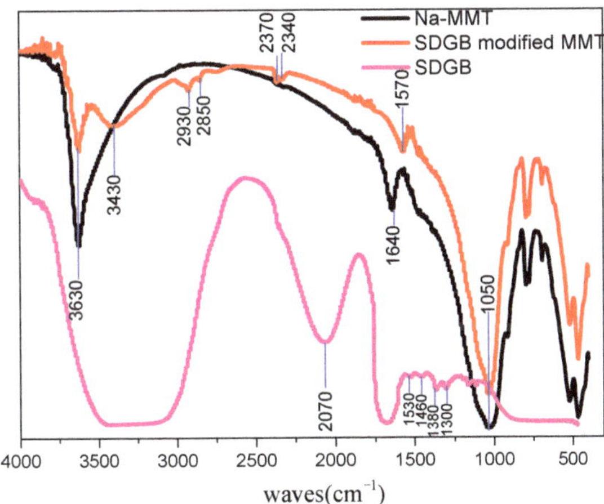

Figure 16. Infrared spectra of sodium montmorillonite before and after modification.

It can be seen from the test results that the main absorption peaks of sodium montmorillonite have not changed significantly before and after modification, indicating that the main molecular structure of sodium montmorillonite has not changed before and after SDGB modification. It can be seen from the infrared spectra of Na-MMT that 3630 cm^{-1} can be attributed to the Tensile vibration of the hydroxyl group, because bound water molecules still exist in the molecular layer of Na-MMT, which is the result of the conjugation of freely bound water and bound water. 1050 cm^{-1} is the vibration absorption peak of silica-silicon in sodium montmorillonite molecular layer. 1640 cm^{-1} can be attributed to the tensile vibration of hydroxyl groups adsorbed in water between different layers of sodium montmorillonite. The results showed that the above main absorption peaks did not change significantly before and after modification, which indicated that the action of SDGB and sodium montmorillonite would not change the main molecular structure of bentonite silicate framework. However, some new infrared absorption peaks can also be found in SDGB-modified sodium montmorillonite. 3430 cm^{-1} can be attributed to the tensile vibration absorption peak formed by the near oxygen atoms connected with benzene ring. 2930 cm^{-1} and 2850 cm^{-1} can be attributed to the symmetric vibration absorption peak of methylene. Meanwhile, the intermolecular hydroxy tensile vibration absorption peak initially appeared at 1640 cm^{-1} in the modified Na-MMT spectrum appeared at 1570 cm^{-1} and moved to low frequency to some extent, indicating that SDGB formed hydrogen bond with sodium montmorillonite.

2.4.4. Summary and Analysis of Mechanism

Traditional high-performance amine shale inhibitors often work through higher amine density. In the presence of a large number of amines, a large number of amines in the inhibitor molecules are positively charged, which can electrostatically absorb with negatively charged bentonite particles, thus neutralizing the negative charges of active clays, compressing their diffusion double layer, driving out water molecules and thus inhibiting shale hydration. The schematic diagram of action mechanism of SDGB is shown in Figure 17. SDGB surface does not contain a large number of strong cationic groups to keep the drilling fluid from functioning (most water-based drilling fluid additives are negatively charged). First, the adsorption of SDGB occurs through the presence of trace metal ions on the shale surface. The structure of catechol in SDGB can chelate with metal ions. The active groups such as catechol adsorb rapidly and firmly on the surface of rock particles on the well wall through hydrogen bond and chelation, encapsulate the active groups, and play a "sticky" role in preventing rock particles from dispersing under the action of drilling fluids and improving rock strength. Second, during the adsorption process between chemical wall fixing agent and the surface of rock particles, the polymer chain can also fill and plug the pores of rock, reduce rock permeability, and improve the stability of drilled rock. Third, the molecules of chemical wall fixing agent contain strong adsorptive groups, which can inhibit hydration when strongly adsorbed on rock surface. In short, the combination of "binder curing to increase strength physical blockage to reduce permeability of rock samples to reduce hydration rejection" improves wellbore stability.

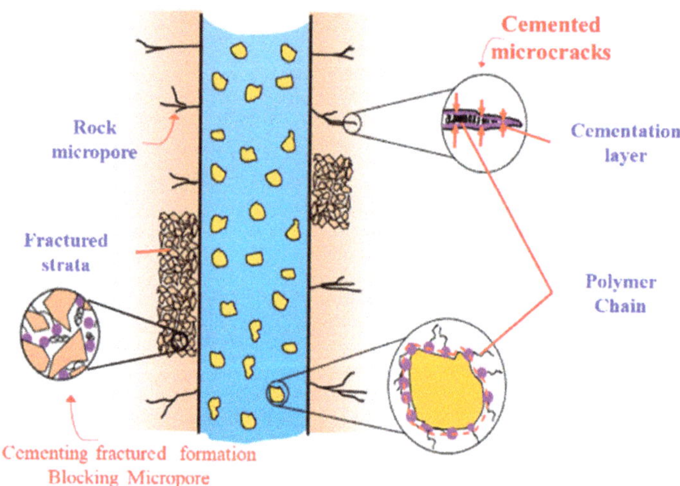

Figure 17. Mechanism diagram of SDGB enhancing wellbore stability.

3. Conclusions

Inspired by mussels, a novel catechol-conjugated-chitosan biopolymer encapsulators SDGB was synthesized. The optimum synthesis conditions are determined in this paper. The structure of the products was characterized by IR, NMR, and TG. The results showed that the catechol reacted successfully with chitosan, and the decomposition temperature was over 180 °C. The experiment shows that SDGB can effectively improve the shear strength and inhibition of shale. The recovery rate of the shale cuttings was still up to 75.6% after 150 °C/16 h hot rolling, and the bearing capacity of the rock can be increased by 24%. A wellbore strengthening drilling fluid was constructed based on SDGB. The rheological and filtration performance of this drilling fluid is stable before and after 130 °C/16 h hot rolling, and shale inhibition behavior is good. Moreover, SDGB could chelate with metal ions, forming a stable covalent bond, which plays an important role in adhesiveness, inhibition, and blockage.

4. Materials and Methods

4.1. Materials

3,4-dihydroxybenzoic acid(catechol), chitosan (degree of deacetylation (>95%), *N*-(3-dimethylaminopropyl)-*N*-ethylcarbodiimide hydrochloride (EDC) and *N*-hydroxysuccinimide (NHS) were commercial products from Aladdin reagent company, Shanghai, China, respectively. Na_2SO_3, NaCl, $CaCl_2$, NaOH, and Na_2SiO_3 were purchased from China Pharmaceutical Reagents Co., Ltd. in Shanghai, China. Sodium-based bentonite and artificial cores for drilling fluid were purchased from Huawei Bentonite Group Co., Ltd. in Weifang City, China and Haian Petroleum Research Instrument Co., Ltd. in Nantong City, China, respectively. The main reagents for the reaction are detailed in Table 4.

4.2. Methods

4.2.1. Synthesis of Catechol-Chitosan Biopolymer Encapsulators

Ferrous chloride was added and fully dissolved in hot ethanol solution (50% wt, 30 mL). A certain amount of polymer containing catechol structure was added into a three-neck flask and stirred with an electric stir bar under an inert N_2 atmosphere. A certain amount of chitosan was dissolved with 20 mL, 1 mol/L hydrochloric acid and them nitrogen is introduced into the reaction system. After the pH was adjusted to a certain value with 5 mol/L NaOH solution, the initiator is dissolved in 10 mL of deionized water and then added dropwise to the reaction solution to trigger the reaction. The mixture was refluxed for a period of time at a certain temperature. The precipitate was filtered and washed with ethanol to get the final product. The molecular structure of SDGB is shown in Figure 18.

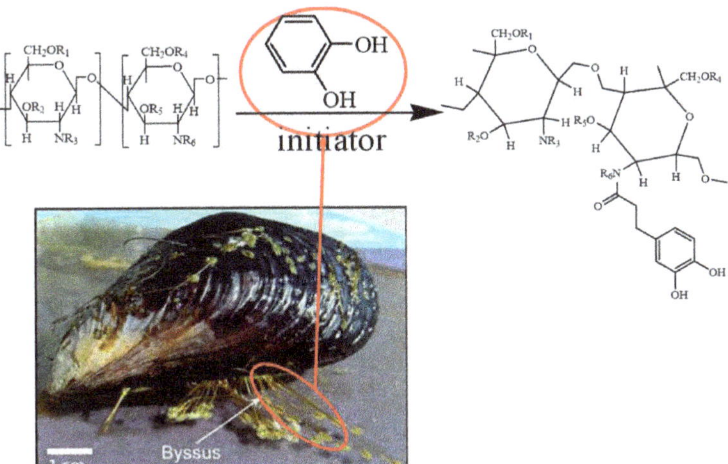

Figure 18. The molecular structure of SDGB.

Orthogonal tests were used to optimize the four main factors to determine the best formulation and these four factors are the mole ratio between catechol and chitosan, reaction time, initiator concentration, and reaction temperature. The influence of different factors was analyzed by testing the tensile shear strength and rolling recovery of the product.

4.2.2. Tensile Shear Strength Test

The mechanical properties of core before and after treated by SDGB are tested by uniaxial mechanical strength test(Bairoe Test Instrument Co., Ltd. in Shanghai, China), which can effectively reflect the improvement of shale by strong wall drilling fluids. In accordance with the Chinese National standard "methods for determination of tensile shear strength of adhesives" (GB7124-1986) and "determination of chemical resistance of adhesives" (GB/t13353-92), the aqueous solution of SDGB is applied to the single lap surface of the sample and the lap joint sample is pressed at 5 MPa for 2 h, then placed in water at 50 °C for 24 h. Then longitudinal tensile force is applied to the single lap joint surface of the sample to test the maximum load that the sample can bear on wdw-100 microcomputer controlled triaxial multifunctional machine (Figure 19).

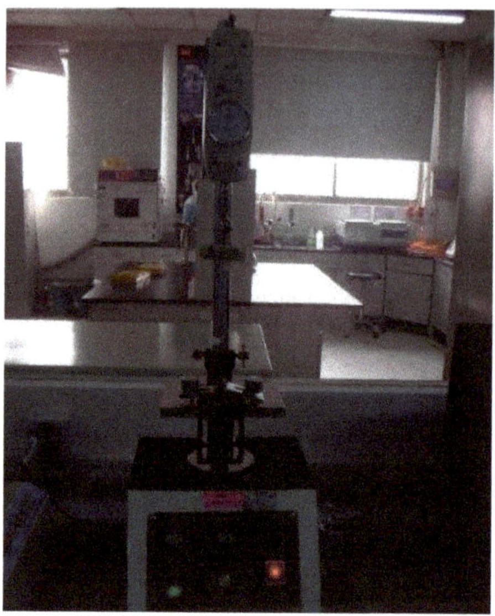

Figure 19. WDW-100 tensile shear strength device.

4.2.3. Cuttings Hot-Rolling Dispersion Test

Shale dispersion recovery experiment is a common method to evaluate the shale inhibition ability of treatment agent. The dispersion performance of shale is related to the stability of wellbore. It is one of the important indexes to evaluate the stability of shale wellbore macroscopically. The dispersion experiment can be used to understand the hydration and dispersion performance of rock samples, and can also be used as a means to evaluate the inhibition of drilling fluid on shale. 350 mL of solution with encapsulators of various concentrations and 50 g of shale cuttings (2–5 mm) obtained from the upper layer of Shahejie formation in Dongying oil field were added into sealed cells. After hot rolling at 77 °C for 16 h, the cuttings were washed with 10% KCl solution and screened through a 40-mesh sieve. The recovered cuttings were dried at 105 °C for 4 h and the percentage of recovery was determined.

4.2.4. Characterization of SDGB

A Bruker AVANCE 400 NMR spectrometer (Brooke Co., Ltd., Zurich, Switzerland) was used to measure the proton nuclear magnetic resonance (^{1}HNMR) spectra. D_2O was used for field frequency lock, and the observed ^{1}H chemical shifts were reported in parts per million (ppm). Before ^{1}H NMR analysis, the pH value of the polymer solution was adjusted to around 9.0 by dilute $NaOH/D_2O$ solution.

A total of 1 mg dry SDGB powder and 20 mg KBr were mixed fully. The mixture was loaded into the mold and compacted with 50 MPa pressure. FTIR of the compacted tableting were obtained on a NEXUS FT-IR spectrometer(Thermo Nicolet Corporation, Waltham, MA, USA).

The thermogravimetric analysis of SDGB was carried out by mettler-teledothermo gravimetric analyzer (NETZSCH-Gerätebau Co., Ltd., Selb, Germany). The temperature range was from room temperature to 1000 °C, the heating rate was 10 K/min, and the atmosphere was nitrogen, and the gas flow rate was 50 mL/min.

The relative molecular mass of SDGB was determined by the SFD gel permeation chromatograph (GPC) (Schambeck Co., Ltd., Berlin, Germany). The mobile phase was phosphate buffer solution. The column was SHODEX (K-806M chloroform system, and the filler was styrene and two vinyl benzene copolymer).

4.2.5. Core Uniaxial Compressive Strength Test

The rock samples were soaked in different wellbore strengthening agent solutions for 4 h, then taken out carefully and dried at room temperature. Then compressive strength was tested on wdw-100 microcomputer controlled triaxial multifunctional machine (Bairoe Test Instrument Co., Ltd., Shanghai, China).

4.2.6. The Hydroscopicity Test

The hydroscopicity test was conducted by self-made core self-priming experimental device. The water absorption and water absorption rate of typical shale in deionized water were tested. At the same time, the change of water absorption of rock samples treated with different encapsulators with time was tested to comprehensively evaluate the hydration and dispersion performance of rock samples. First, the experimental core is dried in an oven at 105 °C for 12 h, and then placed at 25 °C for 24 h. During the experiment, the room temperature is maintained at 25 °C. Then, the shale standard core is suspended on the hook of the electronic balance, so that its lower end just contacts the solution. The total mass of the rock sample changes after water absorption over time is recorded.

4.2.7. Scanning Electron Microscope Test

Core micro morphology was tested by S-4800 field emission scanning electron microscope (HitaChi, Ltd., Tokyo, Japan) after being treated by SDGB, sodium silicate and aluminum humate.

4.2.8. Adsorption Isotherm Test

In order to further study the action mechanism of SDGB in drilling fluids, the adsorption characteristics of SDGB on bentonite particles were studied.

(1). A sufficient amount of 10% hydrogen peroxide was added to the beaker together with 100 g of bentonite and stirred simultaneously on a heating oven to eliminate the possible influence of the residual organics in the bentonite on the adsorption test results. Deionized water was then added so that the contents in the beaker are well dispersed, and then was placed in a centrifuge and centrifuged at 8000 rpm for 15 min. The upper clear liquid was poured out, and deionized water was continued to be centrifuged repeatedly for three times. The precipitate was collected and then placed into a blast drying oven and dried to constant weight, crushed and sieved.

(2). Some SDGB was added into a beaker containing deionized water, stirred well and dissolved; 2.0 g of purified bentonite was added to another beaker containing deionized water, stirred well and dispersed. After the two beakers had been mixed, they were placed in a magnetic stirrer and stirred thoroughly for 24 h until the adsorption equilibrium was reached. The mixed suspension was placed in a centrifuge and centrifuged at a high speed of 10,000 rpm for 15 min. The absorbance in λ = 200–400 nm of the supernatant was measured by a uv-1750 UV VIS spectrophotometer(Yuanxi Instrument Co., Ltd., Shanghai, China).

(3). In λ = 277.8 nm, the adsorption capacity on the surface of bentonite sodium was calculated from the standard curve of polymer solution concentration and light transmittance (as shown in Figure 12a).

4.2.9. Rheological and Filtration Testing of Drilling Fluids

(1) Drilling fluid preparation and hot rolling

A total of 16 g of sodium-based bentonite was added to 400 mL of clear water, and stirred at 8000 rpm in a high-speed blender for 30 min. Then 0.8 g of Na_2CO_3 was added, stirred for 20 min, and then pre-hydrated for 24 h. A certain amount of polymer was then added to the slurry, which was stirred at 8000 rpm for 30 min in a high-speed blender. The composites were aged at set temperature for 16 h by hot rolling and were cooled to room temperature before stirring at high speed for 20 min. Rheological and filtration properties of drilling fluid before and after rolling at a specific temperature/16 h were tested according to the drilling fluid performance evaluation standard SY/t5621-1993.16g.

(2) API Static Filtration Test

The static API filterability of drilling fluid was tested with ZNZ-D3 API medium pressure filter (Qingdao Haitong Instrument Co., Ltd., Qingdao, China). A certain amount of drilling fluid was loaded into the filter kettle and the top was covered with API filter paper and placed under a pressure of 100 psi. The filtered volume (FLAPI) of the drilling fluid is recorded for 30 min, which is recommended by API.

(3). Rheological property test

The rheological parameters of the drilling fluid are tested according to the drilling fluid performance evaluation standard SY/T5621-1993. Apparent viscosity, plastic viscosity, and shear force of drilling fluid were measured with ZNP-M7 6-speed rotating viscometer (Haitong Instrument Co., Ltd., Qingdao, China). Then the apparent viscosity, plastic viscosity, and shear force of drilling fluid with φ600 and φ were measured. The value of 300 is calculated according to the test program recommended by API.

$$AV = \varphi600/2 \quad (1)$$

$$PV = \varphi600 - \varphi300 \quad (2)$$

where:

AV is the apparent viscosity (mPa·s);

PV is the plastic viscosity (mPa·s);

φ600 is the dial reading of 6-speed rotational viscometer at 600 r/min (dia);

φ300 is the dial reading of 6-speed rotational viscometer at 300 r/min (dia).

Author Contributions: Conceptualization, Z.Q. and H.Z.; methodology, H.Z.; investigation, Z.T. and H.Z.; resources, Z.Q. and B.G.; data curation, Z.T. and Y.K.; writing—original draft preparation, Z.T.; writing—review and editing, Z.T. and Z.Q.; supervision, Z.Q. and H.Z.; project administration, Z.Q. All authors have read and agreed to the published version of the manuscript.

Funding: This research received no external funding.

Institutional Review Board Statement: Not applicable.

Informed Consent Statement: Not applicable.

Conflicts of Interest: The authors declare no conflict of interest.

References

1. Lee, H.; Ong, S.H.; Azeemuddin, M.; Goodman, H. A wellbore stability model for formations with anisotropic rock strengths. *J. Pet. Sci. Eng.* **2012**, *96*, 109–119. [CrossRef]
2. Huang, X.; Shen, H.; Sun, J.; Lv, K.; Liu, J.; Dong, X.; Luo, S. Nanoscale laponite as a potential shale inhibitor in water-based drilling fluid for stabilization of wellbore stability and mechanism study. *ACS Appl. Mater. Interfaces* **2018**, *10*, 33252–33259. [CrossRef] [PubMed]
3. Bauder, T.; Barbarick, K.; Ippolito, J.; Shanahan, J.; Ayers, P. Soil properties affecting wheat yields following drilling-fluid application. *J. Environ. Qual.* **2005**, *34*, 1687–1696. [CrossRef]
4. Tianshou, M.; Ping, C. Study of meso-damage characteristics of shale hydration based on CT scanning technology. *Pet. Explor. Dev.* **2014**, *41*, 249–256.
5. Bird, P. Hydration-phase diagrams and friction of montmorillonite under laboratory and geologic conditions, with implications for shale compaction, slope stability, and strength of fault gouge. *Tectonophysics* **1984**, *107*, 235–260. [CrossRef]
6. Heidug, W.; Wong, S.W. Hydration swelling of water-absorbing rocks: A constitutive model. *Int. J. Numer. Anal. Methods Geomech.* **1996**, *20*, 403–430. [CrossRef]
7. Shadizadeh, S.R.; Moslemizadeh, A.; Dezaki, A.S. A novel nonionic surfactant for inhibiting shale hydration. *Appl. Clay Sci.* **2015**, *118*, 74–86. [CrossRef]
8. Liu, H.; Meng, Y.; Li, G.; Li, P.; Deng, Y. Theoretical simulation and experimental evaluation of the effect of hydration on the shale rock strength. *Drill. Prod. Technol.* **2010**, *33*, 18–20.
9. Zhong, H.; Qiu, Z.; Tang, Z.; Zhang, X.; Zhang, D.; Huang, W. Minimization shale hydration with the combination of hydroxyl-terminated PAMAM dendrimers and KCl. *J. Mater. Sci.* **2016**, *51*, 8484–8501. [CrossRef]

0. Gou, S.; Yin, T.; Liu, K.; Guo, Q. Water-soluble complexes of an acrylamide copolymer and ionic liquids for inhibiting shale hydration. *New J. Chem.* **2015**, *39*, 2155–2161. [CrossRef]
1. Zeng, F.; Zhang, Q.; Guo, J.; Zeng, B.; Zhang, Y.; He, S. Mechanisms of shale hydration and water block removal. *Pet. Explor. Dev.* **2021**, *48*, 752–761. [CrossRef]
2. Simpson, J.; Walker, T.; Jiang, G. Environmentally acceptable water-base mud can prevent shale hydration and maintain borehole stability. *SPE Drill. Completion* **1995**, *10*, 242–249. [CrossRef]
3. Liu, Y.; Chen, L.; Tang, Y.; Zhang, X.; Qiu, Z. Synthesis and characterization of nano-SiO_2@ octadecylbisimidazoline quaternary ammonium salt used as acidizing corrosion inhibitor. *Rev. Adv. Mater. Sci.* **2022**, *61*, 186–194. [CrossRef]
4. Baoyou, R.; Xiaolin, P.; Cheng, C.; Chuan, M. Experimental study on improving the formation pressure-bearing capacity by using nano-drilling fluid. *Oil Drill. Prod. Technol.* **2018**, *40*, 179–184.
5. Kang, Y.; Xu, C.; Tang, L.; Li, S.; Li, D. Constructing a tough shield around the wellbore: Theory and method for lost-circulation control. *Pet. Explor. Dev.* **2014**, *41*, 520–527. [CrossRef]
6. Li, J.; Qiu, Z.; Liu, Z.; Yang, Y.; Song, H.; Liang, Y. Application of wellbore strengthening drilling fluid technology in Lingshui gas field. In Proceedings of the IOP Conference Series: Earth and Environmental Science, Chengdu, China, 23–25 April 2021; p. 012190.
7. Qiu, Z.; Bao, D.; Li, J.; Liu, J.; Chen, J. Mechanisms of wellbore strengthening and new advances in lost circulation control with dense pressure bearing zone. *Drill. Fluid Completion Fluid* **2018**, *35*, 1–6.
8. Huang, X.; Sun, J.; Lv, K.; Liu, J.; Shen, H.; Zhang, F. Application of core-shell structural acrylic resin/nano-SiO_2 composite in water based drilling fluid to plug shale pores. *J. Nat. Gas Sci. Eng.* **2018**, *55*, 418–425. [CrossRef]
9. Hoxha, B.B.; van Oort, E.; Daigle, H. How do nanoparticles stabilize shale? *SPE Drill. Completion* **2019**, *34*, 143–158. [CrossRef]
20. Akhtarmanesh, S.; Shahrabi, M.A.; Atashnezhad, A. Improvement of wellbore stability in shale using nanoparticles. *J. Pet. Sci. Eng.* **2013**, *112*, 290–295. [CrossRef]
21. Zhong, H.; Gao, X.; Zhang, X.; Qiu, Z.; Zhao, C.; Zhang, X.; Jin, J. Improving the shale stability with nano-silica grafted with hyperbranched polyethyleneimine in water-based drilling fluid. *J. Nat. Gas Sci. Eng.* **2020**, *83*, 103624. [CrossRef]
22. Ghasemi, A.; Jalalifar, H.; Apourvari, S.N.; Sakebi, M.R. Mechanistic study of improvement of wellbore stability in shale formations using a natural inhibitor. *J. Pet. Sci. Eng.* **2019**, *181*, 106222. [CrossRef]
23. Pourkhalil, H.; Nakhaee, A. Effect of nano ZnO on wellbore stability in shale: An experimental investigation. *J. Pet. Sci. Eng.* **2019**, *173*, 880–888. [CrossRef]
24. Zhang, J.; Li, L.; Wang, S.; Wang, J.; Yang, H.; Zhao, Z.; Zhu, J.; Zhang, Z. Novel micro and nano particle-based drilling fluids: Pioneering approach to overcome the borehole instability problem in shale formations. In Proceedings of the SPE Asia Pacific Unconventional Resources Conference and Exhibition, Brisbane, Australia, 9–11 November 2015.
25. Jung, C.M.; Zhang, R.; Chenevert, M.; Sharma, M. High-performance water-based mud using nanoparticles for shale reservoirs. In Proceedings of the SPE/AAPG/SEG Unconventional Resources Technology Conference, Denver, CO, USA, 12–14 August 2013.
26. Luo, Z.; Wang, L.; Yu, P.; Chen, Z. Experimental study on the application of an ionic liquid as a shale inhibitor and inhibitive mechanism. *Appl. Clay Sci.* **2017**, *150*, 267–274. [CrossRef]
27. Zhong, H.; Qiu, Z.; Huang, W.; Cao, J. Poly (oxypropylene)-amidoamine modified bentonite as potential shale inhibitor in water-based drilling fluids. *Appl. Clay Sci.* **2012**, *67*, 36–43. [CrossRef]
28. Huang, X.; Sun, J.; Jin, J.; Lv, K.; Li, H.; Rong, K.; Zhang, C.; Meng, X. Use of silicone quaternary ammonium salt for hydrophobic surface modification to inhibit shale hydration and mechanism study. *J. Mol. Liq.* **2021**, *341*, 117369. [CrossRef]
29. Zhang, S.; He, Y.; Chen, Z.; Sheng, J.J.; Fu, L. Application of polyether amine, poly alcohol or KCl to maintain the stability of shales containing Na-smectite and Ca-smectiteShifeng Zhang et al. Maintaining stability of shale with Na-, Ca-smectite. *Clay Miner.* **2018**, *53*, 29–39. [CrossRef]
30. Patel, A.D. Design and development of quaternary amine compounds: Shale inhibition with improved environmental profile. In Proceedings of the SPE International Symposium on Oilfield Chemistry, The Woodlands, TX, USA, 20–22 April 2009.
31. Ali, J.A.; Hamadamin, A.B.; Ahmed, S.M.; Mahmood, B.S.; Sajadi, S.M.; Manshad, A.K. Synergistic Effect of Nanoinhibitive Drilling Fluid on the Shale Swelling Performance at High Temperature and High Pressure. *Energy Fuels* **2022**, *36*, 1996–2006. [CrossRef]
32. Schlemmer, R.F. Membrane Forming In-Situ Polymerization for Water Based Drilling Fluids. U.S. Patent 7279445B2, 9 October 2007.
33. Schlemmer, R.; Friedheim, J.; Growcock, F.; Bloys, J.; Headley, J.; Polnaszek, S. Chemical osmosis, shale, and drilling fluids. *SPE Drill. Completion* **2003**, *18*, 318–331. [CrossRef]
34. Schlemmer, R.; Friedheim, J.; Growcock, F.; Bloys, J.; Headley, J.; Polnaszek, S. Membrane efficiency in shale-an empirical evaluation of drilling fluid chemistries and implications for fluid design. In Proceedings of the IADC/SPE Drilling Conference, Dallas, TX, USA, 26–28 February 2002.
35. Albooyeh, M.; Kivi, I.R.; Ameri, M. Promoting wellbore stability in active shale formations by water-based muds: A case study in Pabdeh shale, Southwestern Iran. *J. Nat. Gas Sci. Eng.* **2018**, *56*, 166–174. [CrossRef]
36. Chu, Q.; Luo, P.; Zhao, Q.; Feng, J.; Kuang, X.; Wang, D. Application of a new family of organosilicon quadripolymer as a fluid loss additive for drilling fluid at high temperature. *J. Appl. Polym. Sci.* **2013**, *128*, 28–40. [CrossRef]
37. Fang, W.; Jiang, H.; Li, J.; Li, W.; Li, J.; Zhao, L.; Feng, X. A new experimental methodology to investigate formation damage in clay-bearing reservoirs. *J. Pet. Sci. Eng.* **2016**, *143*, 226–234. [CrossRef]

38. Gomez, S.; He, W. Fighting wellbore instability: Customizing drilling fluids based on laboratory studies of shale-fluid interactions. In Proceedings of the IADC/SPE Asia Pacific Drilling Technology Conference and Exhibition, Tianjin, China, 9–11 July 2012.
39. Tan, C.P.; Richards, B.G.; Rahman, S. Managing physico-chemical wellbore instability in shales with the chemical potential mechanism. In Proceedings of the SPE Asia Pacific Oil and Gas Conference, Adelaide, Australia, 28–31 October 1996.
40. Zhang, Q.; Jia, W.; Fan, X.; Liang, Y.; Yang, Y. A review of the shale wellbore stability mechanism based on mechanical–chemical coupling theories. *Petroleum* **2015**, *1*, 91–96. [CrossRef]
41. Kord Forooshani, P.; Lee, B.P. Recent approaches in designing bioadhesive materials inspired by mussel adhesive protein. *J. Polym. Sci. Part A Polym. Chem.* **2017**, *55*, 9–33. [CrossRef] [PubMed]
42. Lee, H.; Rho, J.; Messersmith, P.B. Facile conjugation of biomolecules onto surfaces via mussel adhesive protein inspired coatings. *Adv. Mater.* **2009**, *21*, 431–434. [CrossRef]
43. Guo, Z.; Ni, K.; Wei, D.; Ren, Y. Fe^{3+}-induced oxidation and coordination cross-linking in catechol–chitosan hydrogels under acidic pH conditions. *RSC Adv.* **2015**, *5*, 37377–37384. [CrossRef]
44. Shu, L.; Chiou, Y.-M.; Orville, A.M.; Miller, M.A.; Lipscomb, J.D.; Que, L., Jr. X-ray absorption spectroscopic studies of the Fe (II) active site of catechol 2,3-dioxygenase. Implications for the extradiol cleavage mechanism. *Biochemistry* **1995**, *34*, 6649–6659. [CrossRef]
45. Deming, T.J. Mussel byssus and biomolecular materials. *Curr. Opin. Chem. Biol.* **1999**, *3*, 100–105. [CrossRef]

 gels

Article

Synthesis and Weak Hydrogelling Properties of a Salt Resistance Copolymer Based on Fumaric Acid Sludge and Its Application in Oil Well Drilling Fluids

Zhongjin Wei, Fengshan Zhou *, Sinan Chen and Wenjun Long

Beijing Key Laboratory of Materials Utilization of Nonmetallic Minerals and Solid Wastes, National Laboratory of Mineral Materials, School of Materials Science and Technology, China University of Geosciences (Beijing), Beijing 100083, China; 3003190032@cugb.edu.cn (Z.W.); 2103190085@cugb.edu.cn (S.C.); longwj@cugb.edu.cn (W.L.)
* Correspondence: zhoufs@cugb.edu.cn

Abstract: Fumaric acid sludge (FAS) by-produced from phthalic anhydride production wastewater treatment contains a large amount of refractory organic compounds with a complex composition, which will cause environmental pollution unless it is treated in a deep, harmless manner. FAS included saturated carboxylic acid, more than 60%, and unsaturated carboxylic acid, close to 30%, which accounted for the total mass of dry sludge. A new oil well drilling fluid filtrate loss reducer, poly(AM-AMPS-FAS) (PAAF), was synthesized by copolymerizing FAS with acrylamide (AM) and 2-acrylamide-2-methyl propane sulfonic acid (AMPS). Without a refining requirement for FAS, it can be used as a polymerizable free radical monomer for the synthesis of PAAF after a simple drying process. The copolymer PAAF synthesis process was studied, and the optimal monomer mass ratio was determined to be AM:AMPS:FAS = 1:1:1. The temperature resistance of the synthesized PAAF was significantly improved when 5% sodium silicate was added as a cross-linking agent. The structural characterization and evaluation of temperature and complex saline resistance performance of PAAF were carried out. The FT-IR results show that the structure of PAAF contained amide groups and sulfonic acid groups. The TGA results show that PAAF has good temperature resistance. As an oilfield filtrate loss reducer, the cost-effective copolymer PAAF not only has excellent temperature and complex saline resistance, the API filtration loss (FL) was only 13.2 mL/30 min after 16 h of hot rolling and aging at 150 °C in the complex saline-based mud, which is smaller compared with other filtrate loss reducer copolymers, but it also has little effect on the rheological properties of drilling fluid.

Keywords: fumaric acid sludge (FAS); free radical copolymerization; filtrate loss reducer; organic wastewater of phthalic anhydride; oil well drilling fluids

Citation: Wei, Z.; Zhou, F.; Chen, S.; Long, W. Synthesis and Weak Hydrogelling Properties of a Salt Resistance Copolymer Based on Fumaric Acid Sludge and Its Application in Oil Well Drilling Fluids. *Gels* **2022**, *8*, 251. https://doi.org/10.3390/gels8050251

Academic Editors: Qing You, Guang Zhao and Xindi Sun

Received: 29 March 2022
Accepted: 18 April 2022
Published: 20 April 2022

1. Introduction

Phthalic anhydride is usually produced by the naphthalene method [1], which is produced by the synthesis of naphthalene and o-xylene through the process of oxidation, isomerization, drying and crystallization, etc. The organic tail gas produced in the production of phthalic anhydride becomes acidic organic wastewater after absorbing washing water, and then the catalyst thiourea was added and the maleic acid in the wastewater was transformed into fumaric acid. After cooling, crystallization, centrifugation, and other operations, the fumaric acid production wastewater is produced.

A large amount of fumaric acid production wastewater is produced during the production of phthalic anhydride, and the COD of these organic wastewater is about 6000–12,000 mg/L [2], which requires the addition of flocculants for flocculation and sedimentation, and the flocculation products are pressed and filtered to obtain press-filtered sludge. Fumaric acid sludge (FAS) can be obtained by drying the press-filtered sludge, as shown in Figure 1. A plant producing about 150,000 tons of phthalic anhydride will produce about 4000 tons of FAS every year, and this FAS sludge may contain hazardous chemicals such as naphthoquinone, naphthalene, maleic anhydride, etc., which will cause environmental pollution unless it is treated in a deep, harmless manner. Therefore, utilized FAS as a high-value resource has important environmental and economic significance.

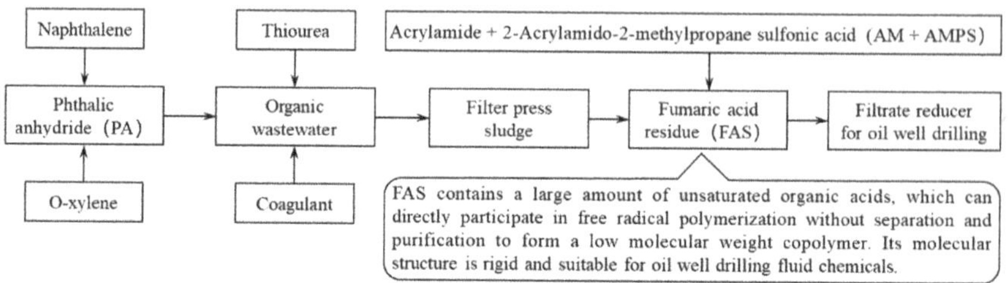

Figure 1. Flowchart of the source of FAS and its high-value utilization method.

The specific composition of FAS is different due to different synthetic process routes, operation process parameters of different manufacturers, organic wastewater treatment process, etc. The substances contained in FAS generally include phthalic acid, benzoic acid, citric acid, maleic acid, fumaric acid and other organic acids, mainly including fumaric acid and maleic acid.

Due to the high value of fumaric acid, many researchers try to separate and purify fumaric acid from FAS [3–5]. However, it is difficult to separate high-purity fumaric acid from FAS, and the process is very complicated, going through the steps of washing and sedimentation, decolorization, filtration, transposition reaction, drying, filtrate treatment, etc. These processing methods require many instruments, high energy consumption, high treatment cost, and low economic efficiency; finding a cost-effective way to treat FAS has become a challenge.

Some researchers have studied the use of FAS to produce polyol unsaturated polyester resin to produce artificial marble/granite and foamed resin for home decoration building materials, but failed to realize industrial application [6]. In addition, only unsaturated carboxylic acids (about one-third) in FAS are utilized in the production of unsaturated polyester resin. Even if the average price of maleic acid and fumaric acid industrial products is 6000 CNY/t, the utilization value is only about 2000 CNY/t. How to realize the utilization of saturated carboxylic acid (about three-fifths) in FAS is more economical.

Unsaturated carboxylic acids account for about one-third of FAS, while saturated carboxylic acids account for two-thirds. This compositional feature determines that the utilization of FAS cannot only consider unsaturated dicarboxylic acids, and the value of saturated carboxylic acids is equally important, or even more important. Citric acid and its aluminum salts are widely used in oilfield water injection scale inhibitors and cross-linking agents for polymer hydrogels [7–10]. Therefore, the direct utilization of FAS may be the best in the field of oilfield chemicals.

Drilling fluid, known as the "blood" in drilling engineering, has the functions to carry cuttings, protect the well-bore wall, balance formation pressure and reduce drilling tool wear [11]. With the continuous low price of oil, the direction of development of drilling fluid additives highlights the utilized resources of wastes to reduce the production cost of the additives [12–14]. If FAS can be directly utilized to prepare drilling fluid additives, it is not only an important process method to solve the clean production of phthalic anhydride, but also an effective way to reduce the production cost of drilling fluid additives.

Oilfield filtrate loss reducer is the core additive of drilling fluid, which can form a thin and dense mud cake at the well wall, reduce the amount of drilling fluid filtrate loss penetration into the formation interior, avoid well collapse, and ensure safe oilfield production. Filtration control is an important property of a drilling fluid, particularly when drilling through permeable formations, where the hydrostatic pressure exceeds the formation pressure. Commonly used filtrate loss reducer includes cellulose derivatives, starch derivatives, humic acids, synthetic resins, and sulfonic acid-based multipolymer [15,16]. Both fumaric acid and maleic acid contain –C=C– in their molecular structures, and it is feasible to utilize FAS as a polymeric monomer for the preparation of multiple copolymers and then as a filtrate loss reducer for oilfield drilling applications. Many researchers have utilized fumaric acid and maleic acid to synthesize oilfield chemical additives in recent years [17–19].

FAS contains a large amount of fumaric anhydride and maleic anhydride; their molecular structure can easily form a rigid ring structure through hydrogen bonding, which results in low polymerization activity (low reactivity rate) and difficulty in forming a large molecular weight in a free radical polymerization reaction (such as hydrolyzed poly-maleic anhydride (HPMA), a polymer-scale inhibitor with excellent performance, the short molecular chain of which is due to the carboxyl group formed by hydrolysis, which makes it easier to form carboxyl salts with metal ions). Therefore, polymer containing FAS has the function of small-molecule viscosity reducer. On the other hand, the ring structure in the molecular structure greatly increases the molecular rigidity of the polymer, which makes the polymer relatively superior in salt resistance and high-temperature resistance to a certain extent. The industrial by-product FAS used in the preparation of polymer filtrate loss reducer can not only reduce the cost of the polymer, but also improve the temperature and salt resistance of filtrate loss reducer. Therefore, FAS used in the synthesis of polymer filtrate loss reducer has a high application value.

In order to realize the efficient utilization of industrial byproduct FAS and prepare a low-cost filtrate loss reducer with temperature and salt resistance, in our work, firstly, a series of qualitative and quantitative analyses of FAS were carried out; and then put forward a method of industrialized utilization of FAS as a high-value resource; then, a temperature and complex saline resistance filtrate loss reducer based on FAS was successfully synthesized; finally, the performance evaluation and structural characterization of the synthesized copolymer were carried out (Figure 1).

2. Results and Discussion

2.1. Component Analyses of FAS by Modern Instrumental Analysis Method

FAS appears as a dark yellow massive solid, which can easily absorb water and agglomerate when exposed to air, turning to light yellow fine powder after drying and crushing, as shown in Figure 2. In order to find out the resource utilization method of FAS, a series of qualitative and quantitative analyses of FAS were carried out: FT-IR is used to detect possible functional groups; ^{1}H-NMR is used to calculate the degree of unsaturation of different substances in FAS; XRD is used to analyze the main phases of FAS; XRF is used to analyze the elements contained in FAS and the content of each element; GC-MS and Py-GC-MS are used to determine possible volatiles and pyrolysis products in FAS; and LC-MS is used to detect the content of water-soluble substances in FAS. Then, the detailed components and ratios of FAS were obtained, as shown in Table 1.

(a) 

(b) 

Figure 2. Appearance comparison of FAS: (**a**) original appearance; (**b**) after drying and crushing.

Table 1. Component composition of FAS.

Component Name	Mass Content/%	CAS#	Carboxylic Acid Saturation	Carboxylic Acid Type
Fumaric acid	19.82	110-17-8	Unsaturation (33.69%)	Diprotic (51.04%)
Fumaric anhydride	0.07	108-30-5		
Maleic acid	13.11	110-16-7		
Maleic anhydride	0.62	108-31-6		
Acrylic acid	0.07	79-10-7		
Phthalic acid	17.10	88-99-3	Saturation (64.21%)	
Phthalic anhydride	0.25	85-44-9		
Benzoic acid	3.71	65-85-0		Mono
Acetic acid	0.15	64-19-7		
Citric acid	43.02	77-92-9		Ternary
Silicate	0.25	-	-	-
Water	1.85	-	-	-

It can be seen from Table 1 that the content of unsaturated carboxylic acid is 33.69%, the content of saturated carboxylic acid is 64.21%, and the content of diprotic carboxylic acid is 51.04%. For oilfield chemicals, the saturated acids in FAS can be applied to oilfield water injection scale inhibitor and cross-linking agent for polymer hydrogels, and unsaturated acids can be used to synthesize filtrate reducer. The application of FAS in the field of oilfield chemicals can effectively improve the utilization rate of FAS.

AM (acrylamide) and AMPS (2-acrylamide-2-methyl propane sulfonic acid) are two commonly used monomers for the synthesis of polyacrylamide, since their molecular structures contain functional groups such as amide and sulfonic acid groups, which are often introduced into the molecular structure of oilfield filtrate loss reducer to improve their temperature and saline resistance [20,21]. In recent years, many researchers have used AM/AMPS in copolymerization to synthesize copolymers and applied them in various fields [22–25].

Therefore, we propose an industrialized utilization of FAS: due to the presence of unsaturated carboxylic acids, their molecular structure contains –C=C–, and the FAS can be utilized as a polymerizable free radical monomer and copolymerized with AM/AMPS to synthesize copolymer as oilfield drilling filtrate loss reducer. In this way, the role of saturated carboxylic acid and unsaturated carboxylic acid in FAS can be fully exerted, and the utilization rate of FAS can reach 97.9%.

2.2. *Structural Characterization of PAAF*

2.2.1. FT-IR

The infrared spectrum of PAAF and FAS are shown in Figure 3a. In the FT-IR spectrum of FAS, the absorption peak at 1678 cm^{-1} corresponds to the characteristic absorption peak of C=O, and 3088 and 934 cm^{-1} correspond to the stretching vibration in =CH and the –OH non-planar wobble vibration, respectively [26]. While in the FT-IR spectrum of PAAF, the absorption peak at 3088 cm^{-1} becomes very flat, which indicates that the stretching vibration of =CH is very weak in the PAAF structure and the C=C in the raw material AM/AMPS/FAS structure has been broken. In addition, in the FT-IR spectrum of PAAF, the absorption peaks at 3341 and 1671 cm^{-1} correspond to the N–H bond and C=O bond stretching vibration peaks, respectively [27]; 2934 cm^{-1} corresponds to the stretching vibration of saturated –CH [27]; the peaks appearing at 1187, 1041, 626, and 513 cm^{-1} are the $–SO_3$ absorption characteristic peak [28]. The characterization results show that the polymer PAAF structure contains groups such as $–CONH_2$, $–SO_3$, etc. These functional groups are introduced by copolymerization through C=C breakage in the raw material AM/AMPS/FAS structure.

Figure 3b gives the FT-IR spectrum of PAAF prepared with the following different monomers: FAS and MA (analytically pure), FA (analytically pure), and MA/FA (analytically pure). It can be seen from the figure that the FT-IR spectrum of this series of PAAF does not differ much, and the positions of the corresponding absorption characteristic peaks of the main functional groups are basically the same. It means that PAAF synthesized with unpurified FAS has the same functional groups as that synthesized with analytically pure compounds, and FAS does not need to be refined and can be utilized directly as a polymeric monomer to copolymerize with other monomers.

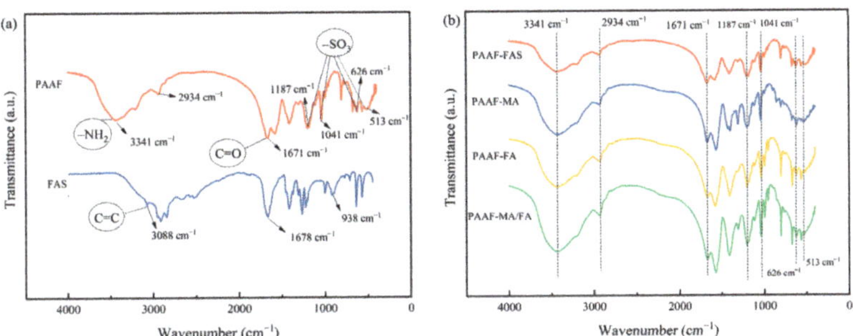

Figure 3. Infrared spectrum comparison: (**a**) PAAF and FAS; (**b**) PAAF-FAS\PAAF-MA\PAAF-FA\PAAF-MA-FA.

2.2.2. TGA

Figure 4 shows the TGA-DTGA curve of PAAF. As can be seen from the figure, the decomposition of PAAF is divided into four stages. The first stage occurs at room temperature to 107 °C, which is the free water evaporation stage. The evaporation of intramolecular and intermolecular water corresponds to a weight loss of 3.0%. The second stage occurs at 107 to 261 °C; at this stage, the amide group undergoes thermal acylamination reaction [29], resulting in a weight loss of 6.4%. The third stage occurs at 261 to 315 °C, the –C–C– bond on the side chain of PAAF begins to break, and the molecular structure on the side chain begins to detach from the main chain of PAAF and undergo carbonization, corresponding to a weight loss of 4.3%. The fourth stage occurs at 315 to 350 °C; in this stage, the –C–C– bond in the main chain of the PAAF molecular structure begins to break, and the macromolecular chain of PAAF begins to break into small molecular chains, corresponding to a weight loss of 9.7%. From the TGA-DTGA results, it can be seen that the molecular structure of PAAF began to decompose from 261 °C, and the general oil well filtrate loss reducer use environment is lower than 200 °C, so as an oil well filtrate loss reducer, PAAF can have a good temperature resistance.

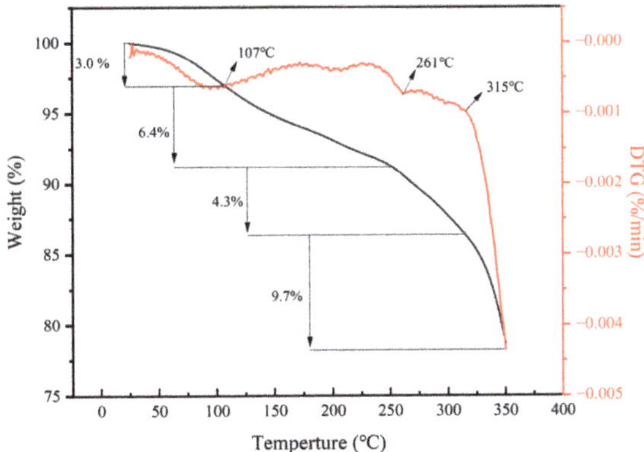

Figure 4. TGA-DTGA curve of PAAF.

2.3. Performance Evaluation of PAAF

2.3.1. Effect of Monomer Mass Ratio

To improve the utilization rate of FAS, a series of PAAF were synthesized by gradually increasing the proportion of FAS in the total monomer (0%, 20%, 25%, 33%, 50%) while keeping the mass ratio AM:AMPS = 1:1 and the total mass of the three monomers AM/AMPS/FAS constant, which were recorded as PAAF110, PAAF221, PAAF332, PAAF111 and PAAF112. Weigh two portions of a certain mass of PAAF and add them to two groups of complex saline-based mud, respectively. One portion was stirred at high speed for 20 min, and then stirred at high speed for 10 min after 24 h of maintenance to determine the room temperature filtration loss FL_{API} and rheological parameters; the other group was stirred at high speed for 20 min, loaded into a high-temperature aging tank, and hot rolled and aged at 150 °C for 16 h. After it cools, take it out and stir at high speed for 10 min, and then determine the room temperature filtration loss FL_{API} and rheological parameters.

The AV, PV, YP, and FL_{API} of the complex saline-based mud were all higher after 24 h of room temperature maintenance than after aging at 150 °C for 16 h (Figure 5), indicating that part of the carbon chain structure of PAAF was destroyed at 150 °C, leading to a decrease in its rheological properties and filtration loss reduction performance. With the increase in FAS proportion, the AV and PV of the complex saline-based mud showed a decreasing trend (Figure 5a,b), but FL_{API} did not increase significantly (Figure 5d), indicating that the addition of a certain proportion of FAS can significantly reduce the viscosity of the complex saline-based mud, and the filtrate loss reduction performance will not be significantly reduced. However, there is a significant increase in FL_{API} as the proportion of FAS increases to 50% (Figure 5d), comparing PAAF111/PAAF112, from 22.0 to 82.0 mL after hot rolling at 150 °C, which can no longer meet the technical requirements of oilfield filtrate loss reducer ($\leq$ 25 mL). It is possible that the proportion of FAS in the total amount of monomers is too high, and FAS contains a large amount of fumaric acid, and the double carboxyl group of fumaric acid is located on the opposite side of the carbon-carbon double bond, which has a large polymerization site resistance and is not conducive to the polymerization between the monomers, making the degree of polymerization between AM/AMPS/FAS lower, resulting in a sharp decrease in the performance of the synthesized PAAF to reduce filtration loss. Therefore, considering the utilization rate of FAS and filtration loss reduction performance, the synthesized PAAF can meet the requirements of oilfield filtrate loss reducer and maximize the utilization of FAS when the mass ratio AM:AMPS:FAS = 1:1:1.

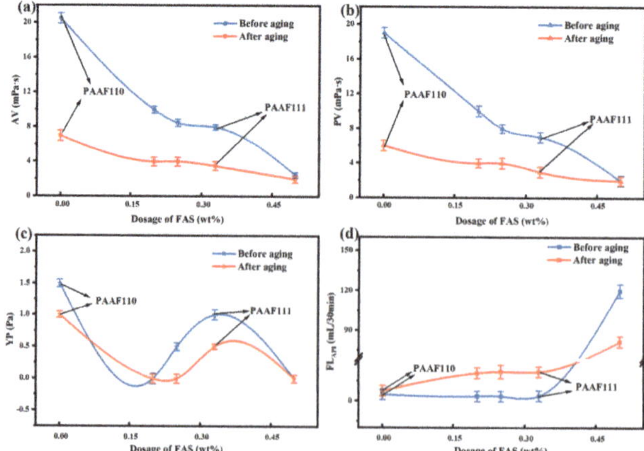

Figure 5. Effect of monomer ratio on the performance of PAAF: (**a**) effect of monomer ratio on AV; (**b**) effect of monomer ratio on PV; (**c**) effect of monomer ratio on YP; (**d**) effect of monomer ratio on FL_{API}.

2.3.2. Effect of Cross-Linking Agent

Unsaturated carboxylic acids account for about one-third of FAS, while saturated carboxylic acids account for about three-fifths. This compositional feature determines that the utilization of FAS cannot only consider unsaturated dicarboxylic acids, and the value of saturated carboxylic acids is equally important, or even more important. Citric acid and its aluminum salts are widely used in oilfield water injection scale inhibitors and cross-linking agents for polymer hydrogels, easy-to-form weak hydrogels. In the reaction of synthesizing copolymers, the saturated polyacids in FAS may be thermally cross-linked with small-molecule polymers to form weak gels with a blocking effect. However, due to the cross-linking formed by covalent bonds, its cross-linking strength is relatively weak. Although it has an effect on improving the fluid loss reduction performance at room temperature, it is relatively weak in improving the high-temperature resistance of the copolymer. We conducted an experimental exploration to improve the high temperature resistance of FAS-based copolymers by using silicates as cross-linking agent.

Sodium silicate readily forms polysilicate at a certain pH range and can be applied in polymerization reactions to strengthen the structure of polymer molecules [30–34]. Under the condition of monomer mass ratio AM:AMPS:FAS = 1:1:1, different masses of sodium silicate were added to the solution as cross-linking agent to improve the temperature resistance of PAAF, and the masses of sodium silicate were 0.0, 1.5, 3.0, 6.0, and 12.0 g. After adding different masses of sodium silicate to synthesize the series of PAAF, the FL_{API} and rheological parameters of the complex saline-based mud were measured based on the previously introduced test method for temperature and salt resistance, and the test results are shown in Table 2.

Table 2. Effect of cross-linking agent on the performance of PAAF.

Mass of Sodium Silicate (g)	AV (mPa·s)		PV (mPa·s)		YP (Pa)		FL_{API} (mL/30 min)	
	Before Aging	After Aging	Before Aging	After Aging	Before Aging	After Aging	Before Aging	After Aging
0.0	10.0	4.0	10.0	4.0	0.0	0.0	3.6	22.0
1.5	8.5	3.0	8.0	3.0	0.5	0.0	6.8	16.0
3.0	7.0	3.5	6.0	3.0	1.0	0.5	6.4	13.2
6.0	7.0	3.0	6.0	3.0	1.0	0.0	9.6	66.0
12.0	7.0	3.5	6.0	3.0	1.0	0.5	12.4	74.0

With the addition of cross-linking agent, the FL_{API} before aging does increase, but it is obvious that the FL_{API} after aging first decreases significantly, and then starts to increase when the addition of cross-linking agent exceeds 6.0 g, which means that the cross-linking agent can improve PAAF's ability to reduce filtration in high temperature, but adding too much cross-linking agent will greatly reduce the hydration ability of PAAF molecules, resulting in a decrease in its ability to reduce filtration. Therefore, there is an optimal mass for the cross-linking agent. Combining the data of FL_{API} before aging and FL_{API} after aging, we believe that PAAF's ability to reduce filtration is the best when the addition of the cross-linking agent is 3 g. On the other hand, the YP value changes significantly with the increase in the crosslinking agent. This is because the PAAF molecules form a certain network structure through the crosslinking reaction, which greatly enhances the yield value. It also happens that when the addition of cross-linking agent is about 3.0 g, the YP before aging and after aging is relatively high, which is good for the cuttings carrying ability of the drilling fluid. Therefore, it is very necessary to optimize the amount of cross-linking agent in general, and it has indeed achieved better results.

As can be seen from Table 2, the rheological parameters AV, PV and YP of the complex saline-based mud did not change much after the addition of a certain mass of sodium silicate, indicating that the addition of sodium silicate hardly affected the rheological properties of PAAF. There was a significant decrease in FL_{API} after hot rolling and aging at

150 °C for 16 h, FL_{API} decreased from 22.0 to 16.0 mL/30 min after adding 1.5 g of sodium silicate, and FL_{API} was lowest at 13.2 mL/30 min when 3.0 g sodium silicate was added, indicating that a certain mass of sodium silicate could improve the temperature resistance of PAAF. However, with the further increase in sodium silicate mass (increased to 6.0 g), the FL_{API} after hot rolling and aging increased sharply instead (from 13.2 mL/30 min to 66.0 mL/30 min), indicating that the addition of too much sodium silicate hinders the polymerization between AM/AMPS/FAS. It makes the polymerization of PAAF decrease, which leads to the carbon chain of PAAF to break easily at high temperatures, thus losing the function of reducing filtration loss.

Therefore, the addition of 3.0 g sodium silicate (5% of the total monomer mass) can make the monomers on the main chain of PAAF cross-link with each other and ensure that the structure of PAAF is not easily destroyed at high temperature; thus, the temperature resistance of PAAF can be improved.

2.3.3. Effect of FAS on Rheological Properties

In order to more scientifically evaluate the rheological properties of the complex saline-based mud used in this study, the Bingham Plasticity Model and Power Law Model were applied to the viscometric data obtained before and after aging. As can be seen from Figure 6, whether it is before or after aging, the Power Law Model is more consistent with the experimental data curve than the Bingham Plastic Model. Therefore, we consider the complex saline-based mud to be fitted with the Power Law Model.

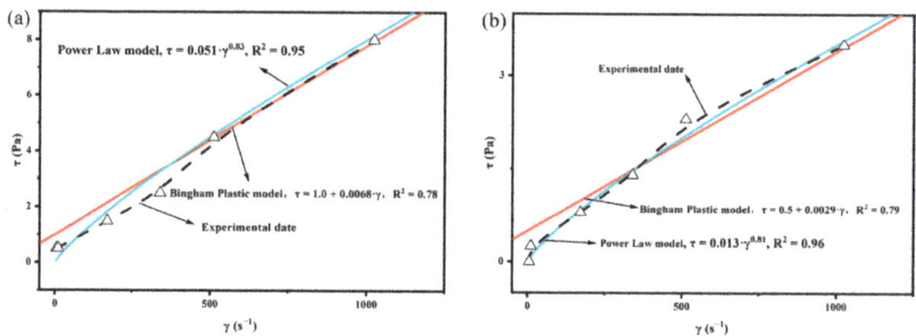

Figure 6. Rheological model cure: (**a**) rheological models applied to viscometric data obtained before aging; (**b**) rheological models applied to viscometric data obtained after aging.

According to Power Law Model: $\tau = K\gamma^n$, where τ is the shear stress, γ is the shear rate, n is the fluidity index, and K is the consistency coefficient. The value of n reflects the shear dilution performance of drilling fluid, and K is related to the viscosity and sheared force of drilling fluid. The K value reflects the pumpability of drilling fluid; the large value of K will make it difficult to repump. In order to study the effect of FAS on the rheological properties of drilling fluid, we added two additives to the complex saline-based mud, PAAF110 (without FAS) and PAAF111 (mass ratio AM:AMPS:FAS = 1:1:1), and tested the changes of rheological parameters of composite brine base mud under different additives, as shown in Table 3.

It can be seen from Table 3 that at 25 °C, the AV and PV values of based mud with PAAF110 are much greater than those of based mud with PAAF111, which indicates that the based mud with PAAF111 has better fluidity; at 150 °C, the RYP value and n value of based mud with PAAF110 are consistent with based mud with PAAF111, indicating that the two additives have the same shear ability, and there is no difference in the ability to carry cuttings. The K value of based mud with PAAF110 is half of that of based mud with PAAF111, which means that based mud with PAAF110 is more viscous, has worse pumpability, and is more difficult to pump the drilling fluid to the ground. Therefore, the

PAAF111 synthesized based on FAS can not only maintain the ability of drilling fluid to carry cuttings, but also make the drilling fluid have better fluidity and facilitate pumping.

Table 3. Rheological property results of the complex saline-based mud with the addition of PAAF110 or PAAF111.

Temperature/ °C	Index	Complex Saline-Based Mud	
		With PAAF110	**With PAAF111**
25	AV/ mPa·s	20.5	8.0
	PV/ mPa·s	19.0	7.0
	RYP/(Pa/mPa·s)	0.0807	0.1486
	n/ Dimensionless	0.8981	0.8301
	K/ Pa·s^n	0.0415	0.0260
150	AV/ mPa·s	7.0	3.5
	PV/ mPa·s	6.0	3.0
	RYP/(Pa/mPa·s)	0.1703	0.1703
	n/ Dimensionless	0.8074	0.8074
	K/ Pa·s^n	0.0266	0.0133

In general, polymer filtrate loss reducer has a positive effect on rheological properties, the value of n is appropriate to the control at 0.4–0.7, and the value RYP is appropriate to the control at 0.36–0.48 Pa/mPa·s. Obviously, the effect of PAAF and general polymer filtrate loss reducer on rheological properties is not the same, because PAAF is a dilution-type filtrate loss reducer with a relatively smaller molecular weight. This can be clearly verified from the results of previous studies on China's most famous polymer diluent, XY-27 (ultra-low-molecular-weight poly(AM-AA)) [35].

After PAAF111 is added, the value of RYP and K is relatively small. On the one hand, the dilution of PAAF weakens the network structure formed between polymer and bentonite particles, which reduces the structural viscosity (intrinsic viscosity) of drilling fluid; On the other hand, a certain amount of saturated organic acids (citric acid, phthalic acid, etc., which will not participate in the polymerization of PAAF) contained in FAS will form small molecular organic salts with metal ions on the end face of bentonite particles under high-temperature conditions, compress the electric double layer on the surface of bentonite particles, and appear with the tendency of high temperature passivation of bentonite particles, which affects the stability of the bentonite colloidal system and leads to poor thixotropy.

However, this paper only studies the effect of a single PAAF in the bentonite-based mud on the rheological properties, in order to understand the influence of the difference of PAAF itself on the rheological properties of the drilling fluid. In practical industrial applications, drilling fluid is composed of many kinds of functional additives (sometimes even more than ten kinds of additives), and the effect of different additives on drilling fluid rheological properties will be affected by their interaction Therefore, the rheological properties of the drilling fluid system is the result of the joint action of various additives, and it is also the result of the coupling effect of various additives complementing each other. It is not difficult to understand that even if the effect of a single additive on the rheological properties is negative, the combined effect in the actual drilling fluid may still be positive. Examples abound in this regard, such as Fe-Cr-lignosulfonate (FCLS), the most effective diluent in brine drilling fluid systems, which has exactly the same effect on the rheological properties of drilling fluid systems [35].

2.3.4. Comparison of By-Product FAS with Analytical Pure MA and FA

Under optimal experimental conditions, FAS, maleic acid (MA, analytical pure), fumaric acid (FA, analytical pure), and MA:FA = 1:3 were used to co-polymerize with AM/AMPS to form a series of PAAF, respectively. The AV and FL_{API} of the complex saline-based mud after adding the series PAAF were tested before and after hot rolling and aging at 150 °C for 16 h according to the performance evaluation method, and the test results are shown in Figure 7.

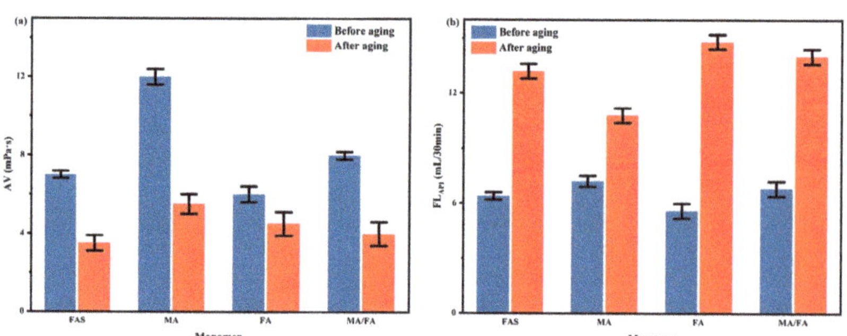

Figure 7. Effect of different monomers on the performance of PAAF: (**a**) effect of different monomers on AV; (**b**) effect of different monomers on FL_{API}.

From the AV values before hot rolling and aging (Figure 7a), the apparent viscosity of MA is the highest (12.0 mPa·s) and the apparent viscosity of FA is the lowest (6.0 mPa·s), which indicates that MA has the highest degree of polymerization with AM/AMPS and the highest molecular weight of the synthesized PAAF compared to FAS and FA. From the FL_{API} after hot rolling and aging (Figure 7b), the PAAF synthesized using MA has the best reduce filtration loss performance (10.8 mL, the double carboxyl group of MA is located on the same side of the carbon-carbon double bond, and the site resistance of polymerization is smaller than FA, which is easier to polymerize). There was little difference in the filtration loss reduction effect between PAAF synthesized directly using FAS and PAAF synthesized at MA:FA = 1:3 (13.2 mL and 14.0 mL, respectively), which indicates that the components of FAS other than fumaric acid and maleic acid do not affect the polymerization between FAS and AM/AMPS. Therefore, there is no complicated refining requirement, and FAS can be used directly as polymerizable free radical monomer for the synthesis of oilfield filtrate loss reducer PAAF.

2.3.5. Filtration Properties of PAAF

According to Darcy's law [35]: $dV_f/dt = KA\triangle p/(\mu h_{mc})$, where dV_f/dt is filtration loss rate, cm^3/s; K is mud cake permeability, μm^2; A is percolation area, cm^2; $\triangle p$ is percolation pressure, 105 Pa; μ is filtration loss viscosity, 0.1 mPa·s, h_{mc} is mud cake thickness, cm; V_f is filtration loss (FL), cm^3; t is percolation time, s. We can infer that FL and $\sqrt{t}$ should be proportional. Figure 8 shows the plot of the relationship between square root of time and FL (After aging at 150 °C). As $\sqrt{t}$ increases, the change trend of FL is almost linear. We take the FL at 7.5 and 30 min, respectively, and fit a straight line; we can find that the values of FL are almost all on this straight line. Therefore, it can be considered that FL is proportional to $\sqrt{t}$, and the filtration loss characteristics of PAAF follow Darcy's law. On the other hand, we can find that this straight line is not at the origin and has a positive intercept, indicating that before the mud cake is formed, there is a sudden spurt of initial filtration loss, that is, spurt loss. When spurt loss occurs, the spurt loss molecules can quickly enter the bottom of the thin cuttings formed by the broken rock of the drill bit, which can help to strip the thin cuttings from the rock and wash away immediately, or make them not closed, which is beneficial to increase the mechanical rotational speed.

The ability of the mud to seal permeable formations exposed by the bit with a thin, low-permeability filter cake is another major requirement for successful completion of the hole. Because the pressure of the mud column must be greater than the formation pore pressure in order to prevent the inflow of formation fluids, the mud would continuously invade permeable formations if a filter cake were not formed.

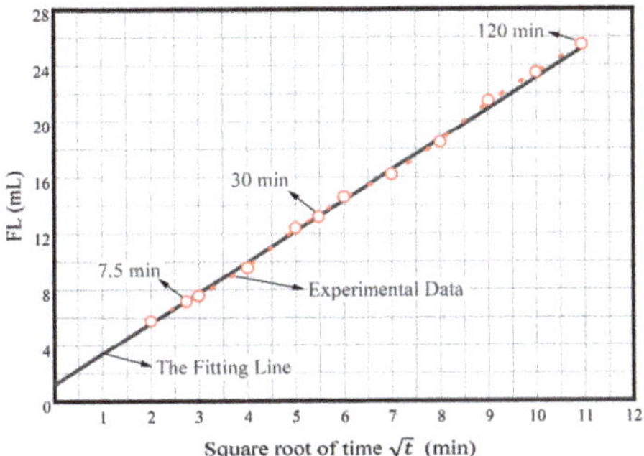

Figure 8. The plot of relationship between square root of time and FL (after aging at 150 °C).

The rate of filtration and the increase in cake thickness depend on whether or not the surface of the cake is being subjected to fluid or mechanical erosion during the filtration process. When the mud is static, the filtrate volume and the cake thickness increase in proportion to the square root of time (hence, at a decreasing rate). Under dynamic conditions, the surface of the cake is subjected to erosion at a constant rate, and when the rate of growth of the filter cake becomes equal to the rate of erosion, the thickness of the cake and the rate of filtration remain constant. In the well, because of erosion by the mud and because of mechanical wear by the drill string, filtration is dynamic while drilling is proceeding; however, it is static during round trips. All routine testing of filtration properties is made under static conditions because dynamic filtration tests and static tests under high temperature and high pressure (HTHP) are time-consuming and require elaborate equipment. Thus, filtration rates and cake thicknesses measured in surface tests correlate only approximately to those prevailing down-hole and can be grossly misleading. The permeability of the filter cake, which may readily be calculated from static test data, is a better criterion because it is the fundamental factor controlling both static and dynamic filtration.

2.3.6. Comparison of Filtration-Loss-Controlling Ability of Popular Copolymers

We selected several popular low-molecular-weight copolymer filtrate loss reducers (Na-PAN, NH_4-PAN and PAMAP) and compared their filtration loss in different based mud. The Na-PAN is hydrolyzed polyacrylonitrile sodium salt, which is a commonly used industrial filtrate loss reducer. It is insensitive to NaCl and has good filtration loss performance for based mud containing NaCl. The NH_4-PAN is also an industrial filtrate loss reducer, owing to the fact that NH_4^+ released by NH_4-PAN in the drilling fluid can be embedded in the shale; it can not only reduce the filtration loss, but also have a certain effect of preventing well collapse. The PAMPAP is a copolymer obtained by copolymerization of AM/AMPS/AA/AP and has a similar molecular structure to PAAF [27]. The FL data for these filtrate loss reducers are listed in Table 4.

Table 4. FL comparison of different types of filtrate loss reducers in different based mud (after aging at 150 °C).

Filtrate Loss Reducer	Saturated Brine-Based Mud		Complex Saline-Based Mud	
	FL_{API}/mL	FL_{HTHP}/mL	FL_{API}/mL	FL_{HTHP}/mL
Standard [36]	≤25.0	-	≤25.0	-
PAAF	14.2	62.4	13.2	126.0
Na-PAN	7.2	25.6	94.0	250.0
NH_4-PAN	32.0	250.0	110.0	250.0
PAMAP	15.8	49.4	-	-

As can be seen from Table 4, compared with Na-PAN, PAAF has worse filtration loss performance in saturated brine-based mud, while PAAF has better filtration performance in complex saline-based mud, which means that although the PAAF has a weaker NaCl resistance performance, its complex saline resistance performance is much greater than Na-PAN. Compared to NH_4-PAN, it is clear that PAAF exhibits stronger filtration loss performance both in saturated brine-based mud and in complex saline-based mud. Compared with PAMAP, which is similar in molecular structure, the filtration loss performance of PAAF at room temperature medium pressure is better, but the filtration loss performance of PAAF is not good enough at high temperature high pressure (HTHP), the reason being that PAAF is a small molecule filtrate loss reducer with a dilution function, meaning the molecular weight of PAAF is too small to block the filtration loss channel. Therefore, even a salt resistance filtrate loss reducer such as PAAF, in the complex saline-based mud used in this paper, does not have a good enough HTHP filtration loss performance (FL_{HTHP} is 126 mL).

The danger of relying on the API filter loss as a criterion for downhole dynamic filtration rates is obvious. A treating agent that was recommended on the basis of API test results might give higher rates downhole than another agent that gave a higher API filtrate loss. Worse still, an agent that reduced the API loss might increase the downhole filtration rate.

The HTHP static loss correlated quite well with the short term (30 min) dynamic loss but had virtually no relationship with the long term (equilibrium) dynamic loss.

Despite its shortcomings, the API static test is the only practical test for control of filtration at the wellsite. However, results should be interpreted in the light of correlations made in the laboratory between API filter loss and dynamic filtration rate.

2.3.7. Mud Cake Properties of PAAF

Figure 9 shows the mud cake produced at room temperature medium pressure and the mud cake produced at high temperature high pressure (HTHP). It can be seen that the mud cake produced at room temperature medium pressure is very thin and dense and is tightly attached to the filter paper, showing a strong toughness, and the thickness of the mud cake is less than 0.5 mm. The mud cake formed at high temperature high pressure is not dense enough, and the thickness of the mud cake is about 1.8 mm. The particles on the surface of the mud cake are evenly distributed, and there is no large particle agglomeration. This shows that PAAF has played a diluting and dispersing role even in high temperature and high pressure (HTHP) condition, so that the bentonite particles do not agglomerate together.

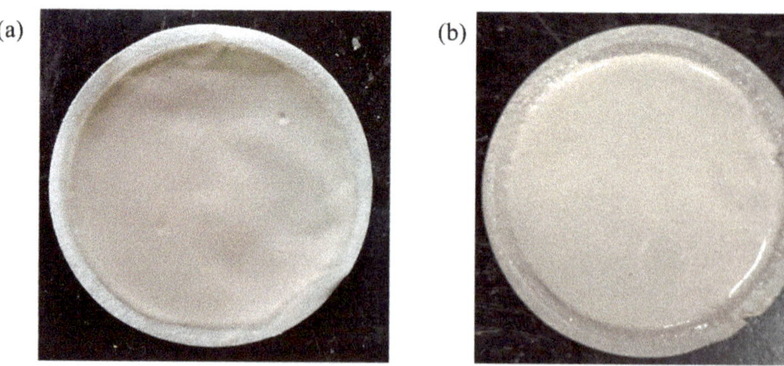

Figure 9. Mud cake produced under different conditions: (**a**) room temperature medium pressure; (**b**) high temperature high pressure (HTHP).

For a mud cake to form, it is essential that the mud contain some particles of a size only slightly smaller than that of the pore openings of the formation. These particles, which are known as bridging particles, are trapped in the surface pores, while the finer particles are, at first, carried deeper into the formation. The bridged zone in the surface pores begins to trap successively smaller particles, and, in a few seconds, only liquid invades the formation. The suspension of fine particles that enters the formation while the cake is being established is known as the "mud spurt". The liquid that enters subsequently is known as the filtrate.

The permeability of the mud cake depends on the particle size distribution in the mud and on the electrochemical conditions. In general, the more particles there are in the colloidal size range, the lower the cake permeability. The presence of soluble salts in clay muds increases the permeability of the filter cake sharply, but certain organic colloids enable low cake permeabilities to be obtained even in saturated salt solutions. Thinners usually decrease cake permeabilities because they disperse clay aggregates to smaller particles.

2.3.8. Salt Resistance of PAAF

Figure 10 shows the comparison of the FL_{API} (before aging) in different based mud. After adding 1wt% PAAF to the based mud, the filtration loss of freshwater-based mud was reduced by 27.8 mL/30 min; the filtration loss of saturated brine-based mud was reduced by 99.2 mL/30 min, and the filtration loss of complex saline-based mud was even lower, 113.6 mL/30 min. This result shows that PAAF exhibits excellent salt resistance. The salt resistance mechanism of PAAF can be explained as follows: On the one hand, the molecular structure of PAAF contains sulfonic acid groups, which are not sensitive to alkaline earth metal ions, so PAAF can also play a good hydration effect in a high-salinity environment, forming a thicker hydration shell, slowing down the loss of water in the mud. On the other hand, a large number of amide groups are distributed in the molecular structure of PAAF, and the amide group has a strong adsorption effect [37], which can eliminate the expansion of the interlayer spacing of bentonite by metal ions, promoting the formation of dense mud cake.

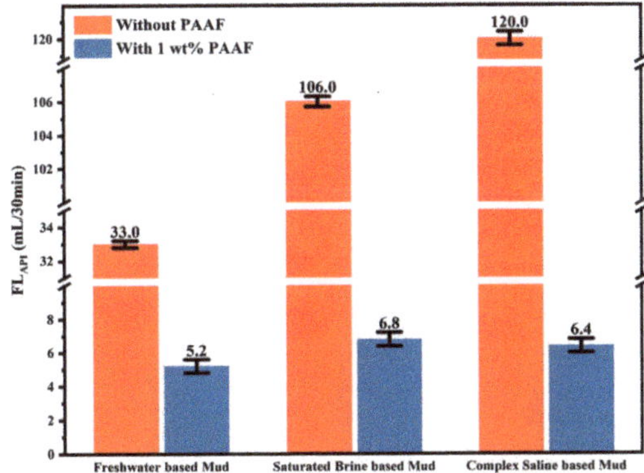

Figure 10. FL_{API} comparison under different based mud (before aging).

3. Conclusions

The FAS is a byproduct of the processing of phthalic anhydride organic wastewater, which is rich in saturated carboxylic acid and unsaturated carboxylic acid. FAS as a free radical monomer can directly participate copolymerization with acrylamide (AM) and 2-acrylamido-2-methylpropane sulfonic acid (AMPS) without separation and purification to form a low-molecular-weight copolymer. Its molecular structure is rigid and suitable for oil well drilling fluid chemicals such as filtrate loss reducer with a very low cost and good comprehensive performance. We achieved the efficient utilization of FAS, and the utilization rate of FAS can reach 97.9%.

In the mass ratio AM:AMPS:FAS = 1:1:1, with the addition of 5 wt% sodium silicate the prepared copolymer PAAF has the best comprehensive performance. As a filtrate loss reducer for oil well drilling fluid, PAAF has excellent temperature and complex saline resistance. Under complex saline-based mud, the FL_{API} is 13.2 mL/30 min after aging at 150 °C for 16 h, reaching the industry standard requirement (≤25 mL/30 min).

Compared with hydrolyzed polyacrylonitrile sodium salt (Na-PAN), which is also a small molecular copolymer filtrate reducer, PAAF has a better resistance in complex salts containing calcium and magnesium ions. Compared with another small molecular copolymer filtrate reducer, hydrolyzed polyacrylonitrile ammonium salt (NH_4-PAN), PAAF has a better filtration-loss-controlling performance. Compared with polycarboxylic acid comb copolymer PAMAP, PAAF has better resistance to saturated brine. Furthermore, PAAF prepared based on low-cost FAS has an advantage in cost and can be used as a low-cost filtrate loss reducer in oil well drilling fluids.

4. Materials and Methods

4.1. Materials

Fumaric acid sludge (FAS) was obtained from Karamay Zhengcheng Co., Ltd. (Xinjiang, Karamay, China). 2-acrylamido-2-methylpropanesulfonic acid (AMPS) was purchased from Jingwen Dongxin Biotechnology Co., Ltd. (Beijing, China). Acrylamide (AM) was purchased from Shandong Duofeng Chemical Co., Ltd. (Shandong, Zibo, China). potassium persulfate, sodium hydroxide, sodium carbonate, sodium bicarbonate, sodium chloride, anhydrous calcium chloride, magnesium chloride, fumaric acid, maleic acid, etc., were purchased from Beijing Yili Fine Chemicals Co., Ltd. (Beijing, China).

4.2. Methods

4.2.1. Preparation of PAAF Based on FAS

FAS pretreatment: FAS was dried in an oven at 80 °C for 24 h, and then crushed into a fine powder with a pulverizer, that is, the fine powder of FAS.

According to the best ratio, 15.0 g NaOH and 5 g Na_2CO_3 were added to 80.0 g deionized water and dissolved fully, then 20.0 g FAS fine powder was added slowly and dissolved fully before use; 20.0 g AM and 20.0 g AMPS were added to 40.0 g deionized water and dissolved fully before use; weigh 0.6 g of $K_2S_2O_8$ initiator and 3.0 g of Na_2SiO_3 cross-linking agent into 10.0 g of deionized water, dissolve fully before use. The FAS solution was mixed with the AM solution and the AMPS solution, and then transferred to a three-necked flask to obtain a mixed solution. The temperature of the mix solution was raised to 55 °C while N_2 was passed into the flask for 30 min. Then, slowly add the $K_2S_2O_8$ initiator mixture dropwise to the three-necked flask to initiate the polymerization. After 5 h of reaction, a brown viscous liquid was obtained, which was completely dried in an oven at 80 °C. Finally, the dried liquid was pulverized into a fine powder with a pulverizer to obtain PAAF.

4.2.2. Performance Evaluation Method of PAAF

The method of GB/T 16783.1 "Field Testing of Drilling Fluids for Oil and Gas Industry Part 1: Water-based Drilling Fluids" was adopted to measure the room temperature filtration loss (FL_{API}), high-temperature and high-pressure filtration loss (FL_{HTHP}) and rheological properties of PAAF [38]. In addition, the temperature and complex saline resistance of PAAF was tested according to the technical requirements of the corporate standard "Salt Resistant filtrate loss reducer FRS for Drilling Fluids", issued by China National Offshore Oil Corporation (CNOOC) [36]. The specific test methods are as follows:

Method of room temperature filtration loss FL_{API}: Pour the drilling fluid sample into the filtration loss meter cup, make the liquid level reach the scale line in the filtration loss meter cup, put the filtrate paper and install the filtration loss meter, and put the dry measuring cylinder underneath to receive the filtrate. Close the pressure relief valve and adjust the pressure regulator to make the pressure reach 690 ± 35 kPa (100 ± 5 psi) in 30 s or less, and start timing while pressurizing. After reaching 30 min, measure the collected filtrate volume, which is the room temperature filtration loss FL_{API}, and the test schematic is shown in Figure 11a.

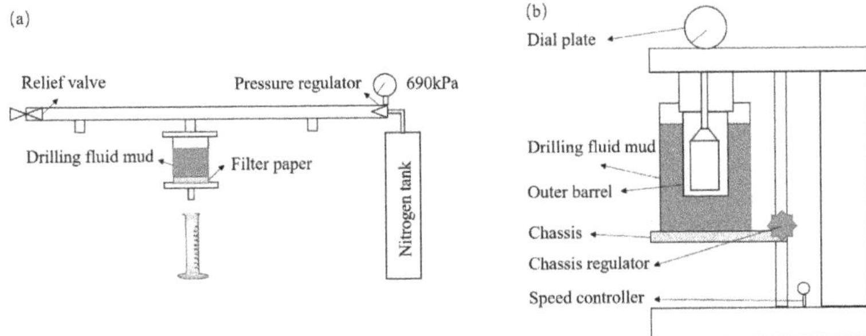

Figure 11. Test schematic: (**a**) FL_{API}; (**b**) rheological properties.

Method of the rheological properties: Pour the drilling fluid sample into the sample cup of the rotational viscometer, make the liquid level reach the scale line in the sample cup of the rotational viscometer, put the sample cup on the bottom frame of the viscometer, move the bottom frame so that the sample liquid level coincides with the scale line on the outer cylinder, and measure and record the temperature of the drilling fluid sample. Adjust the rotational speed of the outer cylinder of the rotational viscometer, and after the dial reading value is stabilized, read and record the dial reading value under different rotational speeds, as shown in Figure 11b. Calculate the apparent viscosity, plastic viscosity and dynamic shear force based on the following equation:

$$AV = R_{600}/2 \quad (1)$$

$$PV = R_{600} - R_{300} \quad (2)$$

$$YP = 0.511(R_{300} - PV) \quad (3)$$

$$RYP = YP/PV \quad (4)$$

$$n = 3.3221\,g(R_{600}/R_{300}) \quad (5)$$

$$K = (0.511\,R_{300})/511^{n} \quad (6)$$

where:

R_{600} = the reading of the viscometer at 600 r/min (dia);
R_{300} = the reading of the viscometer at 300 r/min (dia);
AV = apparent viscosity (mPa·s);
PV = plastic viscosity (mPa·s);
YP = yield point (Pa);
RYP = the ratio of yield point to plastic viscosity (Pa/mPa·s);
n = fluidity index of power law model rheological equation (dimensionless quantity);
K = the consistency coefficient of power law model rheological equation (Pa·s^n).

Method of temperature and complex saline resistance evaluation: Complex saline was prepared by adding 45.0 g of sodium chloride, 5.0 g of anhydrous calcium chloride and 13.0 g of magnesium chloride per liter of distilled water. Add 1.0 g of sodium bicarbonate and 35.0 g of evaluation clay to 350 mL of complex saline and stir at high speed for 20 min, stopping at least twice during the period to scrape off the clay adhering to the walls of the container (maintenance at 25 ± 3 °C 24 h). Two portions of the prepared base mud were added with a certain mass of PAAF and stirred at high speed for 20 min. One portion, after being maintained for 24 h, was taken out and stirred at high speed for 10 min, and then the room temperature filtration loss FL_{API} was determined, along with FL_{HTHP} and rheologica parameters. The other portion was loaded into a high-temperature tank and hot rolled a 150 °C for 16 h; after it had cooled, it was taken out and stirred at high speed for 10 mir and then the room temperature filtration loss FL_{API} was determined, along with FL_{HTH} and rheological parameters.

In this test process, a hot roll furnace XGRL-5 was used for hot rolling and aginş the drilling fluid, a medium-pressure ZNS-2A was used to determine FL_{API}, and high pressure filtration apparatus GGS42-2 was used to determine FL_{HTHP}. The rheologica parameters, such as AV, PV, and YP, were determined by a ZNN-D6 rotating viscomete All of the equipment was made by Qingdao Haitongda Special Instruments Co., Ltd. Qingdao, China.

4.2.3. Fourier Transform Infrared Spectra (FT-IR)

Sample preparation of PAAF was achieved using the KBr pellet method, and the sam ple was scanned by infrared spectrometer (Nicolet IS5; PerkinElmer; Waltham, MA, USA in a wave number range of 400 to 4000 cm^{-1}, a signal-to-noise ratio of 50,000:1. The infrared spectrum of PAAF samples was recorded at a resolution of 4 cm^{-1} with 64 scans.

4.2.4. Thermogravimetric Analysis (TGA)

The thermal decomposition behavior of the PAAF samples was investigated using ther mogravimetric analysis (STA449F5; NETZSCH; Selb, Bavaria, Germany) under nitrogen(N_2 atmosphere, and the N_2 flow rate was 40 mL/min. The PAAF samples were placed in a clean crucible and heated from 25 to 350 °C at a ramp rate of 10 °C/min.

Author Contributions: All authors contributed to the conception and design of the study. Material preparation, data collection, and analyses were performed by Z.W., S.C. and F.Z. Original draft preparation, Z.W. Review and editing, Z.W. and F.Z. Software, Z.W. and W.L. All authors have read and agreed to the published version of the manuscript.

Funding: This research was funded by the Fundamental Research Funds for the Central Universities, grant number No. 2- 9-2019-141.

Institutional Review Board Statement: Not applicable.

Informed Consent Statement: Not applicable.

Data Availability Statement: Data are available from the authors upon request.

Conflicts of Interest: We declare that this is original research that has not been published previously and has not been under consideration for publishing elsewhere. There is no conflict of interest in the submission of this manuscript, and the manuscript is approved by all authors for publication.

References

1. Dong, J.F. Process Study and Optimization of Preparing Phthalic Anhydride from Naphthalene. Master Thesis, Beijing University of Chemical Technology, Beijing, China, 2016. Available online: https://d.wanfangdata.com.cn/thesis/Y3094719 (accessed on 6 July 2021).
2. Tao, W.H. Study on Synthesis of Polyvinyl Pyridine Resin and Its Application in Wastewater Treatment. Doctor Thesis, Nanjing University, Nanjing, China, 2013. Available online: http://cdmd.cnki.com.cn/article/cdmd-10284-1016000517.htm (accessed on 6 July 2021).
3. Zuo, W.X.; Liu, Y.N.; Zhao, Z.Q. Method of Recovering Fumaric Acid from Phthalic Anhydride Wastewater. China Patent CN103553899A, 22 November 2013. Available online: https://d.wanfangdata.com.cn/patent/CN201310592787.0. (accessed on 6 July 2021).
4. Zhu, Y.T.; Li, Q.H.; Lv, R.; Liu, C.Y.; Chen, W.P. The experiment investigation of high purity fumaric acid preparation from phthalic anhydride wastewater. *Guangdong Chem.* **2014**, *41*, 24–25. [CrossRef]
5. Huang, H.T.; Bao, J.; Gong, D.S. A Process for Producing Fumaric Acid from Phthalic Anhydride Wastewater. China Patent CN106631765A, 5 December 2016. Available online: https://d.wanfangdata.com.cn/patent/CN201611105287.X (accessed on 6 July 2021).
6. Cao, Q. Study on Synthesis of Unsaturated Polyester Resin by Waste Residue of Fumaric Acid Production. *Environ. Sci. Surv.* **2017**, *36*, 58–60. Available online: https://d.wanfangdata.com.cn/periodical/ynhjkx201704014 (accessed on 6 July 2021).

7. Wang, Y.B.; Wang, L.; Fan, H.L. Synthesis of dioctadecyl fumarate-vinyl acetate copolymer and its effect on pour point reduction and viscosity reduction. *Appl. Chem. Ind.* **2006**, *4*, 420–424. [CrossRef]
8. Peng, Y.H.; Qi, G.R. Synthesis of n-diphenyl fumarate vinyl acetate copolymers and their effects on pour point depression and viscosity reduction. *J. Zhejiang Univ.* **2011**, *28*, 284–288. [CrossRef]
9. Khan, I.; Tango, C.N.; Miskeen, S.; Oh, D.H. Evaluation of nisin-loaded chitosan-monomethyl fumaric acid nanoparticles as a direct food additive. *Carbohyd. Polym.* **2018**, *184*, 100–107. [CrossRef]
10. Guo, F.; Wu, M.; Dai, Z.X.; Zhang, S.J.; Zhang, W.M.; Dong, W.L.; Zhou, J.; Jiang, M.; Xin, F.X. Current ad-vances on biological production of fumaric acid. *Biochem. Eng. J.* **2020**, *153*, 107397. [CrossRef]
11. Yan, J.N. *Drilling Fluid Technology*; China University of Petroleum Press: Dongying, China, 2001; pp. 12–25.
12. Zhang, Y.H.; Meng, X.H.; Zhou, F.S.; Lu, J.B.; Shang, J.W.; Xing, J.; Chu, P.K. Utilization of recycled chemical residues from sodium hydrosulfite production in solid lubricant for drilling fluids. *Desalin. Water. Treat.* **2016**, *57*, 1804–1813. [CrossRef]
13. Zhang, Z.L.; Zhou, F.S.; Zhang, Y.H.; Huang, H.W.; Shang, J.W.; Yu, L.; Wang, H.Z.; Tong, W.S. A Promising Material by Using Residue Waste from Bisphenol A Manufacturing to Prepare Fluid-Loss-Control Additive in Oil Well Drilling Fluid. *J. Spectrosc.* **2013**, *2013*, 370325. [CrossRef]
14. Long, W.J.; Luo, H.Z.; Yan, Z.; Zhang, C.L.; Hao, W.S.; Wei, Z.J.; Zhu, X.L.; Zhou, F.S.; Cha, R.T. Synthesis of filtrate reducer from biogas residue and its application in drilling fluid. *TAPPI J.* **2020**, *19*, 151–158. [CrossRef]
15. Wang, Z.H. *Practical Manual for Drilling Fluid Treatment Agents*; Sinopec: Beijing, China, 2016; pp. 35–42.
16. Long, W.; Zhu, X.; Zhou, F.; Yan, Z.; Evelina, A.; Liu, J.; Wei, Z.; Ma, L. Preparation and Hydrogelling Performances of a New Drilling Fluid Filtrate Reducer from Plant Press Slag. *Gels* **2022**, *8*, 201. [CrossRef]
17. Kar, Y.; Al-Moajil, A.M.; Nasr-El-Din, H.A.; Al-Bagoury, M.; Steele, C.D. Environmentally Friendly Dispersants for HP/HT Aqueous Drilling Fluids Containing Mn3O4, Contaminated with Cement, Rock Salt, and Clay. In Proceedings of the SPE Middle East Oil and Gas Show and Conference, Manama, Bahrain, 25 September 2011. [CrossRef]
18. Arora, A.; Pandey, S.K. Review on Materials for Corrosion Prevention in Oil Industry. In Proceedings of the SPE International Conference & Workshop on Oilfield Corrosion, Aberdeen, UK, 28 May 2012. [CrossRef]
19. Assanov, A.A.; Tulenbayev, M.S. Obtaining Water-Soluble Polymeric Electrolytes (WSPE), Comprises Hydrolyzing Product of Copolymerization of Fumaric Acid with Acrylonitrile and a Stabilizing Action on Clay Solutions. Derwent Patent KZ26071-A4, 14 September 2012. Available online: https://www.webofscience.com/wos/alldb/full-record/DIIDW:201930516Y (accessed on 10 July 2021).
20. Perricone, A.C.; Enright, D.P.; Lucas, J.M. Vinyl Sulfonate Copolymers for High-Temperature Filtration Control of Water-Based Muds. *SPE Drill. Eng.* **1986**, *1*, 358–364. [CrossRef]
21. Wang, M.G.; Yan, X.; Peng, Z. High temperature salt resistant filter loss reducers for water base drilling fluids: Synthesis with explosive polymerization method and performance evaluation. *Drill. Fluid Completion Fluid* **2019**, *36*, 148–152. [CrossRef]
22. Phetphaisit, C.W.; Yuanyang, S.; Chaiyasith, W.C. Polyacrylamido-2-methyl-1-propane sulfonic ac-id-grafted-natural rubber as bio-adsorbent for heavy metal removal from aqueous standard solution and industrial wastewater. *J. Hazard. Mater.* **2016**, *301*, 163–171. [CrossRef] [PubMed]
23. An, Y.; Zheng, H.; Zheng, X.; Sun, Q.; Zhou, Y. Use of a floating adsorbent to remove dyes from water: A novel efficient surface separation method. *J. Hazard. Mater.* **2019**, *375*, 138–148. [CrossRef] [PubMed]
24. Sharma, G.; Kumar, A.; Naushad, M.; Thakur, B.; Vo, D.V.N.; Gao, B.; Al-Kahtani, A.A.; Stadler, F.J. Adsorp-tional-photocatalytic removal of fast sulphone black dye by using chitin-cl-poly(itaconic acid-co-acrylamide)/zirconium tungstate nanocomposite hydrogel. *J. Hazard. Mater.* **2021**, *416*, 125714. [CrossRef] [PubMed]
25. Yuan, M.; Gu, Z.; Minale, M.; Xia, S.; Zhao, J.; Wang, X. Simultaneous adsorption and oxidation of Sb(III) from water by the pH-sensitive superabsorbent polymer hydrogel incorporated with Fe-Mn binary oxides composite. *J. Hazard. Mater.* **2022**, *423*, 127013. [CrossRef]
26. Guan, X.M.; Wang, Q.L.; Wang, Q.P.; Hu, W.Q.; Li, Q.; Luo, Y. *Modern Analytical Testing Techniques for Materials*, 2nd ed.; China university of Mining and Technology Press: Xuzhou, China, 2018; pp. 194–242.
27. Chen, X.F.; Li, F.; Song, B.T.; Wu, H.Z.; Li, B.Y. Preparation and performance of PAMAP filter loss reducer with high temperature and salt resistant. *Spec. Petrochem.* **2020**, *37*, 36–40. [CrossRef]
28. Ma, X.P.; Li, J.C.; Zhou, Y.Z.; Huang, L.; Liao, M.F. Synthesis and evaluation of zwitterionic polymer fluid loss agent. *Petrochem. Technol.* **2020**, *49*, 75–82. [CrossRef]
29. Demir, Y.K.; Metin, A.Ü.; Şatıroğlu, B.; Solmaz, M.E.; Kayser, V.; Mäder, K. Poly (methyl vinyl ether-co-maleic acid)—Pectin based hydrogel-forming systems: Gel, film, and microneedles. *Eur. J. Pharm. Biopharm.* **2017**, *117*, 182–194. [CrossRef]
30. Krysztafkiewicz, A.; Rager, B.; Maik, M.; Walkowiak, J. Modified sodium aluminium silicate—A highly dispersed polymer filler and a pigment. *Colloids Surf. A Physicochem. Eng. Asp.* **1996**, *113*, 203–214. [CrossRef]
31. Miyaji, F.; Kim, H.; Handa, S.; Kokubo, T.; Nakamura, T. Bonelike apatite coating on organic polymers: Novel nucleation process using sodium silicate solution. *Biomaterials* **1999**, *20*, 913–919. [CrossRef]
32. Zhao, R. Influence of carbon dioxide on the polymerization behavior of sodium silicate-acrylamide solution and products properties. *Chem. J. Chin. Univ.* **2009**, *30*, 596–600. [CrossRef]
33. Annenkov, V.; Danilovtseva, E.; Pal'shin, V.; Zelinskiy, S.; Chebykin, E.; Gak, V.; Shendrik, R. Luminescent siliceous materials based on sodium silicate, organic polymers and silicon analogs. *Mater. Chem. Phys.* **2017**, *185*, 65–72. [CrossRef]

34. Song, L.; Liu, W.; Xin, F.; Li, Y. "Materials Studio" Simulation Study of the Adsorption and Polymerization Mechanism of Sodium Silicate on Active Silica Surface at Different Temperatures. *Int. J. Met.* **2021**, *15*, 1091–1098. [CrossRef]
35. Huang, H. *The Principles and Technology of Drilling Fluids*; Petroleum Industry Press: Beijing, China, 2016.
36. *Q-81SHL 13-2019*; CNPC Enterprise Standard. Salt Resistant Filtrate Loss Reducer FRS for Drilling Fluids. China National Offshore Oil Corporation: Beijing, China, 2019. (In Chinese)
37. Choi, H.; Kim, T.; Kim, S.Y. Poly (Amidehydrazide) Hydrogel Particles for Removal of Cu2+ and Cd2+ Ions from Water. *Gels* **2021**, *7*, 121. [CrossRef] [PubMed]
38. *GB/T 16783.1*; Chinese National Standard. Field Testing of Drilling Fluids for Oil and Gas Industry Part 1: Water-Based Drilling Fluids. The State Bureau of Quality and Technical Supervision of China: Beijing, China, 2014. (In Chinese)

 gels

Article

Preparation and Hydrogelling Performances of a New Drilling Fluid Filtrate Reducer from Plant Press Slag

Wenjun Long [1], Xialei Zhu [1,2], Fengshan Zhou [1,*], Zhen Yan [1,3], Amutenya Evelina [1], Jinliang Liu [1], Zhongjin Wei [1] and Liang Ma [1]

1 Beijing Key Laboratory of Materials Utilization of Nonmetallic Minerals and Solid Wastes, National Laboratory of Mineral Materials, School of Materials Science and Technology, China University of Geosciences (Beijing), No. 29 Xueyuan Road, Haidian District, Beijing 100083, China; longwj@cugb.edu.cn (W.L.); zhuxialei1@163.com (X.Z.); yanzh@bjeea.cn (Z.Y.); evelinametine288@gmail.com (A.E.); 2003190045@cugb.edu.cn (J.L.); weizhongjin@hotmail.com (Z.W.); maxliang@cugb.edu.cn (L.M.)

2 Goldwind Environmental Protection Co., Ltd., No. 26 Kechuang 13th Street, Daxing District, Beijing 100176, China

3 Beijing Education Examination Authority, No. 9 Zhixin East Road, Haidian District, Beijing 100083, China

* Correspondence: zhoufs@cugb.edu.cn

Abstract: Plant press slag (PPS) containing abundant cellulose and starch is a byproduct in the deep processing of fruits, cereals, and tuberous crops products. PPS can be modified by using caustic soda and chloroacetic acid to obtain an inexpensive and environmentally friendly filtrate reducer of drilling fluids. The optimum mass ratio of m_{NaOH}:m_{MCA}:m_{PPS} is 1:1:2, the optimum etherification temperature is 75 °C, and the obtained product is a natural mixture of carboxymethyl cellulose and carboxymethyl starch (CMCS). PPS and CMCS are characterized by using X-ray diffraction, scanning electron microscopy, Fourier transform infrared spectroscopy, thermogravimetric, X-ray photoelectron spectroscopy, and elemental analysis. The filtration loss performance of CMCS is stable before and after hot-rolling aging at 120 °C in 4.00% NaCl and saturated NaCl brine base slurry. The minimum filtration loss value of CMCS is 5.28 mL/30 min at the dosage of 1.50%. Compared with the commercial filtrate reducers with a single component, i.e., carboxymethyl starch (CMS) and low viscosity sodium carboxymethyl cellulose (LV-CMC), CMCS have a better tolerance to high temperature of 120 °C and high concentration of NaCl. The filtration loss performance of low-cost CMCS can reach the standards of LV-CMC and CMS of the specification of water-based drilling fluid materials in petroleum industry.

Keywords: plant press slag (PPS); carboxymethylation; mixture of carboxymethyl cellulose and carboxymethyl starch (CMCS); filtrate reducer; drilling fluids

Citation: Long, W.; Zhu, X.; Zhou, F.; Yan, Z.; Evelina, A.; Liu, J.; Wei, Z.; Ma, L. Preparation and Hydrogelling Performances of a New Drilling Fluid Filtrate Reducer from Plant Press Slag. *Gels* **2022**, *8*, 201. https://doi.org/10.3390/gels8040201

Academic Editor: Georgios Bokias

Received: 22 February 2022
Accepted: 17 March 2022
Published: 23 March 2022

1. Introduction

The comprehensive utilization of agricultural waste materials has gained much popularity in recent years. Agricultural waste materials show numerous advantages such as cost effectiveness and reducing environmental hazards [1–4]. Many agricultural waste materials are produced because of the rapid development of agriculture in China. For example, the corn production was about 2.16×10^8 tons in 2017 in China, and about 3.5×10^7–3.9×10^7 tons corncob could be obtained from the corn production [5]. Agricultural waste material contains starch, cellulose, hemi-cellulose, etc. It could ferment and deteriorate easily and would cause environmental pollution. Some of them have been separated and purified to product feed additive [6], prebiotic food [7], protein feed [8], etc. Additionally, some of them have even been used to produce bioethanol [9]. However, its popularization and application are difficult to realize due to the complex process, huge investment, and low economic effectiveness.

Many studies have addressed the subject of using cellulosic agricultural wastes in oil field applications and a wide variety of cellulosic agricultural waste materials are available for using as drilling fluids additives, including corncob, rice husk, bamboo, and so on [10]. Some of them showed a good performance such as Tapioca starch [11] and data seed powder (DSP) [12]. However, there are still many engineering problems in the direct use of agricultural wastes, such as too much added dosages needed in the drilling fluids [13], some samples might cause a negative impact to the rheological properties of the drilling fluids [14,15]. In a nutshell, the added value of using agricultural wastes directly as drilling fluid additives is low. However, through appropriate low-cost chemical modification techniques, the performances and added value of additives derived based on the agricultural wastes can be greatly improved [16,17]. This is very beneficial for both additive producers and oilfield users.

There is no well drilling without the use of drilling fluids. Drilling fluids have many functions, including carrying and suspending cuttings, balancing stratum pressure, and reducing the filtration loss value, etc. Water-based drilling fluid is a widely used drilling fluid and it is a kind of suspension composed of bentonite, water, chemical additives, etc. Filtration control is an important property of a drilling fluid, particularly when drilling through permeable formations, where the hydrostatic pressure exceeds the formation pressure. It is important for a drilling fluid to quickly form a filter cake to effectively minimize fluid loss. Fluid loss control additives, also called filtrate-reducing agents or filtrate reducer, is crucial to the performance control of drilling fluid, which is mainly reflected in four aspects: firstly, enhancing the borehole stability for maintain normal and safe drilling in water sensitive formation and crushed formation; secondly, protecting the formation damage caused by filtrate loss immersion; thirdly, maintaining the colloid stability of drilling fluid system to suspend and carry drilling cutting particles; and lastly, aiding the bit tool cut rock while water jet drilling to improve the rate of penetration (ROP). The consumption of filtrate reducers accounts for about 10% of the whole drilling fluid additives in amount, but its cost accounts for about 30%, so filtrate reducer is the most important drilling fluid additive. Commonly used filtrate reducers include modified starch, modified cellulose, humic acid derivatives, low molecular polyacrylamide-like copolymers, partially hydrolyzed polyacrylonitrile, synthetic water-soluble resin, and so on [17–20].

Conventional chemical additives used in drilling fluids have been found to have a negative impact on the environment and human health, and the most commercially available chemical additives are not biodegradable [21,22]. Recently, nanomaterials have been used as drilling fluid additives, improving their rheology, lubricity, and filtration property [23–25]. However, due to their poor dispersion ability, low performance stability, and relatively high prices, their usage is limited [26,27]. Therefore, there is a great need for new widely used biodegradable environmentally friendly drilling fluid additives, which can help control the performances of drilling fluids, causing little impact on the environment and to the health of workers [15]. CMC (carboxymethyl cellulose) and CMS (carboxymethyl starch) are typical biodegradable water-soluble polymers. CMC is often called "the monosodium glutamate of industry". CMC and CMS can be used in oil and gas well drilling [28,29], drug release [30–34], food industry [35,36], etc. When used in well drilling, CMC is roughly divided into two classes, i.e., HV-CMC (high-viscosity sodium carboxymethyl cellulose) and LV-CMC (low-viscosity sodium carboxymethyl cellulose). HV-CMC is mainly used as a natural tackifier [37], while LV-CMC is the most widely used filtrate reducer in water-based drilling fluids [38,39]. However, due to the rising price of short linters for CMC production, preparing CMC using non-cotton cellulose has become increasingly popular. Main alternative raw materials include waste disposable paper cup [40], corncob residue [41], cassava residue [42], etc. CMS is close to LV-CMC in almost all performances except for salt resistance and is often used as drilling fluid loss agent to replace LV-CMC. The reaction mechanism of using cellulose and starch to prepare filtrate reducers is the same, but the reaction conditions are different [43,44].

Plant press slag (PPS) is a byproduct in the deep processing of natural products and it contains starch, cellulose, pectin, etc. It was treated by dehydration, drying, and then processed into powdery or granular fodder. The economic effectiveness of dealing with PPS in this way was low. Because of a high starch and cellulose content in PPS, we can synthesize a mixed product of LV-CMC and CMS by alkalizing and etherifying the PPS, which was a natural mixture of carboxymethyl cellulose and carboxymethyl starch (CMCS). We determined the optimal reaction conditions of CMCS, evaluated its applied performances, and characterized its structure. We compared CMCS with CMS and LV-CMC in terms of the filtration loss performance, tolerance to high temperature and salt. We also compared the filtration loss performance of CMCS with the standards of LV-CMC and CMCS of the specification of drilling fluid materials.

2. Results and Discussion

2.1. Structural Characterizations

We used a X-ray diffraction analyzer, scanning electron microscope, infrared spectrometer, and thermogravimeter to test, analyze, and compare PPS and CMCS. CMCS samples were prepared according to the optimal mass ratio of m_{NaOH}:m_{MCA}:m_{PPS} at 1:1:2 and etherification reaction at 75 °C for 1 h.

2.1.1. Elemental Analysis (EA)

The contents of carbon, hydrogen, and oxygen of PPS and CMCS were determined by using an organic element analyzer. According to the molecular formula of cellulose and starch, the theoretical content ratio of C/H is 7.2, and the ratio of C/O is 0.9. Due to the presence of impurities in PPS, the ratio of C/H and C/O were relatively low than the theoretical ratio. The ratio of C/H and C/O of chloroacetic acid is 8.0 and 0.75, respectively. Therefore, the ratio of C/H should increase, while the ratio of C/O should decrease theoretically after modification. As shown in Table 1, the experimental results were in good agreement with theory, indicating the success of the modification reaction of PPS to a certain extent.

Table 1. Elemental analysis results of PPS and CMCS.

Sample	Weight/mg	C/%	H/%	O/%	C/H	C/O
PPS	2.075	38.560	5.667	48.062	6.804	0.802
CMCS	2.139	32.170	4.691	54.241	6.858	0.593

2.1.2. XRD Characterization

An X-ray diffractometer was used to analyze PPS and CMCS, respectively (Figure 1). Starch has three crystal structures of A, B, and C, generating three types of XRD patterns [29]. From the diffractogram of PPS, PPS has C-type diffraction pattern, which is a mixture of A-type and B-type. Obvious diffraction peaks at around 2θ = 16.96°, 21.97°, 25.38°, and 26.61° could be observed for PPS, indicating its semi-crystalline structure [35,36]. The peaks were broad due to the crystallites of the cellulose in PPS [40]. After the alkalization and etherification in the process of carboxymethylation, the peaks at 2θ = 16.96°, 21.97°, 25.38°, and 26.61° in the diffractogram of CMCS disappeared completely, revealing the crystalline structures of cellulose and starch in PPS were destroyed and the crystallinity decreased dramatically [45,46]. Furthermore, new peaks were observed at 2θ = 31.60°, 45.44°, and 56.46°, which were the characteristic diffraction peaks of NaCl crystal. This was attributed to the crystallinity of the remaining NaCl in CMCS [40,47]. The change in the XRD pattern of the sample after the modification was due to the destruction of the hydrogen bond structure of the cellulose and starch in the PPS under the action of strong alkaline, heat, and mechanical forces, resulting in a decrease in its crystallinity, which is conducive to improving the hydrophilicity of the sample and its filtrate reduction performance.

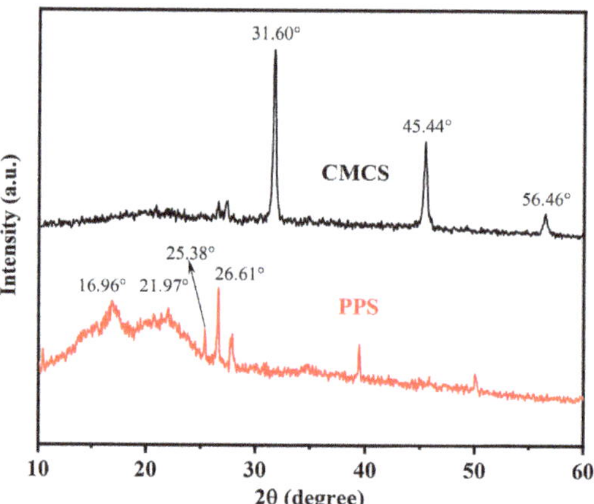

Figure 1. X-ray diffraction spectrograms of PPS and CMCS.

2.1.3. SEM Characterization

As shown in Figure 2, the morphology of PPS particles changed significantly after the modification. The original morphology of PPS particles was in an irregular elliptical egg shape, with a smooth and flat surface, and the diameter of those "eggs" were estimated in the range of approximately 10–30 μm (Figure 2a–c). However, the surface of CMCS became rough, with wrinkles, of which holes and cracks were obviously presented (Figure 2e,f). The hydroxyl group in PPS reacted with sodium hydroxide, part of the hydrogen bonds of PPS were broken, resulting in the decrease in its crystallinity and the formation of holes and cracks under the action of strong alkali, heat, and mechanical forces. Moreover, the diameter of the PPS particles was reduced (Figure 2d). Due to the smaller particles, holes and cracks on its surface, the contact area between PPS and etherification agent was increased, which is beneficial to the etherification reaction [36,43].

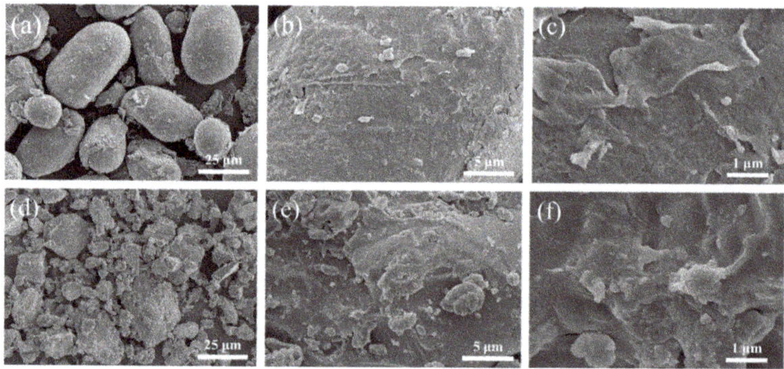

Figure 2. Scanning electron microscope images of PPS and CMCS. (**a**) PPS × 1000; (**b**) PPS × 5000; (**c**) PPS × 20,000; (**d**) CMCS × 1000; (**e**) CMCS × 5000; (**f**) CMCS × 20,000. CMCS was obtained by carboxymethylation when the mass ratio of m_{NaOH}:m_{MCA}:m_{PPS} was 1:1:2 and etherification temperature was 75 °C.

2.1.4. Infrared Spectroscopy Characterization

FTIR spectra was used to analyze the chemical groups of PPS and CMCS (Figure 3). The peaks at 3420.03 cm^{-1} and 2922.92 cm^{-1} in FTIR spectra of PPS were due to the stretching vibration of hydroxyl groups and methylene groups [41]. The peak at 1734.70 cm^{-1} originating from the carbonyl stretching vibrations of hemi-cellulose in PPS [48]. The peaks at 1647.81 cm^{-1}, 1514.43 cm^{-1}, and 1463.91 cm^{-1} were presented in FTIR spectra of PPS, which were due to the aromatic vibrations of the aromatic skeleton of lignin [40,49]. Compared with PPS, the peak at 3416.67 cm^{-1} was also due to the stretching vibration of hydroxyl groups, but the intensity of the signal was slightly reduced, indicating that part of the hydroxyl groups reacted during the modification. The new vibration peaks at 1605.37 cm^{-1} in FTIR spectra of CMCS were attributed to the asymmetrical stretching vibration of C=O in –COO groups. The new peaks at 1421.48 cm^{-1} were the characteristic absorption peak of the long-chain crystalline carboxylate [17,40,49]. Moreover, the peaks at 1324.40 cm^{-1} and 1045.61 cm^{-1} were attributed to the bending vibrations of –OH group and the stretching vibration of CH–O–CH_2 [43]. These results showed that carboxymethyl groups were successfully grafted on the molecular chain of cellulose and starch in PPS by etherification [41,50].

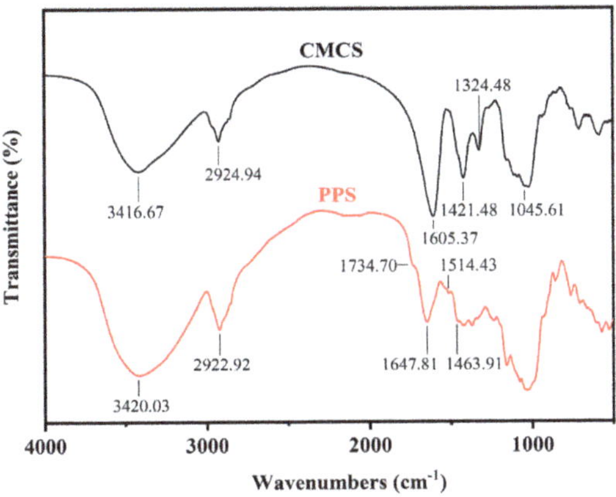

Figure 3. FT-IR spectrograms of PPS and CMCS.

2.1.5. Thermogravimetric Characterization

The weight loss of the samples as a function of temperature (TG curves) is shown in Figure 4a, the derivate weight loss curves (DTG curves) were obtained by calculating the first derivative of TG (Figure 4b). Thermal behavior of PPS and CMCS were examined by the study of TG, DTG thermograms. The lost weight of the samples before 150 °C was attributed to the evaporation of its remaining free water and bound water on its surface [51]. The initial decomposition temperature of PPS was about 184 °C and the weight of PPS at 184 °C was 94.28%. The thermogravimetric curve of PPS gradually became flat at about 401 °C with the weight of 39.15%, losing 55.13% of its weight in this thermal decomposition process and the weight of 33.35% of PPS remained when the temperature reached at 600 °C. However, the initial decomposition temperature of CMCS obtained by carboxymethylation was about 164 °C, which was about 20 °C lower than that of PPS and the weight of CMCS was 94.81% at 164 °C. This is because the crystal structure of cellulose and starch in PPS was destroyed during the preparation of CMCS, the CMCS particles were smaller, and the internal space of CMCS became loose, which led to a poor thermal stability of CMCS [52]. The thermogravimetric curve of CMCS became gradually slower at about 321 °C with the weight of 66.67%, and 28.05% of its weight lost in this decomposition process. The weight of 54.55% of CMCS remained after thermal decomposition (Figure 4a). The maximum thermogravimetric rate of PPS and CMCS was at about 287 °C and 260 °C, respectively. The thermogravimetric interval span of PPS was wider than that of CMCS, and the decomposition of CMCS was relatively slow resulting from the etherification of PPS (Figure 4b).

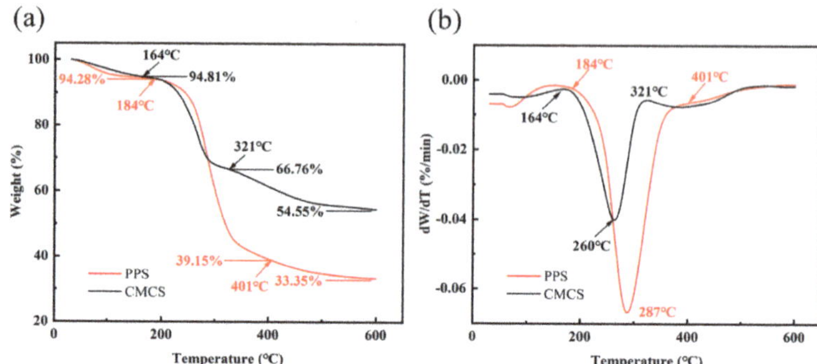

Figure 4. The thermal properties of PPS and CMCS. (**a**) Thermogravimetric (TG) analysis curves of PPS and CMCS; (**b**) Differential thermogravimetric (DTG) curves of PPS and CMCS.

The TG curve indicated the residual mass of CMCS (54.55%) after thermal decomposition was much more than that of PPS (33.35%). There were two reasons for this result. One of the reasons is because after etherification, carboxymethyl groups with high temperature resistance were successfully introduced into the molecular chains of PPS, which significantly improved the thermal stability of CMCS [53,54]. Moreover, the sodium ions introduced during the modification process also attributed to the higher residual mass of CMCS compared with PPS after the thermal decomposition [40].

2.1.6. XPS Analysis

The XPS spectra of PPS before and after modification were investigated to trace and compare the structural differences, the corresponding results are presented in Figure 5. As is shown in Figure 5a. the low-resolution mode spectra of PPS and CMCS indicate that PPS has two sharp peaks (O1s, C1s), whereas CMCS has three peaks (Na1s, O1s, C1s).

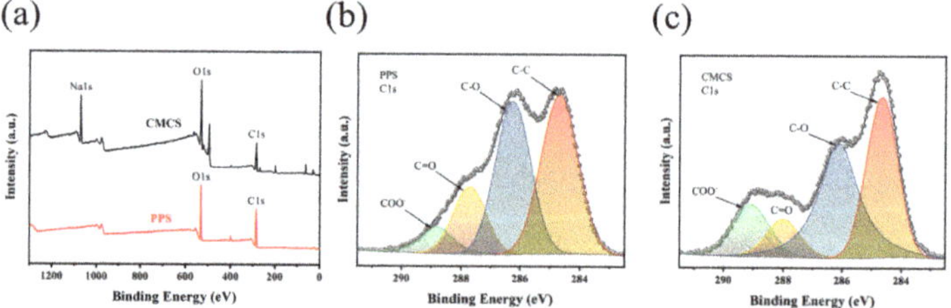

Figure 5. XPS spectra of PPS and CMCS. (**a**) Comparisons of low-resolution mode spectra of PPS and CMCS; (**b**) High-resolution mode XPS spectra of the C1s peaks in PPS; (**c**) High-resolution mode XPS spectra of the C1s peaks in CMCS.

The changes in carbon relative content in different combination states were revealed by the high-resolution C1s peak of PPS and CMCS (Figure 5b,c). The C1s peak was fitted into four distinct peaks that corresponded to C_1 (C–C, 284.7 eV), C_2 (C–O, 286.3 eV), C_3 (C=O, 287.7 eV), and C_4 (O=C–O, 288.8 eV). The presence of C_3, C_4 in PPS was related to its pectin and lignin components. The relative content of C_2 of PPS was decreased, while the composition of C_4 was increased after modification. The above results suggested that the carboxyl functional groups were successfully introduced into the molecular chain of PPS after alkalization and etherification reaction, which is consistent with previous studies [55–58].

2.2. Optimization of Experimental Parameters for Carboxymethylation

The optimized experimental parameters can be obtained according to the minimum value of the filtration loss of the saturated NaCl water-based drilling fluid with the sample dosage of 1.00% (4.00 g sample was added in 400.00 mL brine base slurry) after curing at room temperature for 16 h [59].

2.2.1. Effect of NaOH Dosage on Filtration Loss Performance of CMCS

The alkalized PPS samples were obtained by adding sodium hydroxide solution to the ethanol aqueous solution of PPS, wherein the mass ratio of m_{NaOH}:m_{PPS} was 0.250, 0.375, 0.500, 0.625, and 0.750, respectively. The controlled experimental variables for this parameter were the mass ratio of m_{MCA}:m_{PPS} which was 0.50 and the etherification temperature which was 75 °C. At the sample dosage of 1.00%, the API filtration loss tests of these samples in saturated NaCl brine base slurry were carried out after curing at room temperature for 16 h. As the amount of NaOH increased, the API filtration loss decreased at first, and later reached the minimum value of 6.32 mL when m_{NaOH}:m_{PPS} was 0.50 (Figure 6).

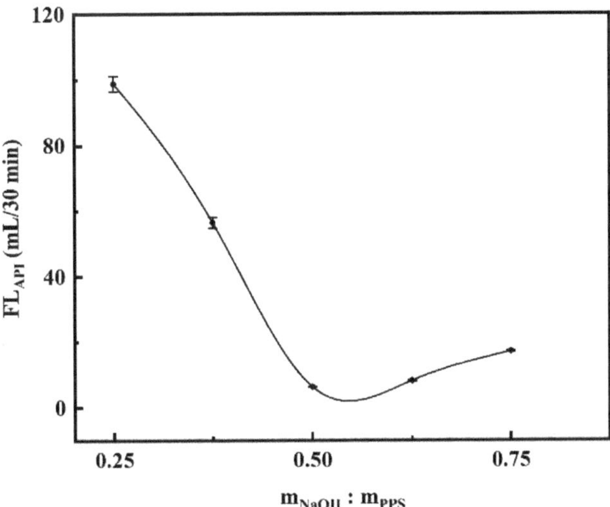

Figure 6. Effect of NaOH on saturated NaCl brine base slurry filtration loss performance of CMCS. (The mass ratio of m_{MCA}:m_{PPS} was 0.50 and the etherification temperature was 75 °C).

When the mass ratio of m_{NaOH}:m_{PPS} was lower than 0.50, the degree of swelling of PPS was insufficient, which was not conducive to the subsequent etherification reaction of MCA and PPS, resulting in lower etherification degree and poor filtration loss performance. When the mass ratio of m_{NaOH}:m_{PPS} was higher than 0.50, the PPS could be degraded by excessive NaOH, and excessive NaOH would react with the etherifying agent (monochloroacetic acid), producing byproducts such as sodium glycolate and lowering the efficient of etherifying agent, which caused a decreased etherification degree and inferior filtration loss performance of the sample [16,43].

2.2.2. Effect of Etherifying Agent Dosage on Filtration Loss Performance of CMCS

The etherified samples were prepared by adding monochloroacetic acid (MCA) to the ethanol aqueous solution of alkalized PPS, wherein the mass ratio of m_{MCA}:m_{PPS} was 0.250, 0.375, 0.500, 0.625, and 0.750, respectively. For this experimental parameter, the controlled conditions were the mass ratio of m_{NaOH}:m_{PPS} which was 0.50 and the etherification temperature which was 75 °C. After curing at room temperature for 16 h, the

API filtration loss tests of these samples in saturated NaCl brine base slurry were executed and the sample dosage was 1.00%. As the amount of MCA increased, the API filtration loss decreased at first and then increased, and it reached the minimum value of 6.32 mL when $m_{MCA}:m_{PPS}$ was 0.50 (Figure 7). When the mass ratio of $m_{MCA}:m_{PPS}$ was lower than 0.50, the inadequate etherification degree resulted in a reduced amount of carboxyl functional groups on CMCS, which led to a poor salt tolerance property and filtrate reduction effect of the sample. The increase in the MCA concentration made it easier to enter the interior of PPS for reaction. However, when the mass ratio of $m_{MCA}:m_{PPS}$ was higher than 0.50, the excessive MCA could affect the PH value of the solution and easily aggravate side reactions, which would reduce the etherification efficiency and filtrate reduction effect [60]

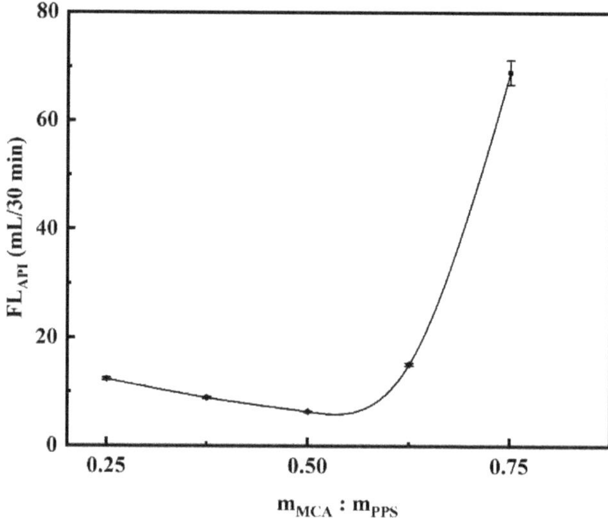

Figure 7. Effect of MCA on saturated NaCl brine base slurry filtration loss performance of CMCS. (The mass ratio of $m_{NaOH}:m_{PPS}$ was 0.50 and the etherification temperature was 75 °C).

2.2.3. Effect of Etherification Temperature on Filtration Loss Performance of CMCS

The etherification was carried out at different temperatures, including 45 °C, 60 °C, 75 °C, 90 °C, and 105 °C. The mass ratio of $m_{NaOH}:m_{MCA}:m_{PPS}$ was kept constant at 1:1:2 for this experimental parameter. At the sample dosage of 1.00%, the API filtration loss tests of these samples in saturated NaCl brine base slurry were implemented after curing at room temperature for 16 h. As the etherification temperature increased, the filtration loss value of sample decreased first and then increased, cause the increasing in temperature between 45 °C and 75 °C improved the swelling of PPS particles and the etherification reaction rate, which enhanced the etherification degree and filtration loss performance. When the etherification temperature was 75 °C, the API filtration loss reached the minimum value of 6.32 mL in saturated NaCl brine base slurry at the sample dosage of 1.00% (Figure 8). However, when the etherification temperature was higher than 75 °C, the excessive temperature would increase the rate constants of the side reaction, resulting in a decline to the filtration loss performance of sample [60]. Considering energy consumption and product performance, 75 °C was selected as the optimum etherification temperature.

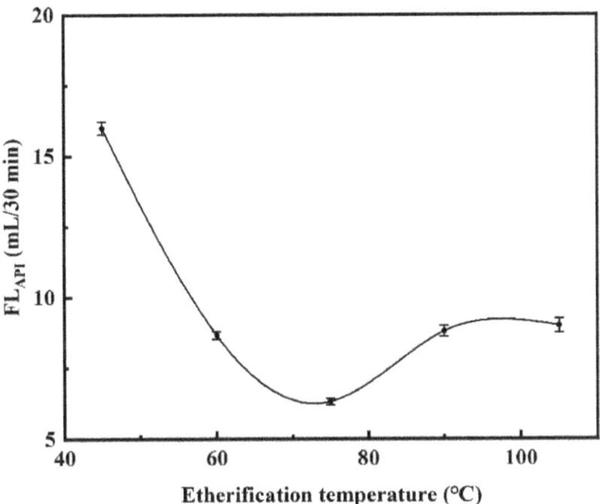

Figure 8. Effect of etherification temperature on saturated NaCl brine base slurry filtration loss performance of CMCS. (The mass ratio of m_{NaOH}:m_{MCA}:m_{PPS} was 1:1:2).

2.3. Comparison between CMCS and Other Fluid Loss Additives

The optimum mass ratio of m_{NaOH}:m_{MCA}:m_{PPS} was 1:1:2 and the optimum etherification temperature was 75 °C (Figures 6–8) The CMCS samples were prepared under this condition. Before and after hot-rolling aging in 4.00% and saturated NaCl brine base slurry for 16 h at 120 °C, the filtration loss tests of PPS, starch, LV-CMC, CMS, and the obtained CMCS were implemented at the sample dosage of 1.50% [59].

By comparing the filtration loss performance of PPS and starch with its corresponding carboxymethylation product CMCS and CMS, the effect of carboxymethylation reaction can be well proved. Figure 9 indicated that PPS and starch had similar fluid loss reduction performance. With the addition of PPS or starch in 4.00% NaCl and saturated NaCl base slurry, the apparent viscosity and fluid loss value of the drilling fluid hardly changed before aging, which revealed PPS and starch did not have any filtration loss effect before aging. However, after aging at 120 °C for 16 h, obvious changes to the apparent viscosity and fluid loss performance of the drilling fluid could be observed with the addition of PPS or starch, which was due to the gelatinization reaction of starch. When starch was heated in water the surface of starch granules would break, resulting in the exudation of internal soluble substances and the increase in the viscosity. At the same time, some amylopectin oozed out, causing the change in the starch structure, which was beneficial to the fluid loss reduction performance of the drilling fluid [61,62]. After aging, the filtration loss performance of PPS greatly improved in 4.00% NaCl and saturated NaCl brine base slurry. However, the degree of change in saturated NaCl brine base slurry was smaller than that in 4.00% NaCl brine base slurry, which was due to the fact that the tolerance to high NaCl brine concentration of PPS was insufficient. The change trend of the starch after aging was similar to PPS (Figure 9). Therefore, modification should be carried out to PPS and starch to improve their fluid loss performance as a fluid loss agent in water-based drilling fluids.

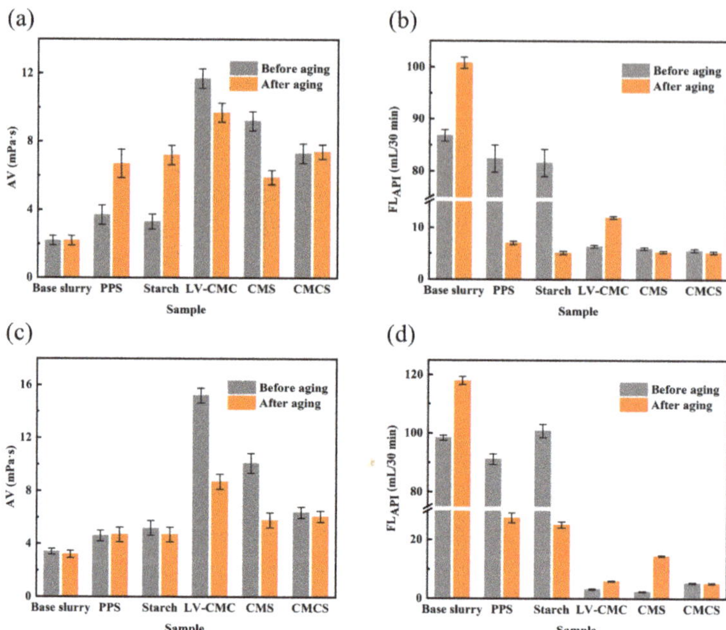

Figure 9. Performances of samples at the sample dosage of 1.50%. (**a**) Apparent viscosity in 4.00% NaCl brine base slurry; (**b**) Filtration loss value in 4.00% NaCl brine base slurry; (**c**) Apparent viscosity in saturated NaCl brine base slurry; (**d**) Filtration loss value in saturated NaCl brine base slurry.

Figure 9 presented that before and after aging, 4.00% NaCl and saturated NaCl brine base slurry with the addition of LV-CMC or CMS showed good rheological properties and fluid loss performance. LV-CMC and CMS could be used as a good filtrate reducer for drilling fluids; however, they all had their own shortcomings. The fluid loss value of 4.00% NaCl brine base slurry with LV-CMC added was 6.40 mL/30 min and 11.96 mL/30 min before and after aging at 120 °C for 16 h, which indicated that the temperature resistance of LV-CMC was inadequate. After aging at 120 °C for 16 h, the filtration loss value of 4.00% NaCl brine base slurry with the addition of CMS was 5.28 mL/30 min, and it increased to 14.48 mL/30 min in saturated NaCl brine base slurry, which indicated that the resistance to high concentration of NaCl of CMS was insufficient.

With the addition of CMCS, the apparent viscosity of the brine base slurry increased significantly, indicating that the water solubility of CMCS was much higher than that of PPS (Figure 9a,c). Furthermore, the filtration loss performance of CMCS at room temperature was significantly improved compared with that of PPS. When the dosage was 1.50%, the filtration loss value of CMCS and PPS was 5.6 mL/30 min and 82.4 mL/30 min in 4.00% NaCl brine base slurry before aging, respectively. It indicated that the carboxymethylation reaction greatly improved the filtration loss performance of PPS at room temperature. Moreover, CMCS showed stable fluid loss reduction performance in different brine base slurries before and after aging at 120 °C for 16 h, indicating that CMCS had a good resistance to temperature and high concentration of NaCl, which depicted that carboxyl group with resistance to temperature and salt had been introduced successfully in the molecular chain of PPS through carboxymethylation reaction (Figure 9b,d).

When the sample dosage was 1.00%, the filtration loss test was carried out for LV-CMC, CMS, and CMCS in the saturated NaCl brine base slurry before and after aging at 120 °C for 16 h. Table 2 showed the obtained CMCS reached the standard of LV-CMC and modified starch in the specification of drilling fluid material [63].

Table 2. Contrast of LV-CMC, CMS, and CMCS according to the standards in the specification of drilling fluid materials.

Filtrate Reducer	FL_{API} in Saturated Brine Base Slurry (mL/30 min)		Market Price (CNY/t)
	Before Aging	After Aging	
LV-CMC	6.0	16.0	11,500
CMS	4.0	34.0	9000
CMCS	6.4	7.0	7000
Standards	≤10.00	-	-

By comparing the fluid loss reduction performance of LV-CMC, CMS, and CMCS, we could conclude that CMCS had more stable temperature resistance than LV-CMC and had more stable resistance to high concentration of NaCl than CMS. After the pilot scale production of CMCS in the cooperative factory, the price of CMCS was calculated to be about CNY 7000 per ton, which was lower than LV-CMC and CMS. Furthermore, LV-CMC, CMS, and CMCS as drilling fluid additives, were compared in terms of thickening, filtrate loss, salt tolerance, temperature resistance, environmental friendliness, and cost (Table 3).

Table 3. Comparison results of LV-CMC, CMS, and CMCS.

Index	LV-CMC	CMS	CMCS
Thickening	ΔΔΔ	ΔΔ	ΔΔ
Filtrate loss	ΔΔΔ	ΔΔΔ	ΔΔΔ
Salt tolerance	ΔΔΔ	Δ	ΔΔΔ
Temperature resistance	ΔΔ	Δ	ΔΔ
Environmental friendliness	ΔΔΔ	ΔΔΔ	ΔΔΔ
Value for money	Δ	ΔΔ	ΔΔΔ

The number of Δ indicates the degree of excellence of an index.

2.4. Rheological Properties of the Drilling Fluid

Drilling fluid is a kind of pseudoplastic fluid, and therefore its rheological properties obey the Power Law equation: $\tau = K\gamma^n$, where τ is the shear stress, γ is shear rate, n is fluidity index, and K is consistency coefficient. K and n are two important parameters of pseudoplastic fluid, the value of n reflect the shear dilution performance of drilling fluid, and k is related to the viscosity and shear force of drilling fluid. In order to ensure carrying cuttings effectively, the n value of drilling fluid is generally required to be around 0.4–0.7. The K value reflects the pumpability of drilling fluid, the large value of K will make it difficult to re-pump, if the K value is too small, it will not be conducive to carrying cuttings. Therefore, the value of K of drilling fluid should also be kept within a proper range. Furthermore, the ratio of yield point to plastic viscosity, which also can reflect the strength of shear dilution, was used to evaluate the rheology behavior of drilling fluid, and is required to be controlled at 0.36–0.48 in drilling fluid technology. Before and after aging, the smaller the changes of the above parameters, the better the high temperature stability of the drilling fluid generally. As is shown in Table 4, the rheological property results of different NaCl concentration brine base slurries with the addition of PPS or CMCS before and after aging were tested and calculated (the dosage was 1.50%). According to the experimental results, different NaCl concentration drilling fluids added with PPS or CMCS have good rheological properties and thermal stability [64–66].

Table 4. Rheological property results of the drilling fluids with the addition of PPS or CMCS.

T/°C	Index	4% NaCl Slurry		Saturated NaCl Slurry	
		with PPS	with CMCS	with PPS	with CMCS
25	AV/mPa·s	4	8	5	7
	PV/mPa·s	3	6	3	5
	RYP/(Pa/mPa·s)	0.3407	0.3407	0.6813	0.4088
	n/Dimensionless	0.6781	0.6781	0.5146	0.6374
	K/Pa·s^n	0.0372	0.0744	0.1445	0.0864
120	AV/mPa·s	7	7.5	4.5	6
	PV/mPa·s	6	5	3	5
	RYP/(Pa/mPa·s)	0.1703	0.5110	0.5110	0.2044
	n/Dimensionless	0.8074	0.5850	0.5850	0.7776
	K/Pa·s^n	0.0266	0.1330	0.0789	0.0280

2.5. Lubricity Performance of CMCS

Before and after aging, the extreme pressure lubrication coefficient reduction rate (Δf) of drilling fluid with different dosages of PPS or CMCS was tested and calculated. Solid lubricants can greatly reduce the friction resistance by changing the sliding friction between drill string and borehole wall into rolling friction, while polymer treatment agents can improve the lubricity of drilling fluid by improving the quality of mud cake and forming adsorption film on drill string and borehole wall [66].

Before aging, the lubricity of PPS was better than that of CMCS with 1% addition, because PPS contains many small water-insoluble matters, the sliding friction between the steel ring and the slider will be changed into rolling friction with the addition of PPS, which greatly reduce the friction resistance. However, the strength of adsorption film formed by 1% addition of CMCS is not enough. When the dosage of samples increased to 2%, the strength of the absorption film formed by CMCS was greatly increased, making its Δf increase to 40%, while excessive PPS addition did not improve the lubricity of drilling fluid (Figure 10).

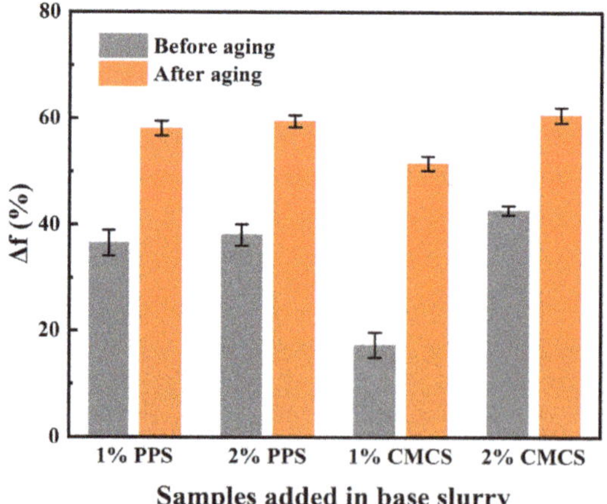

Figure 10. Lubricity performance of PPS and CMCS under different dosages.

After aging, PPS and CMCS had a better lubricity performance. Under the function of high temperature, the bentonite in the base slurry will agglomerate, resulting in increased viscosity and particle size and reduced lubricity [65,66]. After aging, the adsorption film formed by a water-soluble polymer produced by starch in PPS and its other insoluble small size particles improved the lubricity of the drilling fluid together. Due to the presence of

CMCS, the degree of bentonite agglomeration was reduced, and the formed adsorbent film making the drilling fluid had a good lubricity property (Figure 10).

2.6. Biodegradability Performance of CMCS

2.6.1. Determination of BOD_5 of CMCS

The BOD_5 value of CMCS was determined according to the mentioned method, which was 413 mg/L.

2.6.2. Determination of CODcr of CMCS

The CODcr value of CMCS was calculated according to the results in Table 5 and the calculation formula mentioned above. The experiment was carried out for three times using the potassium dichromate method. The average CODcr value of CMCS was 1600 mg/L.

Table 5. The volume of consumed ammonium ferrous sulfate standard solution.

	Dilution Rate	Test 1	Test 2	Test 3	Average Value
V_1/mL	50	19.50	19.20	19.10	19.27
V_2/mL	50	18.50	18.40	18.50	18.47

As a result, the ratio of BOD_5 and CODcr of the CMCS product was 25.81%, which showed that CMCS is an easily degradable chemical additive and causes no harm to the environment.

2.7. Biotoxicity of CMCS

The biological toxicity result of CMCS showed that the EC_{50} value of CMCS was 8.75×10^4 mg/L, which was much higher than the standard requirement value of 2.5×10^4 mg/L. Therefore, CMCS can be used as a non-toxic and eco-friendly filtration reducer for drilling fluids.

3. Conclusions

Plant press slag (PPS) is a byproduct of the deep processing of natural polymer products, which is rich in cellulose and starch. A natural mixture of carboxymethyl cellulose and carboxymethyl starch (CMCS) was prepared by alkalization and etherification through PPS and had the optimal filtration loss performance when the mass ratio of m_{NaOH}:m_{MCA}:m_{PPS} was 1:1:2, and the etherification temperature was 75 °C. After carboxymethylation, the crystallinity of PPS was reduced, and the carboxyl function group was successfully introduced into the molecular chain of PPS. CMCS had a stable filtration loss performance in both 4.00% NaCl and saturated NaCl brine base slurry before and after aging at 120 °C for 16 h. Before aging, the filtration loss value of the obtained CMCS at the dosage of 1.50% in 4.00% NaCl and saturated NaCl brine base slurry was 5.60 mL/30 min and 5.28 mL/30 min, and after aging, the result was 5.20 mL/30 min and 5.28 mL/30 min. Compared with CMS and LV-CMC, CMCS has a better filtration loss performance under high concentration of NaCl and high temperature of 120 °C. The filtration loss performance of CMCS can reach the standards of modified starch and carboxymethyl cellulose for drilling fluids. We can solve the pollution problem of the agricultural waste materials and make full use of them through this chemical modification method. Furthermore, the product CMCS has a good filtration loss performance and it can be used as an environmentally friendly and low-cost filtrate reducer in oilfields.

4. Materials and Methods

4.1. Materials

Plant press slag (PPS) used in this experiment was a byproduct in the deep processing of natural products and its main components include starch (37.00%), hemi-cellulose (14.26%), cellulose (14.77%), pectin (14.97%), lignin (3.98%), etc., (Figure 11a). The composition analysis of PPS was provided by the Qingdao Institute of Bioenergy and Bioprocess Technology, Chinese Academy of Sciences. PPS was obtained from Inner Mongolia Huaou Starch Industry Co., Ltd. (Hohhot, China).

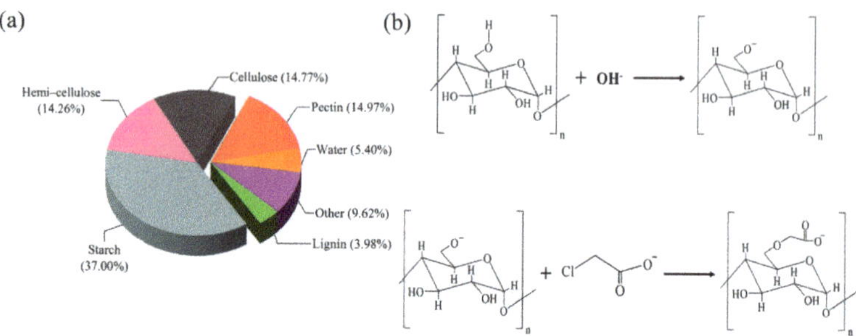

Figure 11. Composition and carboxymethyl reaction mechanism of PPS. (**a**) Compositions of PPS. (**b**) The reaction mechanism of PPS.

Technical-grade cotton fiber was obtained from Chenxiang Mining Co., Ltd. (Shijiazhuang, China), and corn starch was obtained from Hexinrong New Materials Co. Ltd. (Renqiu, China). Technical-grade monochloroacetic acid (MCA) was purchased from Yancheng Jinbiao Chemical Industry Co., Ltd. (Yancheng, China). Technical-grade low-viscosity carboxymethyl cellulose (LV-CMC) was obtained from Chongqing Lihong Fine Chemicals Co., Ltd. (Chongqing, China). Sodium carboxymethyl starch (CMS) was obtained from Zhongke Rising Co., Ltd. (Beijing, China). Analytical grade ammonium ferrous sulfate was purchased from Tianjin Guangfu Technology Development Co., Ltd. (Tianjin, China). Other reagents used were analytical grade, which were purchased from Beijing Chemical Factory (Beijing, China) and used without further purification

4.2. Methods

4.2.1. Synthesis of CMCS

The PPS was alkalized in the NaOH-ethanol solution of a certain concentration to obtain the alkalized PPS, which included cellulose-Na and starch-Na. CMCS was obtained after etherifying the alkalized PPS by monochloroacetic acid (MCA). Figure 11b showed the reaction mechanism of PPS.

Specific procedures: 20.00 g of PPS and 200.00 mL of 90% ethanol-water solution were added into a three-necked flask with a reflux condenser, a thermometer, and a mechanical stirring device with constant stirring speed. After mixing uniformly, 10.00 g of NaOH (solved in 10 mL distilled water) was added to alkalize PPS for 40 min at room temperature. After alkalization, 10.00 g of MCA (solved in 10 mL distilled water) was added to the container, and the etherification was carried out at 75 °C for 60 min. Finally, the sample was washed 2–3 times with 90% ethanol solution, dried in the oven at 60 °C for 12 h, and smashed to obtain the CMCS.

4.2.2. Elemental Analysis (EA)

The organic elements contents of samples before and after modification were determined by using the organic elemental analyzer (Vario EL cube, Elementar Trading Co., Ltd.; Shanghai, China). The detection limit of the sample was 50 ppm and the accuracy was 0.1 percent.

4.2.3. X-ray Diffraction (XRD)

In order to analyze the crystalline structures, samples were characterized by X-ray Diffraction with Cu Kα radiation (λ = 1.5418 Å), ranging from 10° to 70° (D8 Advance X-ray diffractometer, Bruker Scientific Instruments Hongkong Co., Ltd.; Hongkong, China).

4.2.4. Scanning Electron Microscope (SEM)

A dry sample was mounted on an aluminum holder and an ion sputtering device (Model E-1010, Hitachi; Tokyo, Japan) was used for gold sputter coating to make it be conductive. The morphology of the samples was characterized by a scanning electron microscope (SU8020, Hitachi; Tokyo, Japan) operating at an accelerating voltage of 5 kV.

4.2.5. Fourier Transform Infrared Spectra (FT-IR)

Samples were tested using FT-IR spectrometer (Spectrum 100, PerkinElmer; Waltham, MA, USA) in the range of 4000 cm^{-1} to 500 cm^{-1} with the resolution of 4 cm^{-1} and the signal-noise ratio is 50,000:1. All spectrums were obtained by accumulating 64 scans.

4.2.6. Thermogravimetric (TG)

The thermal decomposition behavior of the samples was investigated using thermogravimetric analysis (STA449F3; Netzsch; Selb, Germany) under the nitrogen (N_2) atmosphere and the nitrogen flow rate was 40 mL/min. The samples were placed in a clean crucible and heated from 30 °C to 600 °C at a heating rate of 10 °C/min. Differential thermogravimetric (DTG) analysis curves were obtained with the TGA data by numerically differentiating the latter with respect to temperature.

4.2.7. X-ray Photoelectron Spectroscopy (XPS)

The elements' binding energies of samples before and after modification were carried out using X-ray photoelectron spectroscopy (XPS, Escalab 250Xi, Thermo Fisher Scientific; Waltham, MA, USA). A deconvolution curve fitting was performed for the C_{1s} peaks, and the spectra were fitted using the Gaussian peak profiles and a linear background.

4.2.8. Performance Tests of the Drilling Fluids

Tests of drilling fluid performances included API filtration loss, temperature tolerance, lubricity, and rheological properties. Evaluation methods on drilling fluid performances were carried out in accordance with API and Sinopec group recommended standard method [59]. The experimental apparatus used are shown in Figure 12.

Figure 12. Instruments used during the experiment. (**a**) High-speed mixer; (**b**) Six-speed rotational viscometer; (**c**) Medium-pressure filter press; (**d**) Extreme pressure and lubricity tester; (**e**) High-temperature aging tank; (**f**) High-temperature hot-rolling furnace.

Preparation of Base Slurry

(1) In total, 4.00% NaCl brine base slurry: the base slurry was prepared following the ratio of 100.00 g of England Evaluation Clay (a famous standard kaolin clay from Britain), 40.00 g of NaCl and 2.80 g of $NaHCO_3$ in 1 L of distilled water. The mixture was stirred for 20 min at high speed, and at least two stops were needed during the stirring to scrape

the clay adhered on the container wall. Finally, sealed and conserved at 25 ± 3 °C for 24 h (2) Saturated NaCl brine base slurry: the base slurry was prepared following the ratio of 100.00 g of England Evaluation Clay, 365.00 g of NaCl and 2.80 g of $NaHCO_3$ in 1 L of distilled water. The mixture was stirred for 20 min at high speed, and at least two stops were needed during the stirring to scrape the clay adhered on the container wall. Finally, sealed and conserved at 25 ± 3 °C for 24 h.

Determination of API Filtration Loss

The API filtration loss was determined on a medium-pressure filter press (SD3, Qingdao Tongchun Oil Instrument Co., Ltd.; Qingdao, China). A certain amount of sample was added into 400 mL of brine base slurry and stirred at high speed. After 20 min, the mixture was poured into a sealed container and cured at room temperature. After 16 h, the drilling fluid was stirred at a high speed for 5 min, and poured into the filtration instrument cup up to the scale mark. The filtrate was collected under 0.69 MPa in 30 min. The volume of the obtained filtrate represents the API filtration loss (FL_{API}) of the drilling fluid.

Determination of Temperature Tolerance

The mixture of a certain amount of sample and 400 mL of brine base slurry were stirred at high speed for 20 min, poured into a high-temperature aging tank and placed it in high-temperature hot-rolling furnace (Hot roll furnace, XGRL-5, Qingdao Haitongda Special Instrument Co., Ltd.; Qingdao, China) to age under a certain temperature for 16 h. After 16 h, the drilling fluid was cooled down to room temperature and stirred for 5 min at high speed, and then the filtration loss and rheological property were tested.

Determination of Rheological Property

A certain amount of sample was added into 400 mL of the brine base slurry, which was closely conserved for at least 24 h. The mixture was stirred for 20 min at high speed and poured into the sample cup of Model Fanns 35 6-speed rotational viscometer (ZNN-D6, Qingdao Haitongda Special Instrument Co., Ltd.; Qingdao, China). The liquid level of the sample cup is tangent to the scale mark of drum of the 6-speed rotational viscometer and the stabilized values at different speeds were recorded. The apparent viscosity (AV), plastic viscosity (PV), yield point (YP), ration of yield point to plastic viscosity (RYP), the fluidity index (n), and consistency coefficient (K) of power law model fluid equation were calculated according to the following formulas:

$$\mathrm{AV} = 0.5 \times \Phi_{600} \tag{1}$$

$$\mathrm{PV} = \Phi_{600} - \Phi_{300} \tag{2}$$

$$\mathrm{YP} = 0.511(\Phi_{300} - \mathrm{PV}) \tag{3}$$

$$\mathrm{RYP} = \mathrm{YP}/\mathrm{PV} \tag{4}$$

$$n = 3.322lg(\Phi_{600}/\Phi_{300}) \tag{5}$$

$$\mathrm{K} = (0.511\Phi_{300})/511^{n} \tag{6}$$

where:

AV is the apparent viscosity (mPa·s);
PV is the plastic viscosity (mPa·s);
YP is the yield point (Pa);
RYP is the ratio of yield point to plastic viscosity (mPa·s);
Φ_{600} is the dial reading of 6-speed rotational viscometer at 600 r/min (dia);
Φ_{300} is the dial reading of 6-speed rotational viscometer at 300 r/min (dia);
n is the fluidity index of power law model rheological equation (dimensionless quantity);
K is the consistency coefficient of power law model rheological equation (Pa·s^n).

Determination of Lubricity Performance

The evaluation methods are according to the Sinopec enterprise standard, Q/SH1020 1879–2016 [67]. The extreme pressure lubrication coefficient reduction rate (Δf) was used to evaluate the lubrication performance with extreme pressure (E-P) and lubricity tester (Fann Model 212, Fann Instrument Company; Houston, TX, USA). The test steps were as follows: (1) The extreme pressure lubricator was turned on and preheated for 15 min. (2) The test slider was secured on the bracket and attached to the test ring. (3) The rotation speed of the test ring was adjusted to 60 rpm and torque was adjusted to zero. (4) The cured 400 mL brine base slurry was stirred at high speed for 5 min, and then poured into the test tray to the scale mark. Then, the torque was adjusted to 16.95 N·m, and the reading was recorded as T_1 after testing for 10 min. (5) The cured mixture of a certain amount of sample and 400 mL brine base slurry was stirred at high speed for 5 min. After cleaning the test slider and ring, the mixture was carried out using the same operation in step 4, and the reading was recorded as T_2 after testing for 10 min. Then, Δf was calculated according to the following formula:

$$\Delta f = \frac{(T_1 - T_2)}{T_1} \times 100 \quad (7)$$

where:

Δf is the extreme pressure lubrication coefficient reduction rate of the samples (%);

T_1 is the extreme pressure lubrication coefficient of the base slurry;

T_2 is the extreme pressure lubrication coefficient of the slurry added with samples.

4.2.9. Biodegradability Test

The biodegradable performance of the sample is evaluated by the ratio of Biological Oxygen Demand (BOD_5) and Chemical Oxygen Demand (CODcr).

Determination of BOD_5

The BOD_5 value was determined by measuring the consumed oxygen in the sample decomposition process under certain experimental conditions. The sample was pretreated according to Chinese enterprise standard, HJ 557-2010 [68]. Then, the sample solution was analyzed with a BOD5 tester (JC-870H, Qingdao Juchuang Times Environmental Protection Technology Co., Ltd.; Qingdao, China) and the instrument reading was recorded.

DDetermination of CODcr

The CODcr of sample was determined with potassium dichromate method according to the Chinese national standard, GB/11914-89 [69]. A certain amount of sample was added into a conical flask and oxidized by appropriate amount of potassium bichromate standard solution under the catalysis of sulfuric acid-silver sulfate solution. After heated and reflux for a period of time, the ammonium ferrous sulfate standard solution was used for titration, with 1,10-Phenanthroline as an indicator, and the volume of consumed ammonium ferrous sulfate standard solution was recorded as V_2. Take the same amount of distilled water from the above experimental conditions, and the amount of consumed ammonium ferrous sulfate standard solution of the blank group was recorded as V_1. Then, the CODcr (mg/L) was calculated according to the following formula:

$$\text{COD(mg/L)} = \frac{c(V_1 - V_2) \times 8000}{V_0} \quad (8)$$

where:

c is the concentration of ammonium ferrous sulfate standard solution (mol/L);

V_1 is the volume of consumed ammonium ferrous sulfate standard solution of the blank group (mL);

V_2 is the volume of consumed ammonium ferrous sulfate standard solution of the sample (mL);

V_0 is the volume of sample solution (mL).

4.2.10. Biological Toxicity Test

The biological toxicity tests were carried out according to the CNPC enterprise standard, Q/SY 111-2007 [70]. The clear sample solution was added into the biotoxicity tester (Microtox LX, Modern Water, Shanghai, China), and when the strength of luminescent bacterial luminous became 50%, the concentration of the solution was determined. The value of EC_{50} was calculated according to the standard method by the instrument display value

Author Contributions: All authors contributed to the conception and design of the study. Material preparation, data collection, and analyses were performed by W.L., X.Z. and F.Z. Original draft preparation, W.L., X.Z. and Z.Y. Review and editing, W.L., A.E., J.L. and F.Z. Software, W.L., Z.W. and L.M. Funding, F.Z. All authors have read and agreed to the published version of the manuscript.

Funding: This work was supported by the National Major Science and Technology Projects of China in the 13th Five-Year Plan (Grants 2016ZX05040-005), and the Key Science and Technology Projects of Sinopec Group (P18048-2).

Institutional Review Board Statement: Not applicable.

Informed Consent Statement: Not applicable.

Data Availability Statement: Data are available from the authors upon request.

Conflicts of Interest: We declare that this is original research that has not been published previously and has not been under consideration for publishing elsewhere. There is no conflict of interest in the submission of this manuscript, and the manuscript is approved by all authors for publication.

References

1. Junpen, A.; Pansuk, J.; Kamnoet, O.; Cheewaphongphan, P.; Garivait, S. Emission of Air Pollutants from Rice Residue Open Burning in Thailand, 2018. *Atmosphere* **2018**, *9*, 449. [CrossRef]
2. Zhang, H.; Hu, D.; Chen, J.; Ye, X.; Wang, S.; Hao, J.; Wang, L.; Zhang, R.; An, Z. Particle size distribution and polycyclic aromatic hydrocarbons emissions from agricultural crop residue burning. *Environ. Sci. Technol.* **2011**, *45*, 5477–5482. [CrossRef] [PubMed]
3. Jin, H.; Reed, D.W.; Thompson, V.S.; Fujita, Y.; Jiao, Y.; Crain-Zamora, M.; Fisher, J.; Scalzone, K.; Griffel, M.; Hartley, D.; et al. Sustainable Bioleaching of Rare Earth Elements from Industrial Waste Materials Using Agricultural Wastes. *ACS Sustain. Chem. Eng.* **2019**, *7*, 15311–15319. [CrossRef]
4. Hsu, E. Cost-benefit analysis for recycling of agricultural wastes in Taiwan. *Waste Manag.* **2021**, *120*, 424–432. [CrossRef] [PubMed]
5. Zhang, L.; Wang, Y.; Liu, W.; Ni, Y.; Hou, Q. Corncob residues as carbon quantum dots sources and their application in detection of metal ions. *Ind. Crops Prod.* **2019**, *133*, 18–25. [CrossRef]
6. Jiang, K.; Tang, B.; Wang, Q.; Xu, Z.; Sun, L.; Ma, J.; Li, S.; Xu, H.; Lei, P. The bio-processing of soybean dregs by solid state fermentation using a poly gamma-glutamic acid producing strain and its effect as feed additive. *Bioresour. Technol.* **2019**, *291*, 121841. [CrossRef]
7. Zhou, R.; Ren, Z.; Ye, J.; Fan, Y.; Liu, X.; Yang, J.; Deng, Z.; Li, J. Fermented Soybean Dregs by Neurospora crassa: A Traditional Prebiotic Food. *Appl. Biochem. Biotechnol.* **2019**, *189*, 608–625. [CrossRef]
8. Mo, W.; Man, Y.; Wong, M. Soybean dreg pre-digested by enzymes can effectively replace part of the fishmeal included in feed pellets for rearing gold-lined seabream. *Sci. Total Environ.* **2020**, *704*, 135266. [CrossRef]
9. Zhao, C.; Zou, Z.; Li, J.; Jia, H.; Liesche, J.; Chen, S.; Fang, H. Efficient bioethanol production from sodium hydroxide pretreated corn stover and rice straw in the context of on-site cellulase production. *Renew. Energy* **2018**, *118*, 14–24. [CrossRef]
10. Agwu, O.E.; Akpabio, J.U. Using agro-waste materials as possible filter loss control agents in drilling muds: A review. *J. Pet. Sci. Eng.* **2018**, *163*, 185–198. [CrossRef]
11. Zoveidavianpoor, M.; Samsuri, A. The use of nano-sized Tapioca starch as a natural water-soluble polymer for filtration control in water-based drilling muds. *J. Nat. Gas Sci. Eng.* **2016**, *34*, 832–840. [CrossRef]
12. Amanullah, M.; Ramasamy, J.; Al-Arfaj, M.K.; Aramco, S. Application of an indigenous eco-friendly raw material as fluid loss additive. *J. Pet. Sci. Eng.* **2016**, *139*, 191–197. [CrossRef]
13. Igwilo, K.C.; Anawe, P.A.L.; Okolie, S.T.A.; Uzorchukwu, I.; Charles, O. Evaluation of the Fluid Loss Property of Annona muricata and Carica papaya. *Open J. Yangtze Oil Gas* **2017**, *2*, 144–150. [CrossRef]
14. Hossain, M.E.; Wajheeuddin, M. The use of grass as an environmentally friendly additive in water-based drilling fluids. *Pet. Sci.* **2016**, *13*, 292–303. [CrossRef]
15. Al-Hameedi, A.T.T.; Alkinani, H.H.; Dunn-Norman, S.; Al-Alwani, M.A.; Al-Bazzaz, W.H.; Alshammari, A.F.; Albazzaz, H.W.; Mutar, R.A. Experimental investigation of bio-enhancer drilling fluid additive: Can palm tree leaves be utilized as a supportive eco-friendly additive in water-based drilling fluid system? *J. Pet. Explor. Prod. Technol.* **2019**, *10*, 595–603. [CrossRef]

16. Haleem, N.; Arshad, M.; Shahid, M.; Tahir, M.A. Synthesis of carboxymethyl cellulose from waste of cotton ginning industry. *Carbohydr Polym.* **2014**, *113*, 249–255. [CrossRef]
17. Long, W.; Luo, H.; Yan, Z.; Zhang, C.; Hao, W.; Wei, Z.; Zhu, X.; Zhou, F.; Cha, R. Synthesis of filtrate reducer from biogas residue and its application in drilling fluid. *Tappi J.* **2020**, *19*, 151–158. [CrossRef]
18. Huang, H. *The Principles and Technology of Drilling Fluids*; Petroleum Industry Press: Beijing, China, 2016.
19. Darley, H.C.H.; Gray, G.R. *Composition and Properties of Drilling and Completion Fluids*, 6th ed.; Gulf Professional Publishing: Houston, TX, USA, 2011.
20. Wang, Z. Present Situation and Development Direction of Modified Natural Material Additives for Drilling Fluid in China. *Sino-Glob. Energy* **2018**, *23*, 28–35. (In Chinese)
21. Al-Hameedi, A.T.T.; Alkinani, H.H.; Dunn-Norman, S.; Al-Alwani, M.A.; Alshammari, A.F.; Alkhamis, M.M.; Mutar, R.A.; Al-Bazzaz, W.H. Experimental investigation of environmentally friendly drilling fluid additives (mandarin peels powder) to substitute the conventional chemicals used in water-based drilling fluid. *J. Pet. Explor. Prod. Technol.* **2020**, *10*, 407–417. [CrossRef]
22. Al-Hameedi, A.T.T.; Alkinani, H.H.; Alkhamis, M.M.; Dunn-Norman, S. Utilizing a new eco-friendly drilling mud additive generated from wastes to minimize the use of the conventional chemical additives. *J. Pet. Explor. Prod. Technol.* **2020**, *10*, 3467–3481. [CrossRef]
23. Mohamadian, N.; Ghorbani, H.; Wood, D.A.; Khoshmardan, M.A. A hybrid nanocomposite of poly(styrene-methyl methacrylate-acrylic acid)/clay as a novel rheology-improvement additive for drilling fluids. *J. Polym. Res.* **2019**, *26*, 33. [CrossRef]
24. Mohamadian, N.; Ghorbani, H.; Wood, D.; Hormozi, H.K. Rheological and filtration characteristics of drilling fluids enhanced by nanoparticles with selected additives: An experimental study. *Adv. Geo-Energy Res.* **2018**, *2*, 228–236. [CrossRef]
25. Aftab, A.; Ali, M.; Sahito, M.F.; Mohanty, U.S.; Jha, N.K.; Akhondzadeh, H.; Azhar, M.R.; Ismail, A.R.; Keshavarz, A.; Iglauer, S. Environmental Friendliness and High Performance of Multifunctional Tween 80/ZnO-Nanoparticles-Added Water-Based Drilling Fluid: An Experimental Approach. *ACS Sustain. Chem. Eng.* **2020**, *8*, 11224–11243. [CrossRef]
26. Ghanbari, S.; Kazemzadeh, E.; Soleymani, M.; Naderifar, A. A facile method for synthesis and dispersion of silica nanoparticles in water-based drilling fluid. *Colloid Polym. Sci.* **2016**, *294*, 381–388. [CrossRef]
27. Yang, X.; Yue, Y.; Cai, J.; Liu, Y.; Wu, X. Experimental Study and Stabilization Mechanisms of Silica Nanoparticles Based Brine Mud with High Temperature Resistance for Horizontal Shale Gas Wells. *J. Nanomater.* **2015**, *2015*, 745312. [CrossRef]
28. Xu, J.; Hu, Y.; Chen, Q.; Chen, D.; Lin, J.; Bai, X. Preparation of hydrophobic carboxymethyl starches and analysis of their properties as fluid loss additives in drilling fluids. *Starch-Stärke* **2017**, *69*, 1600153. [CrossRef]
29. Jiang, G.; Hou, X.; Zeng, X.; Zhang, C.; Wu, H.; Shen, G.; Li, S.; Luo, Q.; Li, M.; Liu, X.; et al. Preparation and characterization of indicator films from carboxymethyl-cellulose/starch and purple sweet potato (*Ipomoea batatas* (L.) lam) anthocyanins for monitoring fish freshness. *Int. J. Biol. Macromol.* **2020**, *143*, 359–372. [CrossRef]
30. Pooresmaeil, M.; Javanbakht, S.; Behzadi Nia, S.; Namazi, H. Carboxymethyl cellulose/mesoporous magnetic graphene oxide as a safe and sustained ibuprofen delivery bio-system: Synthesis, characterization, and study of drug release kinetic. *Colloids Surf. A Physicochem. Eng. Asp.* **2020**, *594*, 124662. [CrossRef]
31. Nordin, N.A.; Rahman, N.A.; Talip, N.; Yacob, N. Citric Acid Cross-Linking of Carboxymethyl Sago Starch Based Hydrogel for Controlled Release Application. *Macromol. Symp.* **2018**, *382*, 1800086. [CrossRef]
32. Uva, M.; Mencuccini, L.; Atrei, A.; Innocenti, C.; Fantechi, E.; Sangregorio, C.; Maglio, M.; Fini, M.; Barbucci, R. On the Mechanism of Drug Release from Polysaccharide Hydrogels Cross-Linked with Magnetite Nanoparticles by Applying Alternating Magnetic Fields: The Case of DOXO Delivery. *Gels* **2015**, *1*, 24–43. [CrossRef]
33. Tiwari, N.; Nawale, L.; Sarkar, D.; Badiger, M.V. Carboxymethyl Cellulose-Grafted Mesoporous Silica Hybrid Nanogels for Enhanced Cellular Uptake and Release of Curcumin. *Gels* **2017**, *3*, 8. [CrossRef] [PubMed]
34. Camponeschi, F.; Atrei, A.; Rocchigiani, G.; Mencuccini, L.; Uva, M.; Barbucci, R. New Formulations of Polysaccharide-Based Hydrogels for Drug Release and Tissue Engineering. *Gels* **2015**, *1*, 3–23. [CrossRef]
35. Tavares, K.M.; Campos, A.d.; Mitsuyuki, M.C.; Luchesi, B.R.; Marconcini, J.M. Corn and cassava starch with carboxymethyl cellulose films and its mechanical and hydrophobic properties. *Carbohydr. Polym.* **2019**, *223*, 115055. [CrossRef]
36. Tuan Mohamood, N.F.A.; Zainuddin, N.; Ahmad Ayob, M.; Tan, S.W. Preparation, optimization and swelling study of carboxymethyl sago starch (CMSS)-acid hydrogel. *Chem. Cent. J.* **2018**, *12*, 133. [CrossRef] [PubMed]
37. Tiwari, R.; Kumar, S.; Husein, M.M.; Rane, P.M.; Kumar, N. Environmentally benign invert emulsion mud with optimized performance for shale drilling. *J. Pet. Sci. Eng.* **2020**, *186*, 106791. [CrossRef]
38. Li, F.; Jiang, G.; Wang, Z.; Cui, M. Drilling Fluid from Natural Vegetable Gum. *Pet. Sci. Technol.* **2014**, *32*, 738–744. [CrossRef]
39. Wei, M.; Wu, C.; Zhou, Y. Study on Wellbore Temperature and Pressure Distribution in Process of Gas Hydrate Mined by Polymer Additive CO_2 Jet. *Adv. Polym. Technol.* **2020**, *2020*, 2914375. [CrossRef]
40. Chen, J.; Li, H.; Fang, C.; Cheng, Y.; Tan, T.; Han, H. Synthesis and structure of carboxymethylcellulose with a high degree of substitution derived from waste disposable paper cups. *Carbohydr. Polym.* **2020**, *237*, 116040. [CrossRef]
41. Gao, S.; Liu, Y.; Wang, C.; Chu, F.; Xu, F.; Zhang, D. Structures, Properties and Potential Applications of Corncob Residue Modified by Carboxymethylation. *Polymers* **2020**, *12*, 638. [CrossRef]
42. Yun, T.; Pang, B.; Lu, J.; Lv, Y.; Cheng, Y.; Wang, H. Study on the derivation of cassava residue and its application in surface sizing. *Int. J. Biol. Macromol.* **2019**, *128*, 80–84. [CrossRef]

43. Joshi, G.; Naithani, S.; Varshney, V.K.; Bisht, S.S.; Rana, V.; Gupta, P.K. Synthesis and characterization of carboxymethyl cellulose from office waste paper: A greener approach towards waste management. *Waste Manag.* **2015**, *38*, 33–40. [CrossRef] [PubMed]
44. Shi, H.; Yin, Y.; Jiao, S. Preparation and characterization of carboxymethyl starch under ultrasound-microwave synergistic interaction. *J. Appl. Polym. Sci.* **2014**, *131*, 40906. [CrossRef]
45. Muthukumarana, C.; Kanmani, B.R.; Sharmila, G.; Kumar, N.M. Carboxymethylation of pectin: Optimization, characterization and in-vitro drug release studies. *Carbohydr. Polym.* **2018**, *194*, 311–318. [CrossRef]
46. Santos, M.B.; Dos Santos, C.H.C.; de Carvalho, M.G.; de Carvalho, C.W.P.; Garcia-Rojas, E.E. Physicochemical, thermal and rheological properties of synthesized carboxymethyl tara gum (Caesalpinia spinosa). *Int. J. Biol. Macromol.* **2019**, *134*, 595–603. [CrossRef]
47. Li, H.; Zhang, H.; Xiong, L.; Chen, X.; Wang, C.; Huang, C.; Chen, X. Isolation of Cellulose from Wheat Straw and Its Utilization for the Preparation of Carboxymethyl Cellulose. *Fibers Polym.* **2019**, *20*, 975–981. [CrossRef]
48. He, T.; Jiang, Z.; Wu, P.; Yi, J.; Li, J.; Hu, C. Fractionation for further conversion: From raw corn stover to lactic acid. *Sci. Rep.* **2016**, *6*, 38623. [CrossRef]
49. Yang, H.; Zhang, Y.; Kato, R.; Rowan, S.J. Preparation of cellulose nanofibers from Miscanthus x. Giganteus by ammonium persulfate oxidation. *Carbohydr. Polym.* **2019**, *212*, 30–39. [CrossRef]
50. Mondal, M.I.H.; Ahmed, F. Cellulosic fibres modified by chitosan and synthesized ecofriendly carboxymethyl chitosan from prawn shell waste. *J. Text. Inst.* **2019**, *111*, 49–59. [CrossRef]
51. Mohamad Haafiz, M.K.; Eichhorn, S.J.; Hassan, A.; Jawaid, M. Isolation and characterization of microcrystalline cellulose from oil palm biomass residue. *Carbohydr. Polym.* **2013**, *93*, 628–634. [CrossRef]
52. Moussa, I.; Khiari, R.; Moussa, A.; Belgacem, M.N.; Mhenni, M.F. Preparation and Characterization of Carboxymethyl Cellulose with a High Degree of Substitution from Agricultural Wastes. *Fibers Polym.* **2019**, *20*, 933–943. [CrossRef]
53. Abdullah, C.I.; Azzahari, A.D.; Rahman, N.M.M.A.; Hassan, A.; Yahya, R. Optimizing Treatment of Oil Palm-Empty Fruit Bunch (OP-EFB) Fiber: Chemical, Thermal and Physical Properties of Alkalized Fibers. *Fibers Polym.* **2019**, *20*, 527–537. [CrossRef]
54. Naknaen, P. Physicochemical, Thermal, Pasting and Microstructure Properties of Hydroxypropylated Jackfruit Seed Starch Prepared by Etherification with Propylene Oxide. *Recent Adv. Food Sci.* **2014**, *9*, 249–259. [CrossRef]
55. Peng, X.; Su, S.; Xia, M.; Lou, K.; Yang, F.; Peng, S.; Cai, Y. Fabrication of carboxymethyl-functionalized porous ramie microspheres as effective adsorbents for the removal of cadmium ions. *Cellulose* **2018**, *25*, 1921–1938. [CrossRef]
56. Wu, L.; Lin, X.; Du, X.; Luo, X. Biosorption of uranium(VI) from aqueous solution using microsphere adsorbents of carboxymethyl cellulose loaded with aluminum(III). *J. Radioanal. Nucl. Chem.* **2016**, *310*, 611–622. [CrossRef]
57. Zhou, Z.; Liu, X.; Hu, B.; Wang, J.; Xin, D.; Wang, Z.; Qiu, Y. Hydrophobic surface modification of ramie fibers with ethanol pretreatment and atmospheric pressure plasma treatment. *Surf. Coat. Technol.* **2011**, *205*, 4205–4210. [CrossRef]
58. Zong, Y.; Zhang, Y.; Lin, X.; Ye, D.; Luo, X.; Wang, J. Preparation of a novel microsphere adsorbent of prussian blue capsulated in carboxymethyl cellulose sodium for Cs(I) removal from contaminated water. *J. Radioanal. Nucl. Chem.* **2017**, *311*, 1577–1591. [CrossRef]
59. American Petroleum Institute. *Specification for Drilling Fluids Materials*, 18th ed.; API Specification 13A: Washington, DC, USA, 2009.
60. Wu, Z.; Zhou, P.; Yang, J.; Li, J. Determination of the optimal reaction conditions for the preparation of highly substituted carboxymethyl Cassia tora gum. *Carbohydr. Polym.* **2017**, *157*, 527–532. [CrossRef]
61. Vamadevan, V.; Bertoft, E. Structure-function relationships of starch components. *Starch-Stärke* **2015**, *67*, 55–68. [CrossRef]
62. Hou, C.; Zhao, X.; Tian, M.; Zhou, Y.; Yang, R.; Gu, Z.; Wang, P. Impact of water extractable arabinoxylan with different molecular weight on the gelatinization and retrogradation behavior of wheat starch. *Food Chem.* **2020**, *318*, 126477. [CrossRef]
63. *GB/T 5005-2010*; Chinese National Standard. Specification of Drilling Fluid Materials. National Standards of the People's Republic of China: Beijing, China, 2010. (In Chinese)
64. Liu, J.; Zhou, F.; Deng, F.; Zhao, H.; Wei, Z.; Long, W.; Evelina, L.M.A.; Ma, C.; Chen, S.; Ma, L. Improving the rheological properties of water-based calcium bentonite drilling fluids using water-soluble polymers in high temperature applications. *J. Polym. Eng.* **2021**, *42*, 129–139. [CrossRef]
65. Liu, J.; Cheng, Y.; Zhou, F.; Evelina, L.M.A.; Long, W.; Chen, S.; He, L.; Yi, X.; Yang, X. Evaluation method of thermal stability of bentonite for water-based drilling fluids. *J. Pet. Sci. Eng.* **2022**, *208*, 109239. [CrossRef]
66. Yan, J. *Drilling Fluids Technology*, 1st ed.; China University of Petroleum Press: Shandong, China, 2006.
67. *Q/SH1020 1879-2016*; Sinopec Enterprise Standard. Solid Lubricant Technology Requirements for Drilling Fluid. Sinopec Shengli Petroleum Administration Bureau: Shandong, China, 2016. (In Chinese)
68. *HJ 557-2010*; Chinese Enterprise Standard. 'Solid Waste-Extraction Procedure for Leaching Toxicity-Horizontal Vibration Method. Ministry of Ecology and Environment of the People's Republic of China: Beijing, China, 2010. (In Chinese)
69. *GB/11914-89*; Chinese National Standard. Water Quality-Determination of the Chemical Oxygen Demand-Dichromate Method. The State Bureau of Quality and Technical Supervision of China: Beijing, China, 1990. (In Chinese)
70. *Q/SY 111-2007*; CNPC Enterprise Standard. Grading and Determination of Biotoxicity of Chemicals and Drilling Fluids-Luminescent Bacteria Test. China National Petroleum Corporation: Beijing, China, 2007. (In Chinese)

gels

Review

Novel Trends in the Development of Surfactant-Based Hydraulic Fracturing Fluids: A Review

Andrey V. Shibaev [1,*], Andrei A. Osiptsov [2] and Olga E. Philippova [1]

1 Physics Department, Moscow State University, 119991 Moscow, Russia; phil@polly.phys.msu.ru
2 Skolkovo Institute of Science and Technology (Skoltech), 121205 Moscow, Russia; a.Osiptsov@skoltech.ru
* Correspondence: shibaev@polly.phys.msu.ru

Abstract: Viscoelastic surfactants (VES) are amphiphilic molecules which self-assemble into long polymer-like aggregates—wormlike micelles. Such micellar chains form an entangled network, imparting high viscosity and viscoelasticity to aqueous solutions. VES are currently attracting great attention as the main components of clean hydraulic fracturing fluids used for enhanced oil recovery (EOR). Fracturing fluids consist of proppant particles suspended in a viscoelastic medium. They are pumped into a wellbore under high pressure to create fractures, through which the oil can flow into the well. Polymer gels have been used most often for fracturing operations; however, VES solutions are advantageous as they usually require no breakers other than reservoir hydrocarbons to be cleaned from the well. Many attempts have recently been made to improve the viscoelastic properties, temperature, and salt resistance of VES fluids to make them a cost-effective alternative to polymer gels. This review aims at describing the novel concepts and advancements in the fundamental science of VES-based fracturing fluids reported in the last few years, which have not yet been widely industrially implemented, but are significant for prospective future applications. Recent achievements, reviewed in this paper, include the use of oligomeric surfactants, surfactant mixtures, hybrid nanoparticle/VES, or polymer/VES fluids. The advantages and limitations of the different VES fluids are discussed. The fundamental reasons for the different ways of improvement of VES performance for fracturing are described.

Keywords: viscoelastic surfactants; wormlike micelles; gels; viscoelasticity; oil recovery; hydraulic fracturing; clean fracturing fluids; responsiveness to hydrocarbons; oligomeric surfactants; nanoparticle/VES fluids; polymer/VES fluids

Citation: Shibaev, A.V.; Osiptsov, A.A.; Philippova, O.E. Novel Trends in the Development of Surfactant-Based Hydraulic Fracturing Fluids: A Review. *Gels* **2021**, *7*, 258. https://doi.org/10.3390/gels7040258

Academic Editors: Qing You, Guang Zhao and Xindi Sun

Received: 14 November 2021
Accepted: 10 December 2021
Published: 12 December 2021

1. Introduction

Hydraulic fracturing was first introduced in late 1940s as an enhanced oil recovery (EOR) technique [1], and since then, it has become one of the most widely used and effective methods for the intensification of oil and gas inflow to the wells [2,3]. Hydraulic fracturing increases the well debit, e.g., the volume of liquid or gas extracted from the well per unit of time. Currently, it is used both for resuming the exploitation of depleted wells and for bringing new, especially low-permeable, reservoirs into operation. The fracturing fluids can be water-based or non-aqueous [4].

Water-based fracturing fluids consist of a viscoelastic liquid or a gel in which proppant particles (sand or ceramic particles up to a few millimeters in size) are suspended [5,6]. At the first stage of a fracturing operation, the fluid is injected into the wellbore under high pressure and creates fractures in the oil-bearing layer that propagate in the accumulations of oil located far from the well. At the next stage, the pressure is removed, but the fracture remains open and filled with proppant particles with a large number of pores/channels between them. The fracturing fluid remaining between the particles must be broken and cleaned out from the pores. After that, the hydrocarbons can flow to the well through the pores between the proppant particles. The fracturing operation enlarges the area from which the hydrocarbons are produced, allows the extraction of the hydrocarbons from the accumulations located far from the well and, thus, substantially increases the rate of oil extraction.

For a successful fracturing operation, the fluids should meet the following criteria [7] and should:

(1) Form a stable proppant suspension, prevent its sedimentation, and efficiently transport it into the fracture, which is achieved by high viscosity and elasticity;
(2) Be easily pumped into the wellbore with a minimal pressure drop, for which shear thinning behavior is important;
(3) Pass through perforations without degradation of the mechanical properties, for which recovery after high shear is necessary;
(4) Have sufficient temperature resistance, e.g., not degrade viscosity upon heating to reservoir temperatures;
(5) Show sufficient salt tolerance, e.g., dissolve in water in the presence of salts (e.g. sodium, calcium and magnesium chlorides, or sulfates);
(6) Be easily cleaned out from fracture after it is created, for which the fluids should decrease the viscosity and elasticity under the action of an external stimulus (breaker) or upon contact with reservoir hydrocarbons;
(7) Not damage the formation;
(8) Be compatible with the formation;
(9) Have minimal environmental impact.

An absolute majority of the water-based fracturing fluids currently use polymers as thickening agents [8,9]. Both natural and synthetic polymers are employed, and they may be linear (un-cross-linked) or gelled (cross-linked). Among the natural polymers, polysaccharides are used [10], including guar gum and its derivatives (hydroxypropyl guar-HPG and carboxymethyl hydroxypropyl guar-CMHPG); xanthan gum; cellulose derivatives (hydroxyethyl cellulose, carboxymethyl cellulose, and carboxymethyl hydroxyethyl cellulose); etc. Cross-linkers may include various multivalent metal ions (Zr^{4+}, Ti^{4+}, Cr^{3+}, Al^{3+}, Fe^{3+}, and Fe^{2+}) [10] or borate ions for guar-based fluids [11]. Among the synthetic polymers, polyacrylamide and its derivatives [12], including hydrophobically modified polyacrylamide (HM PAAm) [13,14], are mainly employed.

However, though the technological processes for polymer-based fracturing fluids are well-established, they possess some disadvantages which are difficult to overcome. One of the important problems is the destruction and removal of residual liquid from the fracture (clean-up). For polymer-based fluids, specially introduced substances—breakers—are used, which cause the destruction of the network structure of the gel by breaking up the polymer chains [15]. At present, mainly enzyme or oxidative breakers are employed. Enzyme breakers cleave bonds in the main or side chains of the polysaccharides, leading to their shortening. However, the activity of enzymes can be greatly reduced by changing the external conditions; so, their use is limited. Oxidative breakers (persulfates for example) form free radicals, which also cause the destruction of bonds in the polymer backbone. This involves the use of highly concentrated solutions of oxidants, which may not be ecologically and technologically safe. Moreover, the use of all types of breakers that cut the polymer backbone does not allow complete destruction of the network structure and removal of the gel from the proppant pack.

The exploitation of unconventional reservoirs, including low-permeable, high-temperature, and high-salinity wells [16], has stimulated an extensive search for novel fracturing fluids with enhanced properties. The development of clean and cost-effective fracturing fluids is key for the advancement of the hydraulic fracturing technology. Currently, viscoelastic surfactants (VES) [17–19] are the main alternative to conventional polymer-based fluids.

Though commercial, VES-based fracturing fluids already exist and are industrially employed, new basic concepts and fundamental scientific advancements have been proposed in the last few years and are the focus of this article. Recent efforts to improve the properties of VES-based fluids to extend their use in hydraulic fracturing are reviewed, with particular attention given to the oligomeric surfactants and mixed (including nanocomposite) systems. These new VES-based fluids have not yet found vast application in the oil industry but are of great potential if their enhanced properties can be combined with a moderate cost and availability of the chemicals used (which, among other issues, implies simpler methods of synthesis of new surfactants, polymers, etc.). The fundamental scientific basis of the different ways to improve the VES performance in fracturing operations is discussed.

2. General Concepts of VES-Based Fluids

Since their introduction in 1997 [20,21], polymer-free, surfactant-based fracturing fluids have been attracting constantly increasing attention due to their adequate viscoelastic properties and sand-carrying capacity combined with almost complete clean-up from fracture and low formation damage [22–24]. Such fluids contain VES which are able to thicken aqueous solutions due to the formation of long self-assembled aggregates—wormlike micelles (WLMs) [25].

Surfactant molecules are amphiphilic and contain two parts—a polar hydrophilic head and a non-polar hydrophobic tail. Above the critical micelle concentration (CMC), they self-assemble and form aggregates of various shapes—spherical or cylindrical micelles, lamellae, bicontinuous structures, etc. [26]. The cylinders can grow and form very long chains—WLMs—which resemble polymers [27] and can attain several microns in length. Above an overlap concentration (C*), WLMs form an entangled network, imparting high viscosity and viscoelastic properties to water solutions [28]. Zero-shear viscosities up to 5–7 magnitudes higher than the viscosity of water are attained at moderate surfactant concentrations in the range of a few wt %, making VES highly efficient thickeners for fracturing fluids and comparable to polymer-based fluids [29].

WLMs may be formed by ionic or non-ionic surfactants; however, most of the examples of fracturing fluids currently employed contain ionic VES. Normally, WLMs are formed by ionic surfactants in the case of sufficient screening of electrostatic repulsions between similarly charged head groups, which allows them to pack closer than in the spherical micelles [25]. Screening of electrostatic forces may be achieved by the addition of low-molecular-weight salts (such as KCl) [30–32], oppositely charged surfactants [33], or hydrotrope ions [34–36].

Surfactants containing long (C18–C22) alkyl chains are currently regarded as the best thickening agents for fracturing fluids. Examples of such surfactants include a cationic surfactant with C22 mono-unsaturated tail - erucyl bis-(hydroxyethyl)methylammonium chloride (EHAC) [30] and an anionic surfactant with C18 mono-unsaturated tail – sodium (or potassium) oleate [37].

Another possibility for forming WLMs consists in the use of zwitterionic surfactants, which contain both anionic and cationic groups in the polar head. Such VES are generally more expensive than cationic or anionic species, but they provide stronger thickening effect at lower concentrations. Betaine surfactants with long (C22) tails are among the best thickening agents [29,38–40].

Examples of non-ionic surfactants forming WLMs include mixtures of polyoxyethylene alkyl ethers [41], single ultra-long chain (C18–C24) alkyl ethers [42], sugar-based surfactants [43], cholesterol-based surfactants [44], etc.

3. Advantages and Limitations of VES-Based Fluids

VES solutions have a number of advantages over polymer fracturing fluids [24]. They arise from the fact that, in contrast to polymer chains, the wormlike surfactant micelles are labile structures formed by the self-assembly of small molecules due to weak non-covalent interactions. Therefore, they are dynamic and are able to break and recombine [45]. This imparts reversible shear-thinning to the VES solutions: their viscosity decreases by several orders of magnitude under shear, which is explained by the alignment and breaking of the WLMs in flow, but completely restores at rest. This property is very important in hydraulic fracturing operations to facilitate the pumping of the fluid into the well.

Another advantage of the self-assembled nature of WLMs consists in their high responsiveness to external factors, e.g., to the presence of hydrocarbons [46,47]. Hydrocarbons are solubilized inside hydrophobic micellar cores, which leads to the change of surfactant molecular packing and the transformation of the WLMs to small (several nm in size) microemulsion droplets [48,49]. This is accompanied by a drastic drop of viscosity to the values close to the viscosity of water and by a complete disappearance of the viscoelastic properties. Due to this, VES solutions are called "clean fluids" as in most cases they require no breakers to be removed from the well after fracture formation. Reservoir hydrocarbons serve as a breaker and completely destroy the WLMs, leading to an almost complete fracture clean-up.

However, the labile structure of the micellar chains also imparts some limitations, which include weaker viscoelastic properties than those of the polymer fluids and poor temperature and salt stability.

Therefore, recent efforts in the research on VES fluids have been directed at solving these problems. The main approaches consist in the use of oligomeric surfactants or of mixtures of several surfactants or the addition of nanoparticles (NPs) or polymers to the VES fluids.

4. Oligomeric Surfactants

One of the approaches to increasing the strength of micellar networks for hydraulic fracturing applications is the use of oligomeric surfactants [50]. These surfactants consist of several hydrophobic tails (n) and hydrophilic heads (m) covalently linked to each other [51]. Such molecules exhibit stronger hydrophobic interactions, which enhance micellization, resulting in lower CMC values and smaller amounts of non-aggregated free surfactant, the lower surfactant concentrations necessary to form WLMs, and higher micellar scission energies and more stable micelles.

With regard to the example of cationic surfactant oligomers with quaternary ammonium heads, it was demonstrated that with an increase in the number of units in the oligomer (n), the CMC decreases, and the packing of surfactant molecules in the micelles becomes denser [52]. With an increase in n from 2 to 4, a transition occurs first from the spherical to the cylindrical micelles, which is accompanied by a significant increase in viscosity, and then to the ring-like micelles. Longer alkyl spacer—$(CH_2)_s$—between head groups of quaternary ammonium gemini surfactants also decreases the CMC [53]. An increase in the concentration of dimeric or trimeric surfactants leads to a transition from spherical aggregates to cylindrical and then to long WLMs, as shown by molecular dynamics simulations [54,55].

A promising approach consists in the synthesis of oligomeric surfactants with very long (C18–C22) alkyl tails in order to enhance the hydrophobic interactions. For instance, gemini cationic surfactants EHAB 22:1-s-22:1 (based on two erucyl dimethylammonium bromide moieties linked by an alkyl spacer) with C22 unsaturated tails and different spacers (s = 2, 6) were synthesized, and it was shown that the solutions of gemini EHAB exhibit higher viscosities at lower surfactant concentrations than the monomeric EHAB, which is due to the easier formation of longer WLMs [56]. Addition of a hydrotrope ion (sodium salicylate) to the gemini EHAB 22:1-6-22:1 allows the reaching of the zero-shear viscosity of ca. 30,000 Pa·s.

Recently, a series of dimeric and trimeric VES with long tails have been proposed, aiming at the increased temperature and salt resistance of WLMs [57,58]. The use of a trimeric cationic surfactant with long hydrophobic tails based on erucamidopropyldimethylamine allows the obtaining of WLMs with good thermal stability—due to very strong hydrophobic interactions the micelles do not break down when heated to at least 90 °C, and a stability of the solutions under ultra-high temperatures ranging from 140 to 180 °C is observed [59]. High thermal stability (up to 110–150 °C) was achieved for long-chain (C22) gemini surfactants with various spacers between the polar heads (C25-alkyl(C6)-C25, C25-alkyl(C8)-C25, C25-ester-C25, and C25-amide-C25) [60,61]. The incorporation of a sulfonate group into the spacer of a gemini zwitterionic VES with two unsaturated C22 tails resulted in a significant increase in the salt tolerance as compared to the previously reported cases [62]. Addition of a benzene sulfonate group in the polar head of a gemini zwitterionic surfactant with C22 tails (EDBS) allowed the obtaining of a viscoelastic WLM solution with combined good temperature and salt resistance: it withstood shearing at 170 s^{-1} and 120 °C for 120 min in the 25% standard brine solution [63].

Thus, a variety of gemini and trimeric surfactants with C22 erucyl tails and different chemical structures of spacers between polar groups were proposed (Table 1). However, the variation of the tail length may also be of importance in obtaining an optimized composition for the fracturing fluids. A gemini surfactant with 18:1 oleyl tails was proposed and showed sufficient viscoelastic properties and shear resistance at 80 °C, low proppant settling velocity, and low permeability and conductivity loss rates [64]. However, the behavior of the fluid at higher temperatures (above 80 °C) was not studied. The viscosity of the fluid at 80 °C reached 70.2 mPa·s, and should be lower at higher temperatures, which may be insufficient for high-temperature applications of the fluid.

Table 1. Reported gemini and trimeric surfactants and their characteristics for hydraulic fracturing applications. Reproduced with adaption from Zhang, Y.; Mao, J.; Zhao, J.; Zhang, W.; Liao, Z.; Xu, T.; Du, A.; Zhang, Z.; Yang, X.; and Ni, Y. Preparation of a novel sulfonic gemini zwitterionic, viscoelastic surfactant with superior heat and salt resistance using a rigid–soft combined strategy, *J. Mol. Liq. 318*, 114057, Copyright (2020), with permission from Elsevier.

Chemical Structure	Name	Advantages	Limitations	Refs.
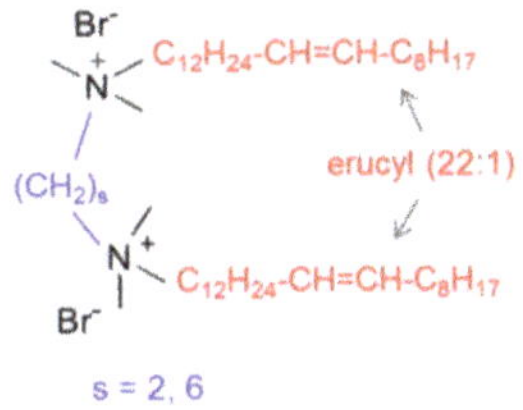 Br⁻, $C_{12}H_{24}$-CH=CH-C_8H_{17}, erucyl (22:1), $(CH_2)_s$, Br⁻; s = 2, 6	EHAB 22:1-s-22:1	Best characteristics are observed for s = 6: highly viscoelastic or gel-like solutions at high temperatures in the presence of salt (viscosity in the presence of NaSal at 110 °C is greater than 107 mPa·s)	no data	[56]
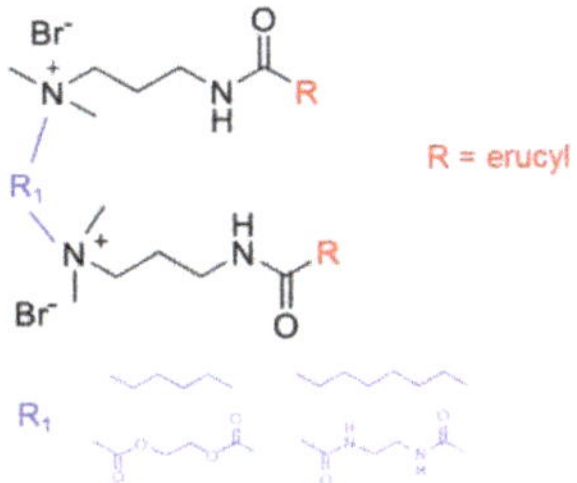 Br⁻, R, R_1, Br⁻; R = erucyl	C25-R1-C25	High temperature resistance (up to 110–150 °C)	sensitivity to high salinity	[60,61]
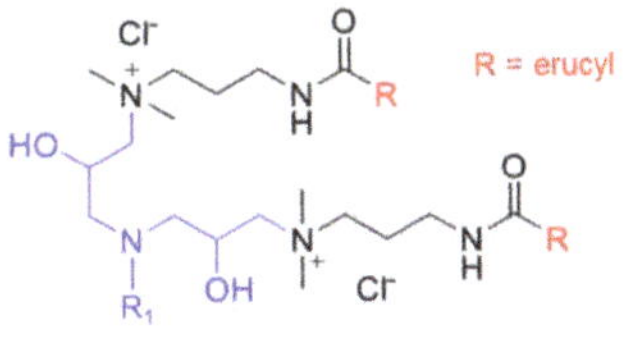 Cl⁻, HO, R, R_1, OH, Cl⁻; R = erucyl	VES-M (R_1 = methyl) VES-E (R_1 = ethyl) VES-P (R_1 = *n*-propyl)	Best characteristics are observed for VES-M with smallest alkyl group in the spacer: - stability up to 139 °C; - no proppant settling for 180 min at ambient temperature; - complete gel breaking by kerosene	sensitivity to high salinity	[57]

Table 1. *Cont.*

Chemical Structure	Name	Advantages	Limitations	Refs.
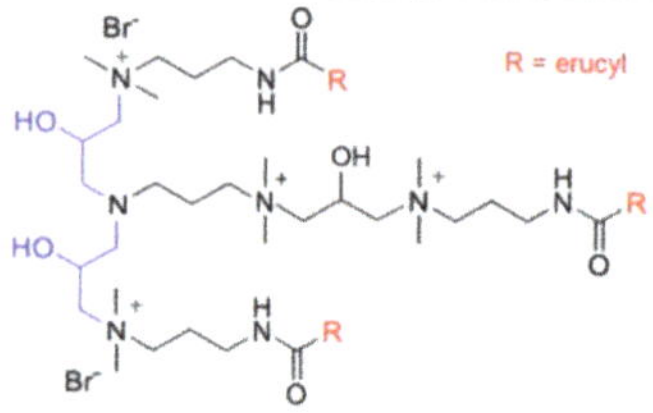 R = erucyl	VES-T	- excellent thermal stabilities under ultra-high temperatures ranging from 140 to 180 °C; - gel broken by standard brines within 2 h	no data	[59]
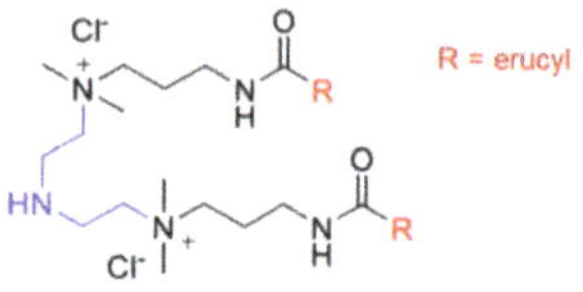 R = erucyl	JS-N-JS	- temperature and shear resistance; - gel breaking into a water-like fluid by hydrocarbons; - low damage to the tight sand reservoirs	sensitivity to high salinity	[58]
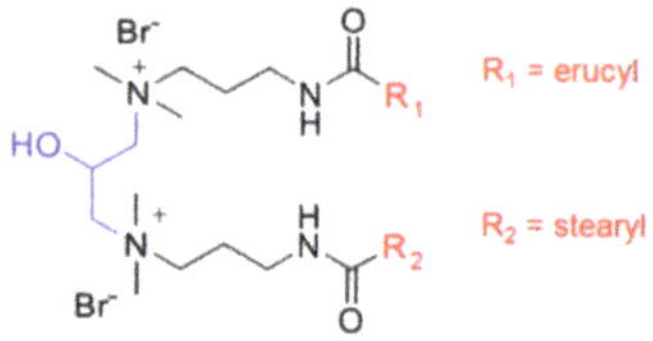 R_1 = erucyl R_2 = stearyl	YS-18	- better temperature and salt tolerance than corresponding symmetric surfactants (stable under a share rate of 170 s^{-1} at 110 °C for 60 min); - gel breaking by reservoir brine and kerosene	no data	[65]
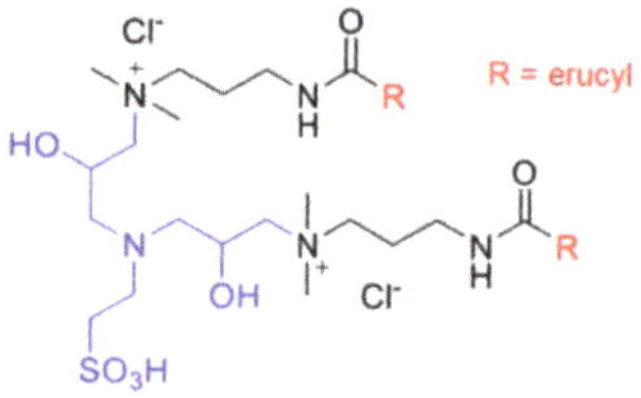 R = erucyl	VES-S	- good salt tolerance (up to 8 wt % NaCl), fracturing fluid may be prepared with seawater; - sufficient viscoelastic properties and proppant-carrying capacity at high salinity; - gel breaking by kerosene	high temperature tolerance not studied in the work	[62]
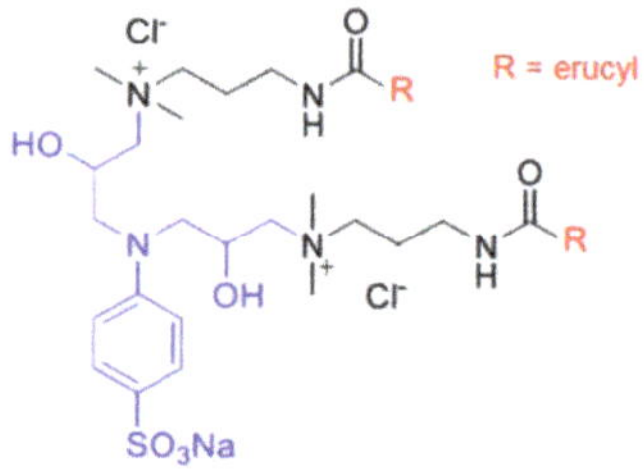 R = erucyl	EDBS	- temperature, shear and salt resistance (stability under a share rate of 170 s^{-1} at 120 °C for 120 min in the 25% standard brine solution); - gel breaking by kerosene	no data	[63]
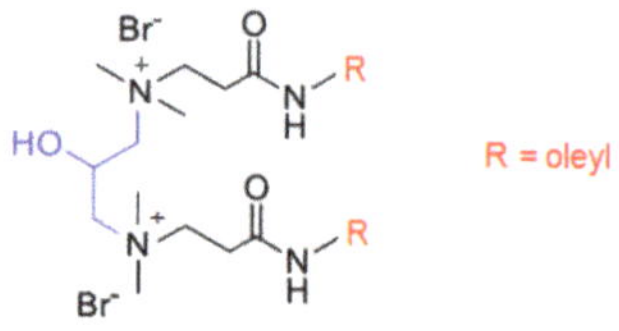 R = oleyl	GLO	- sufficient viscoelastic properties; - shear resistance at 80 °C; - low proppant settling velocity; - low permeability and conductivity loss rates	high temperature tolerance not studied in the work	[64]

Table 1. *Cont.*

Chemical Structure	Name	Advantages	Limitations	Refs.
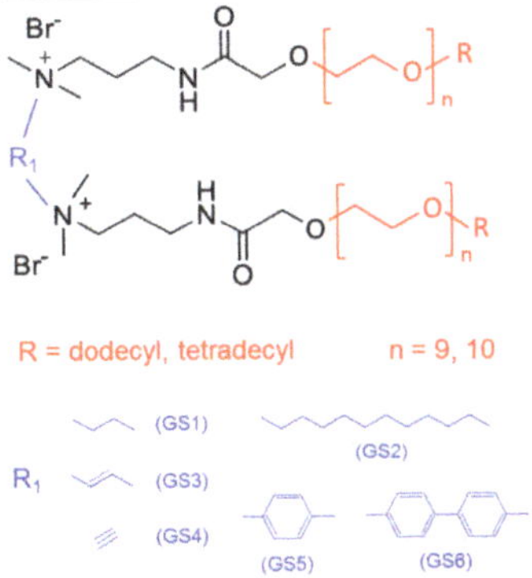 Br⁻ N⁺ … R; R₁; R = dodecyl, tetradecyl; n = 9, 10; (GS1) (GS2) (GS3) (GS4) (GS5) (GS6)	GS	- surfactant inhibits clay swelling; - sandstone retains its permeability at much higher level than for conventional clay stabilizers such as NaCl or KCl; - no increase in rock breakdown pressure	no influence of surfactants on the rheology of fluids reported	[66–69]

A promising improvement of gemini surfactants consists in the synthesis of dissymmetric surfactants containing different alkyl tails. For instance, a heterogemini cationic surfactant, YS-18, containing one saturated (18:0) and one unsaturated (18:1) tail, has much better salt tolerance and thermostability than the corresponding symmetrical surfactants. YS-18/KCl fluid viscosity remains stable under a share rate of 170 s^{-1} at 110 °C for 60 min and can easily be broken by reservoir brine and hydrocarbon [65].

Tariq et al. [67–69] showed that the use of a solution of cationic gemini surfactants with different spacers resulted in the reduction in the breakdown pressure in unconventional tight sandstones compared to that of the deionized water, which was attributed to the clay stabilization against swelling provided by the surfactant. In addition to the lower breakdown pressure, the use of gemini surfactants allows a reduction in the volume of fluid required to fracture the formation. The coreflooding experiment used to evaluate the formation damage demonstrated that the gemini surfactants do not cause any permeability impairment.

Therefore, the use of oligomeric surfactants with long hydrophobic tails is a promising way of improving the properties of WLMs. Incorporation of a sulfonate group or/and a benzene ring in the spacer increases the temperature and salinity stability of the VES fluids. Cationic gemini surfactants reduce the breakdown pressure and do not cause any permeability impairment. Up to now, mostly dimeric (n = 2) and trimeric (n = 3) surfactants have been employed due to the difficulties associated with the synthesis of higher oligomers. Development of new oligomeric surfactants with $n \geq 4$ at a reasonable cost may further advance the fracturing fluid applications of oligomeric surfactants.

5. Mixtures

5.1. Single-Chain Surfactants

A common method of WLM preparation consists in mixing different surfactants—anionic/cationic [70], zwitterionic/anionic [71,72], anionic/non-ionic [73], cationic/non-ionic [74,75], or two non-ionic surfactants [44]. In many cases, mixing results in the synergistic enhancement of the viscoelastic properties.

Strong synergy is observed for many anionic/cationic surfactant systems, even in the absence of added low-molecular-weight salt, and is a result of two processes: (1) the anionic and cationic polar heads reside in close proximity due to the co-assembly of two surfactants, resulting in a very efficient screening of the electrostatic repulsions at the micellar surface and tighter packing of the surfactant molecules, and (2) the counter ions of both surfactants are released to the solution, giving additional salt, which also weakens the electrostatic repulsions. For instance, in the mixtures of sodium or potassium oleate with different alkyltrimethylammonium or alkylpyridinium surfactants (n-octyltrimethylammonium bromide [33], *n*-octylpyridinium chloride [76], 1-dodecylpyridinium chloride [77]), the formation of a viscoelastic network of WLMs with viscosities up to ca. 1000 Pa·s was observed, which was not inherent in the solutions of each surfactant taken separately. Many other viscoelastic mixed systems of anionic/cationic surfactants have been described: cetyltrimethylammonium tosylate/sodium dodecyl benzyl sulfonate [78,79]; hexadecyl- or dodecyltrimethylammonium bromide/N-dodecylglutamic acid salt [70]; cetyltrimethylammonium bromide with sodium laurate [80–82]; sodium dodecyl sulfonate [83] or sodium N-alkylmaleimidepimaric carboxylate [84]; and cetylpyridinium chloride/sodium deoxycholate [85] systems. If the association between the anionic and cationic species is rather strong, such a co-assembly may be treated as a formation of a "pseudo-gemini" betaine surfactant. At a non-stoichiometric ratio of anionic and cationic surfactants, the system to some extent resembles mixtures of a betaine and a single-chain ionic surfactant.

A great advantage of mixed anionic/cationic WLMs is a strong rheological synergy, due to which sufficient viscoelastic properties are attained at a reduced concentration of surfactants. However, these systems are usually prepared without additional low-molecular-weight salts and, thus, are salt-sensitive. Positive cases of viscosity and elasticity increase upon the addition of monovalent salts (KCl) have been reported [33]; however, due to the presence of an anionic surfactant, anionic/cationic WLMs may be intolerant to divalent cations (Ca^{2+}, Mg^{2+}, etc.).

Therefore, other mixed surfactant systems may be advantageous for potential applications, and examples of recently studied systems include mixtures comprising oligomeric surfactants.

5.2. *Oligomeric and Single-Chain Surfactants*

One of the approaches for obtaining long WLMs consists in the mixing of oligomeric surfactants with single-chain species [86,87]. A synergistic effect has been observed for some short-chain (C12) cationic oligomeric surfactants mixed with anionic single-chain species. For instance, the addition of small amounts (5–20 mol%) of single-chain anionic surfactants (sodium dodecyl sulfate, sodium dodecyl benzene sulfonate, and sodium laurate) to a cationic gemini surfactant 12-3(OH)-12 (2-hydroxyl-propanediyl-α,ω-bis (dimethyldodecylammonium bromide) induces an increase in viscosity by 2 orders of magnitude, up to 60–120 Pa·s, and an enhancement of the viscoelastic properties of the micellar network [88]. In the mixtures of a trimeric cationic surfactant DDTPA (penta sodium N,N′,N′′-dodecyl diethylene triamine pentaacetic acid) and sodium dodecyl sulfate in the presence of KCl, a network of WLMs is formed, which imparts high zero-shear viscosity (~400 Pa·s) and viscoelastic behavior [86].

5.3. *Multiple Oligomeric Surfactants*

Very recently, the mixing of two different long-chain (C22) sulfobetaine zwitterionic gemini surfactants, EDABS (based on erucyl dimethyl amidopropyl benzenesulfonic acid) and EDAES (based on erucyl dimethyl amidopropylethanesulfonic acid), which differ only slightly by the presence of a benzene ring in the polar head, was proposed to obtain a combination of viscoelasticity, good shear, and temperature and salt tolerance [89].

5.4. *"Pseudo-Oligomeric" Surfactants*

As stated above (Section 5.1), a strong association of anionic and cationic single-chain surfactants may be treated as a formation of a "pseudo-gemini" betaine surfactant species comprising anionic and cationic polar groups and two tails from both surfactants. However, such associations are transient and usually sensitive to salts.

A recently proposed interesting approach consists in the non-covalent self-assembly of "ordinary" single-chain cationic surfactant cetyltrimethylammonium bromide (CTAB) into a pseudo-gemini species by an interaction with di- or tricarboxylic acids (citric or maleic acid) [90]. The fluid possesses high viscosity (1000 Pa·s), strong viscoelasticity, and temperature and shear resistance. Such an approach eliminates the complex and expensive synthesis of oligomeric surfactants as cheap surfactants and acids may be used.

Therefore, the mixing of surfactants is an efficient way of obtaining viscoelastic networks of long WLMs with enhanced rheological performance. Promising approaches consist in using mixtures of single-chain and oligomeric surfactants or in the formation of "pseudo-oligomeric" surfactants.

6. Nanoparticle-Enhanced VES Fluids

The rheological properties of VES can be significantly enhanced by added nanoparticles (NPs), leading to the extended stability of the VES fluids in withstanding high shear rate and temperature during the fracturing process, minimization of fluid loss, and improvement of sand-carrying ability [17,91].

The enhancement of the rheological properties results from the binding of micellar end-caps to the surface of the NPs, leading to elongation or cross-linking of the micelles by the NPs (Figure 1) [92]. WLMs can interact with a large variety of NPs regardless of their size, shape, and surface functionality [93]. When the NPs are oppositely charged with respect to the surfactant ions, the electrostatic interactions govern the adsorption, and the surfactant is attached to the NPs via its charged head-groups, leaving the hydrophobic tails outside. To exclude the contact of these hydrophobes with water, a second surfactant layer is attached in such a way that the hydrophobic tails face the tails of the first layer, leaving the hydrophilic heads in contact with the water. Thus, in the case of NPs oppositely charged with respect to the surfactant ions, the surfactant double layers are formed. The formation of the double layer results in the charge reversal of NPs, as was demonstrated by the ζ-potential measurements [94]. When the NPs are hydrophobic, the hydrophobic interactions govern the adsorption [95], and the surfactant is attached to the NPs by its hydrophobic tails, forming a monolayer with the polar groups exposed to water. The surfactant can adsorb on the NPs even if they are similarly charged, due to the hydrophobic interactions; in this case, hemimicelles are formed [96].

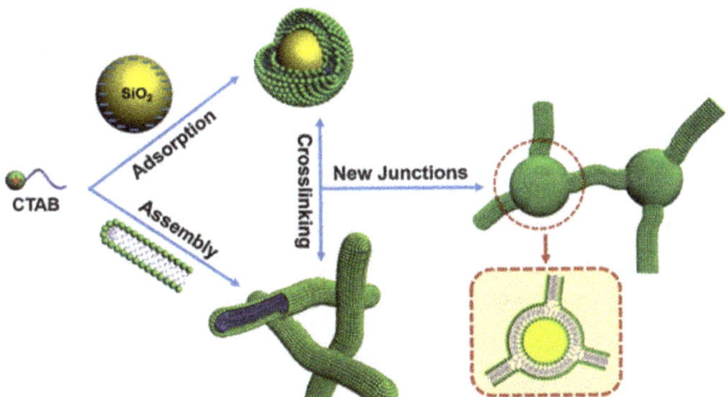

Figure 1. Schematic representation of the mechanism of nanoparticle-enhanced VES fracturing system on an example of a cationic surfactant cetyltrimethylammonium bromide (CTAB) WLMs interacting with anionic silica NPs. Reproduced from Zhang Y., Dai C., Qian Y., Fan X., Jiang J., Wu Y., Wu X., Huang Y., and Zhao M. Rheological properties and formation of dynamic filtration damage evaluation of a novel nanoparticle-enhanced VES fracturing system constructed with wormlike micelles. *Colloids and Surfaces A: Physicochemical and Engineering Aspects* 553, 244–252, Copyright (2018), with permission from Elsevier.

The NPs covered by surfactant can further interact with the WLMs. According to the theoretical studies [97], the interaction proceeds via the fusion of the semi-spherical end-caps of the WLMs with the surfactant aggregates adsorbed on the surface of the NP (Figure 1). The end-caps are mainly involved in the interaction because they represent the energetically unfavorable parts of the micelles where the head-groups are located rather far from each other, allowing penetration of some water molecules into the micelle. Direct evidence of the linking between the WLMs and the NPs was provided by cryo-transmission [96,98] and freeze-fracture transmission electron microscopy [99].

The NPs can increase the zero-shear viscosity η_0, relaxation time τ_{rel}, and the plateau modulus G_0 of VES fluids (Figure 2) [94,96,98–111]. The most pronounced is the effect of NPs on the viscosity, which can augment by up to 3 orders of magnitude [94,96,98–100,104–108,110,111]. It can be attributed to the hindered reptation of WLMs as their motion slows down in the vicinity of NPs. The largest enhancement of viscosity is observed for VES fluids in the vicinity of the overlap concentration C* of WLMs. In these conditions, the micelles are rather short, and therefore, in the system, there is a large quantity of end-caps. When the NPs are added, they interact with the micellar end-caps, thereby building a network. The increase in viscosity by 3 orders of magnitude, leading to the transition of a VES solution into an elastic gel, was observed for anionic WLMs of fatty acid methyl ester sodium sulfonate (MES) and 20–40 nm barium titanate NPs [99]. In semi-dilute solutions of WLMs the enhancement of the viscosity by added NPs usually does not exceed 1.5 orders of magnitude [94,104–107]. It was shown that for ionic VES the increase in viscosity by NPs is more pronounced at lower salt concentrations because the micelles become shorter and the quantity of end-caps in the system increases. For instance, for a 0.6 wt % solution of C22-tailed cationic surfactant erucyl bis-(hydroxyethyl) methylammonium chloride, the enhancement of viscosity by 5 wt % 300 nm magnetite particles is 12.3-fold and 7-fold in 1 and 1.5 wt % KCl, respectively [104].

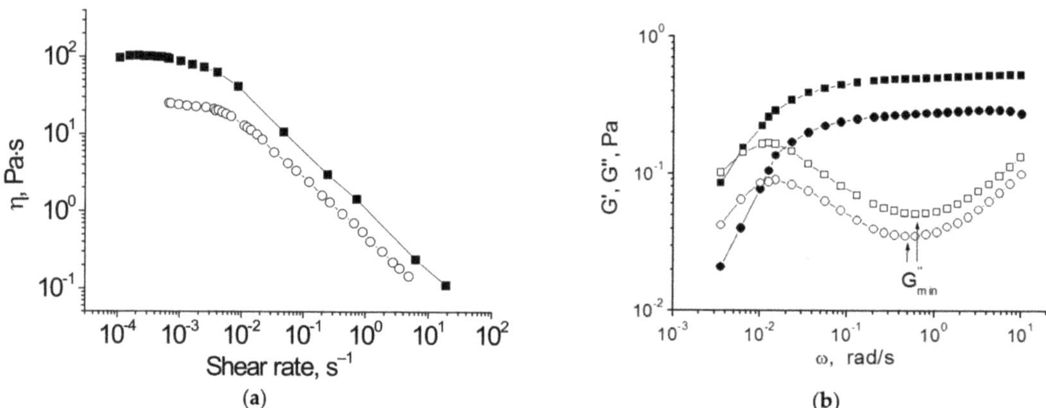

Figure 2. Viscosity as a function of shear rate (**a**) and frequency dependences of storage modulus G′ (filled symbols) and loss modulus G″ (open symbols) (**b**) for 0.6 wt % erucyl bis-(hydroxyethyl) methylammonium chloride solutions with 1.5 wt % 300 nm magnetite particles (squares) and without nanoparticles (circles) at 20 °C. Solvent: 1.5 wt % KCl in water, pH 11. Reprinted with permission from Pletneva V.A., Molchanov V.S., Philippova O.E. *Langmuir* 2015, *31*(1), 110–119. Copyright (2015) American Chemical Society.

In some papers [94,105,107,111], it was observed that upon the gradual increase in the amount of added NPs (at a constant concentration of surfactant) the zero-shear viscosity first increases, reaches a maximum, and then starts to decline. For instance, in VES fluid composed of a C22-tailed cationic surfactant N-erucamidopropyl-N,N-dimethyl-N-allylammonium bromide (EDAA) and negatively charged SiO_2 particles, the maximum value of viscosity was observed at as low an amount of NPs as 0.01 wt %, and the further increase in the NP concentration resulted in the decrease in viscosity, which became even lower than the viscosity of the initial VES solution [94]. In VES fluid consisting of a 3 wt % sodium oleoyl methyl taurate (SOMT), 6 wt % NaCl, and 22 nm SiO_2 particles, the maximum value of viscosity was observed at 0.9 wt % NPs [105]. The drop in viscosity was attributed to the increased electrostatic repulsion between the WLMs and the NPs covered by the same surfactant [94] or to interparticle aggregation leading to the disruption of the network structure [107]. One cannot exclude [93] the fact that the adsorption of the surfactant molecules on the surface of the added NPs can reduce the amount of surfactant involved in the formation of WLMs, which should also significantly contribute to the lowering of the viscosity. In these systems, the relaxation time τ_{rel} passes through the maximum simultaneously with viscosity [105]. The τ_{rel} value in the network of entangled WLMs is determined by the breaking time of the micelles τ_{br} and the reptation time τ_{rep} [29]. It was shown that NPs do not appreciably affect the breaking time of the micelles τ_{br} [105] or even decline it [104]. Therefore, the increase of τ_{rel} is caused by the increase of reptation time τ_{rep} provided by the hindered reptation of WLMs in the presence of NPs.

As to the plateau modulus G_0, it either monotonically increases with the increasing amount of added NPs [96] or first increases and then reaches a constant value [104,105]. The increase of G_0 was attributed to the cross-linking of WLMs by NPs adsorbing the micellar end-caps. When all the micellar end-caps available in the system are linked to NPs, G_0 reaches a constant value. Further addition of particles leads only to a redistribution of micellar end-caps between the NPs, which does not produce new elastically active sub-chains [104]. The increase of G_0 induced by NPs is usually rather small: less than two-fold [96,99,104,105]. This behavior can be explained as follows [96]. If before the addition of NPs each micelle had n elastically active sub-chains due to entanglements with other micelles, the binding of its two end-caps to NPs would give two more elastically active subchains, thereby providing a ca. $(n + 2)/n$-fold increase of the plateau modulus. The lower the number of initial entanglements between the WLMs, the higher the input of NPs in G_0 value. Thus, being taken in a proper amount, the NPs can increase the viscosity η_0, the relaxation time τ_{rel}, and, to a lesser extent, the plateau modulus G_0 of the VES fluids due to the attachment of the energetically unfavorable end-caps of the micellar chains.

NPs can also enhance temperature and shear resistance, which is quite important for fracturing fluids. These parameters can directly influence the fracture-making and sand-carrying capabilities of the fluid [112]. Wu and co-workers [94] examined the temperature and shear resistance of VES fluids containing 1 wt % cationic surfactant EDAA and 0.01 wt % negatively charged SiO_2. The experiments were performed at a shear rate of 170 s^{-1} and at 70 °C. It was shown that for 2 h the fluid with the NPs keeps a rather high shear viscosity of 33 mPa·s, whereas without NPs the shear viscosity was only 24 mPa·s. Note that the increase in viscosity in the presence of NPs was due to higher longest relaxation time τ_{rel} (0.15 s) exceeding that of the VES without NPs (0.054 s). The static particle settling test showed that the EDAA/silica system has a lower settlement rate (0.0021 cm/s) as compared to the pure EDAA solution (0.012 cm/s) and other fracturing fluids, such as guar gum (0.72 cm/s), thereby demonstrating a good proppant-carrying performance.

Zhang et al. [110] studied the effect of NPs on the shear restoration of VES fracturing fluid, demonstrating its ability to recover viscoelasticity after high-speed shearing during pumping. For the fluid composed of 30 mM of cetyltrimethylammonium bromide and 40 mM of sodium salicylate NaSal, it was shown that the addition of 0.1 wt % 140 nm silica NPs increases the shear recovery rate from 83.4% to 92.3%. Therefore, NPs favor faster reconstruction of micellar networks at low shear rates.

Huang and Crews found that NPs added to VES fracturing fluid produce a pseudo-filter cake due to the aggregation with the micelles, thereby significantly reducing the rate of fluid loss and improving fluid efficiency [102]. At the same time, the same fluid without NPs was non-wall-building and had very high fluid leak-off over time [102]. Similar filter cake formation in the presence of NPs was observed by Zhang et al. [110]. It is important to note that the filter cake thus formed can be easily and completely removed at the flow of oil, leaving very little residue or production damage [102].

For enhancement of the performance of VES fluids, different types of NPs were used, including silica SiO_2 [94,96,100,105,110,111], barium titanate $BaTiO_3$ [99], magnetite Fe_3O_4 [104,109,113], MnO [114], ZnO [103,114], TiO_2 [115], and carbon nanotubes [108]. It was shown that among these NPs, long nanotubes induce a smaller increase in viscosity than spherical NPs [108]. This can be due to the sliding of the WLM-nanotube junctions along the nanotube, thus releasing the stress and reducing the viscosity.

In all the discussed papers above, the NPs were added to the WLMs of the surfactants At the same time, the addition of NPs to the VES systems with lamellar structure was also examined [87]. For instance, Baruah et al. studied VES-based fracturing fluids composed of zwitterionic (cocamidopropyl betaine) and anionic (sodium oleate) surfactant mixtures, iso-amyl alcohol as a cosurfactant, clove oil, and water. It was shown that the addition of 0.1% SiO_2 and 0.1% NaOH leads to the increase in apparent viscosity by 5–45%, depending on the temperature and surfactants ratio. The authors explained this behavior by the penetration of the SiO_2 nanoparticles within the lamellae structure of the VES fluids, which helps the development of a 3D network and enhances the interconnections between the mixed surfactant monomers constituting the lamellar liquid crystal phase. The NPs act as a swelling agent that increases the thickness of the water layers in the lamellar structure, making it more stable; in its turn, the lamellar structure stabilizes the dispersion of the NPs. It is interesting that the system was highly pressure-sensitive. At atmospheric pressure, the fluid was unable to maintain an apparent viscosity higher than 90 mPa·s for more than 3 min. However, under higher pressures (2068, 4137, and 6205 kPa), it kept an apparent viscosity higher than 600 mPa·s for 120 min at 103 °C and 100 s^{-1} (this shear rate is associated with the shear which the fluid should experience during pumping in the well), which is suitable for conducing the fracturing job and to suspend proppants.

Thus, the addition of NPs seems to be an effective way to improve various properties of VES-based fracturing fluids.

7. Hybrid Polymer–VES Fluids

7.1. Interaction of Polymers with VES

One of the new directions in the research on hydraulic fracturing fluids is aimed at combining the advantages of polymer- and surfactant-based fluids, which may be achieved by creating mixed (or "hybrid") gels of polymer chains and WLMs. A key factor determining the structure and properties of such fluids is the interaction of the polymer and surfactant. The tuning of this interaction allows the obtaining of homogeneous mixtures and the preserving of the structure of micellar chains.

The addition of polymers may either destroy or preserve WLMs, depending on the hydrophobicity of the polymer. For instance, weakly hydrophobic but water-soluble polymers, such as poly (vinyl methyl ether) or poly (propylene oxide), break WLMs [116–118]. This happens due to the "wrapping" of the polymer chains around the micelle, which is favorable because it reduces the contact of both the micellar and the polymer hydrophobic parts with water. As a result, the WLMs are transformed into spherical [119], ellipsoidal [116], or disc-like aggregates [120].

Some polymer molecules containing hydrophobic groups along the backbone, such as partially sulfonated polystyrene [121], poly(methyl methacrylate-co-sodium styrene sulfonate) copolymer [122], or sodium poly(p-vinylbezoate) [123], can embed into WLMs without their disruption, forming stronger "hybrid" micelles armed by polymer: a more hydrophobic polymer backbone is solubilized in the micellar core, while charged polymer units reside closer to the oppositely charged surfactant head groups, screening electrostatic interactions at the micellar surface. The formation of "hybrid" polymer-surfactant cylindrical micelles was also observed for water-insoluble polymer poly(4-vinylpyridine) [124–126].

However, the most practically important cases comprise water-soluble polymers which do not break WLMs. Two types of such systems have been reported: hydrophobically modified (HM) polymers or fully hydrophilic polymers which do not interact with the micelles.

7.2. HM-Polymers/VES

HM-polymers usually have a main hydrophilic water-soluble backbone and some amount of grafted hydrophobic water-insoluble groups. Such polymers mainly interact with WLMs via their hydrophobic moieties, which can embed into the micelles without their disruption. In this case, common networks of HM-polymer and WLMs can be formed (Figure 3a), and polymer hydrophobic groups serve as "physical" cross-links between the macromolecules and the WLMs. This results in the increase in viscosity and the synergistic enhancement of the viscoelastic properties and temperature stability.

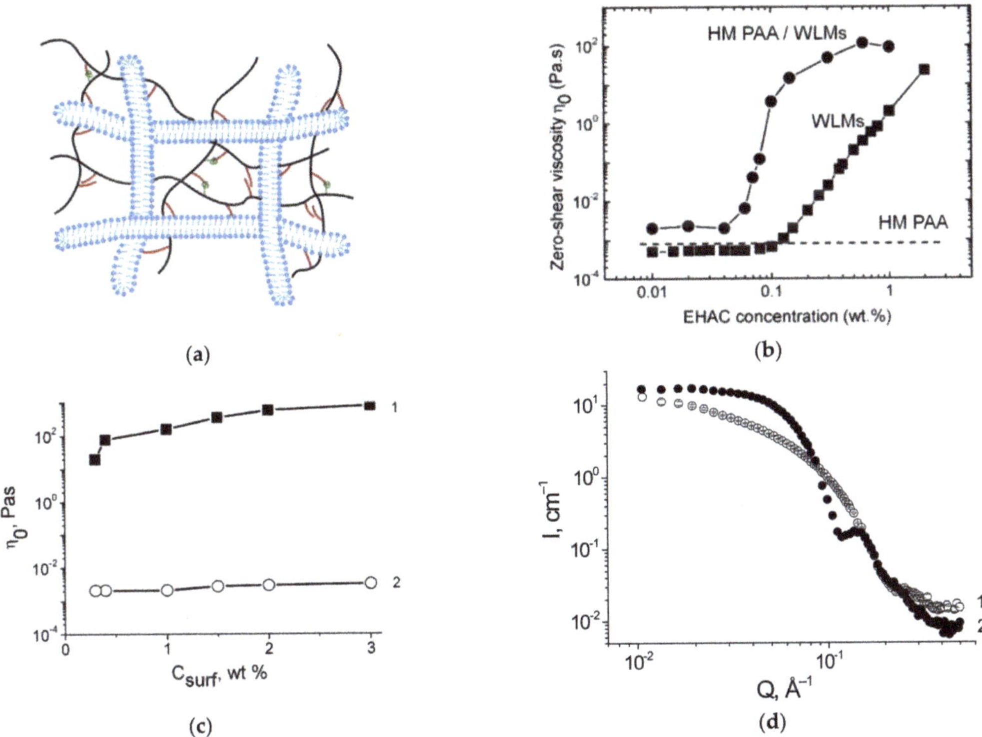

Figure 3. (**a**) Schematic representation of a common network of wormlike surfactant micelles (WLMs) and hydrophobically (HM) modified polymer; Reproduced from Pu, W.; Du, D.; and Liu, R. Preparation and evaluation of supramolecular fracturing fluid of hydrophobically associative polymer and viscoelastic surfactant. *J. Petrol. Sci. Eng. 167*, 568–576, Copyright (2018), with permission from Elsevier; (**b**) zero-shear viscosity as a function of cationic surfactant (EHAC) concentration for the mixtures of WLMs and 0.5 wt % HM PAAm and components (HM PAAm and WLMs) taken separately; solvent: 3 wt % KCl in water, temperature: 60 °C; reprinted and adapted with permission from Shashkina, J.A.; Philippova, O.E.; Zaroslov, Y.D.; Khokhlov, A.R.; Pryakhina, T.A.; and Blagodatskikh, I. Rheology of viscoelastic solutions of cationic surfactant. Effect of added associating polymer. *Langmuir 21*(4), 1524–1530, Copyright (2005) American Chemical Society; (**c**,**d**) zero-shear viscosity as a function of anionic surfactant (potassium oleate) concentration (**c**) and small-angle neutron scattering curves (**d**) for the mixtures of WLMs and 0.5 wt % HM PAA before (1) and after (2) addition of n-dodecane; solvent: 6 wt % KCl in water, temperature: 20 °C; reprinted with permission from Molchanov, V.S.; Philippova, O.E.; Khokhlov, A.R.; Kovalev, Y.A.; and Kuklin, A.I. Self-assembled networks highly responsive to hydrocarbons. *Langmuir 23*(1), 105–111, Copyright (2007) American Chemical Society.

7.2.1. HM PAAm/Surfactants

In most of the published works, synthetic polymers, such as HM PAAm [127–131], hydrophobically modified polyacrylic acid (HM PAA) [132–135], their copolymers [136–138], copolymers of PAAm, and other monomers [139–141] are used. Such polymers are usually synthesized by radical co-polymerization of a hydrophilic (acrylamide/acrylic acid) and hydrophobic (e.g., alkylacrylates) monomers in a common solvent [142]. Otherwise, micellar co-polymerization is employed [143].

In a pioneering work [128], HM PAAm with *n*-dodecyl side chains was mixed with C22-tailed cationic surfactant EHAC, and a huge synergistic increase in viscosity by 4 orders of magnitude was observed, and a rise of viscosity began at lower surfactant concentrations than in the absence of polymer (Figure 3b). This behavior is observed at the rather low HM PAAm concentration of 0.5 wt %, close to C*, at which polymer solutions without surfactant have low viscosities close to water. Such a strong synergistic effect was explained by the formation of a common polymer-micellar network with hydrophobic polymer groups serving as junctions to WLMs.

It is shown that the presence of hydrophobic groups is important for the phase compatibility of PAAm and WLMs [128], and the one-phase region widens with an increase in the length and number of grafted alkyl chains. It confirms that polymer hydrophobic groups are responsible for the interaction with WLMs. The distribution of hydrophobic groups along the polymer chain affects the viscoelastic properties of the mixtures [128]. at a fixed total content of hydrophobic groups, statistical HM PAAm induces a stronger increase in the rheological parameters (viscosity, elastic modulus, and relaxation time) than HM PAAm with a higher degree of blockiness, which is due to the formation of more cross-links between the statistical polymer and WLMs. It should be noted that the characteristic lifetime of such cross-links is mainly determined by the WLM breaking time and not by the time when the polymer hydrophobic group resides inside the micelle.

Thus, the main practically important effect of mixing HM PAAm/HM PAA and WLMs consists in the synergistic increase in the viscoelastic properties. Many works showing a similar effect for the mixtures of HM PAAm or its copolymers and anionic [136,137], zwitterionic [138,140,141], zwitterionic and anionic [139], gemini cationic [144], and trimeric zwitterionic [145] surfactants have been published.

SANS data show that the WLM local structure is preserved in the presence of the HM polymer: the SANS curve of mixed WLM/HM PAAm systems is described by a cylinder model with a radius equal to the radius of WLMs in the absence of HM PAAm [136]. An increase in viscoelasticity was observed both for the C8-C12 polymer hydrophobic groups and for a specific case of HM PAAm modified by ultra-long C22 alkyl groups [135].

The second important effect of mixing HM PAAm and WLMs is the increased temperature stability [128,144]. Upon heating, the viscosity of the polymer/VES mixtures decreases to a lesser extent than that of a VES solution in the absence of polymer. This is explained by the fact that, in contrast to the micellar chains, macromolecules do not break upon heating and contribute to the heating resistance of the fluids.

The third advantage of HM PAAm/WLM fluids is their high responsiveness to hydrocarbons [128,136,144]. Upon contact with aliphatic oils, the fluids completely lose viscoelastic properties, and their viscosity drops by several orders of magnitude, reaching the viscosity of water (Figure 3c) [128,136]. The responsiveness of mixed polymer/WLM fluids is due to the hydrocarbon-induced breaking of the WLM sub-chains into spherical microemulsion droplets (which is proven by SANS, Figure 3d), leading to the disruption of the common network. This effect is analogous to the case of WLMs without polymers. In this regard, mixed polymer/WLM solutions are clean fracturing fluids which do not require additional breakers for their removal from the fracture.

An interesting particular case reported in the literature consists in the replacement of the polymer alkyl hydrophobic moieties by different groups. In the work [145], β-cyclodextrin-functionalized HM PAAm was mixed with a trimeric zwitterionic surfactant (DTPAN), and a pronounced increase in viscosity was observed as compared to the components (polymer or VES) taken separately. The fluid was characterized by a favorable temperature tolerance and shearing resistance and an affordable proppant-carrying capacity and was completely broken by ammonium persulfate.

CO_2-responsive common networks of HM PAAm and WLMs of sodium dodecyl sulfate (SDS)–N,N,N′,N′-tetramethyl-1,3-propanediamine (TMPDA) were reported [146]. The responsiveness of the common networks is explained by the transition of the spherical to the wormlike micelles of SDS–TMPDA mixtures, which is accompanied by a 400-fold increase in viscosity. Sol-gel transition is reversible for at least five cycles of bubbling and removing of CO_2.

Therefore, HM PAAm/VES fluids combine the advantages of both components: increased viscoelasticity and temperature stability due to the polymer component, and responsiveness to hydrocarbons inherent in surfactant WLMs.

7.2.2. Other Synthetic HM Polymers/Surfactants

The formation of the common networks with surfactant WLMs was also observed for other synthetic HM polymers. For instance, water-soluble telechelic poly (ethylene glycol) polymers end-modified by hydrophobic fragments have been reported to form common networks with cetyltrimethylammonium toluene sulfonate [147] or cetylpyridinium chloride in the presence of sodium salicylate [148–150]. It is shown that the hydrophobic end groups of the polymer embed into WLMs, while its main hydrophilic part resides in water and links two micelles together. As a result, the mixtures show viscoelastic behavior which is not observed for the components taken separately [147].

7.2.3. Natural HM Polymers/Surfactants

The common networks of WLMs and natural HM polymers are much less described than the synthetic macromolecules, though natural polymers are preferable for hydraulic fracturing applications due to their better biodegradability, lower environmental impact and lack of need for polymerization processes.

Hydroxypropyl guar, modified by long C22 alkyl groups (HM HPG), was reported to form common viscoelastic networks with cationic surfactant EHAC in the presence of KCl [151], which resulted in a significant synergistic effect in a wide range of concentrations and at elevated temperatures of up to 60 °C. It is important to note that a rather low degree of hydrophobic substitution (10 hydrophobic groups per macromolecule) was enough to achieve a pronounced synergy in rheological properties.

A similar synergistic effect on viscoelastic properties was observed upon the bridging of cationic cetyltrimethylammonium tosylate WLMs with HM chitosan with C12 alkyl side chains [152], or of cetyltrimethylammonium *p*-toluenesulfonate WLMs with HM hydroxyethyl cellulose [153], and, in the latter case, the HM polymer had a much stronger effect on the solution viscosity than a similar unmodified hydroxyethyl cellulose, proving the incorporation of hydrophobic side groups of macromolecules into WLMs.

An absolute majority of WLM/HM polymer (either synthetic or natural) common networks employed alkyl hydrophobic groups attached to polymer. An interesting variation of the hydrophobic modification approach was proposed [154], in which cholesterol-modified gellan gum was used to strengthen a network of cylindrical micelles formed by a non-ionic surfactant polyoxyethylene cholesteryl ester ($ChEO_{10}$), with the hydrophobic tail of the same chemical structure as the polymer hydrophobic groups. At the same time, cholesterol-modified short-chained dextran was shown to break longer WLMs of $ChEO_{10}$ into shorter ones, leading to a more liquid-like behavior [155].

Therefore, the enhancement of the viscoelastic properties of the WLM networks by HM-polymers is a general effect observed for both the synthetic and the natural polymers bearing alkyl groups of different length (C8–C22) or hydrophobic groups of other chemical natures. This effect is due to the bridging of WLMs by polymer chains and the formation of a common polymer/micellar network. The practical application of this approach may be greatly advanced by the easier and cheaper synthesis methods of HM-polymers.

7.2.4. HM Polymers/Surfactants/Nanoparticles

A ternary viscoelastic fracturing fluid containing HM HPG with C14 alkyl side chains cationic gemini surfactant 12:0-3-OH-12:0 and 7–40 nm hydrophobic silica NPs with improved characteristics was described [156]. The viscosity of the fluid was increased by the addition of NPs from 37 mPa·s to 154 mPa·s at 80 °C and 170 s^{-1}; the presence of the surfactant and polymer improved the interfacial properties (oil–water interfacial tension, hydrophilicity of the model core slices aged in the fluid filtrate); the fluid application showed a good oil permeability loss rate (9.4%) and a fracture conductivity retainment rate (95%); and a significant increase in oil production was obtained in the oilfield test. This new approach combines the effect of NPs, which induce elongation of the WLMs and the synergistic effect of HM polymer and the WLMs and shows that a combined synergy may be achieved by using a multi-component fluid.

7.3. Hydrophilic Polymers/VES

A recently described novel approach for the development of viscoelastic polymer/surfactant fluids consists in the use of fully hydrophilic polymers which do not interact with WLMs and do not destroy them.

One of the first reported cases is the mixing of potassium oleate WLMs with a similarly charged polyelectrolyte-sodium polystyrenesulfonate (PSS) [37]. However, in this work, no increase in viscosity or elasticity caused by the polymer was reported. The WLMs of a cationic surfactant EHAC are preserved in the presence of an oppositely charged natural polyelectrolyte–xanthan [157]. One-phase homogeneous mixtures are obtained when a significant amount (4–4.75 wt %) of low-molecular-weight salt KCl is added, which screens the electrostatic attraction between the components and prevents the precipitation of the polymer/surfactant complex. A synergistic increase of the plateau storage modulus is seen for the mixture, which is explained by the formation of a common semi-interpenetrating network of polymer and micellar chains with entanglements between the components (Figure 4a).

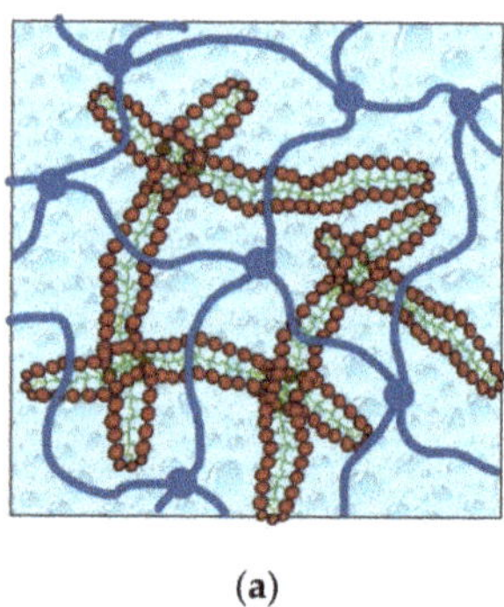

(a)

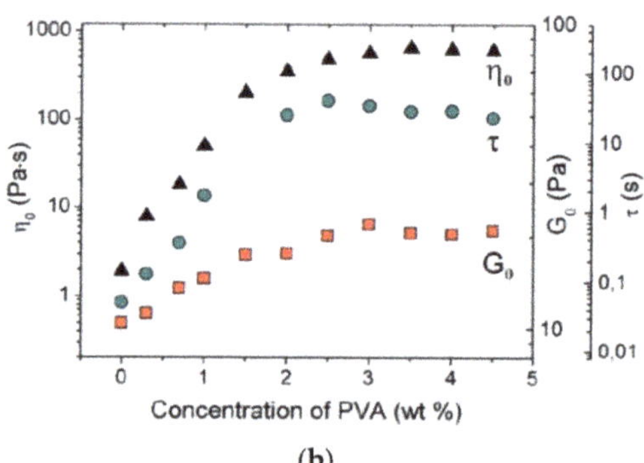

(b)

Figure 4. (**a**) Schematic representation of a common network of wormlike surfactant micelles (WLMs) and hydrophilic polymer, which may be un-cross-linked or cross-linked; (**b**) zero-shear viscosity η_0 (triangles), terminal relaxation time τ (circles), and plateau modulus G_0 (squares) as a function of concentration of added poly(vinyl alcohol) PVA in 3.3 wt % potassium oleate/C8TAB aqueous solutions at 20 °C, molar ratio [potassium oleate]/[C8TAB] = 2.5; reprinted with permission from Shibaev, A.V.; Abrashitova, K.A.; Kuklin, A.I.; Orekhov, A.S.; Vasiliev, A.L.; Iliopoulos, I.; and Philippova, O.E. Viscoelastic synergy and microstructure formation in aqueous mixtures of nonionic hydrophilic polymer and charged wormlike surfactant micelles. *Macromolecules 50*(1), 339–348, Copyright (2017) American Chemical Society.

Synergy in viscoelastic properties has been observed for the mixtures of a synthetic polymer poly(vinyl alcohol) and mixed anionic/cationic WLMs [158,159]. A nearly 3-magnitude increase in viscosity is observed, which is accompanied by a significant enhancement of the elastic modulus and relaxation time (Figure 4b) and is explained by two factors: (1) the formation of entanglements between the polymer and micellar chains, (2) the microphase separation with the formation of polymer-rich and micellar-rich microdomains, which arises from weak repulsive interaction between the polymer and micelles. Microphase separation results in the concentration of both components in their domains, leading to the augmentation of the rheological parameters. Later, the same effect was observed for a natural polymer HPG mixed with anionic/cationic WLMs [160]. The use of a weakly repulsive polymer and WLMs, which form one-phase but microscopically segregated solutions, allows the significant reduction in the concentrations of the components in the fluid in order to obtain viscosity and elasticity sufficient for fracturing operations.

A very recently reported novel approach consists in using the hydrophilic polymer/WLM system and cross-linking the hydrophilic polymer chains into their own network (for instance, by borate ions), which leads to the formation of a double dynamic network, consisting of an entangled network of WLMs interpenetrating with a polymer network with labile cross-links [161]. Such polymer/WLM systems are characterized by significantly enhanced rheological properties as compared to their components taken separately—their zero-shear viscosity and plateau storage modulus are increased by factors of 3400 and 27, respectively. Due to the dynamic nature of the bonds in both networks, they recover mechanical properties after strong shear and may be responsive to different factors.

It was proposed to mix the WLM network and cellulose nanofibrils (CNF). Sodium oleate/KCl/CNF fluid was shown to have higher zero-shear viscosity than a simple oleate/KCl fluid, and was able to form a dense filter cake, which reduced filtration of the fluid into the core [162]. A pseudo-interpenetrating network of sodium oleate/KCl WLMs and a CNF surface modified by carboxyl, sulfonic, or hydrophobic groups was prepared and was characterized by higher viscosity than for pure WLMs (at moderate CNF content which does not break the micellar network) and by improved sand-carrying capacity [163].

Thus, the interaction of polymers with WLMs may result in the disruption of the micellar chains, the obtaining of hybrid cylindrical micelles with embedded polymer, or the formation of the common polymer-micellar networks. Among these various structures, common networks of HM polymers and WLMs show the best rheological characteristics (high viscosity and elastic modulus exceeding the values for individual components by several orders of magnitude) and temperature stability combined with a pronounced responsiveness to hydrocarbons. The best properties of mixed fluids are observed when HM polymer contains rather long (C12-C22) alkyl tails or other bulky hydrophobic groups (for instance, β-cyclodextrin, or cholesterol) providing a rather strong hydrophobic interaction with WLMs. The synergistic effect of mixing HM polymers and surfactants has already been observed at a rather low amount of hydrophobic substituents in the polymer chain, which may be only a few per macromolecule. However, the application of HM polymer/VES fluids in hydraulic fracturing is limited by a rather high cost and the limited availability of HM polymers. An alternative approach may consist in the use of fully hydrophilic polymers weakly interacting with the micelles.

8. Conclusions and Outlook

In this review, several new approaches are described for improving the properties of VES-based fluids, aiming at their application as a cost-effective alternative to polymer-based hydraulic fracturing fluids.

Gemini and oligomeric surfactants show enhanced rheological properties and temperature resistance compared to their single-chain counterparts. Promising ways to obtain better performance of such fluids consist in the use of longer alkyl tails and higher oligomer numbers ($n \geq 4$), providing stronger hydrophobic interactions within WLMs. Other possibilities include the variation of the polar head chemical substituents, or the use of dissymmetric surfactants, combining alkyl tails of different length within one molecule. A significant advancement in this field would consist in the development of easier synthetic methods, leading to a reduction in the oligomeric VES cost.

A rather simple way for optimizing the properties of VES fluids is the use of surfactant mixtures. One of the prospective routes consists in mixing gemini or oligomeric surfactants with single-chain surfactants or together, which allows the reduction in the concentration of a more expensive oligomeric component while exploiting the synergy between different surfactant species. New interesting solutions may include obtaining "pseudo-gemini" surfactants by self-assembly of multiple cheaper components.

Addition of NPs is an effective and rather inexpensive way for improving the performance of VES fracturing fluids, including the rheological properties (viscosity, plateau shear modulus, and relaxation time), the temperature and shear resistance, and the proppant carrying capacity. An attractive pathway lies in combining the addition of NPs with oligomeric surfactants, polymer/surfactant mixtures, etc.

The development of hybrid polymer/VES fluids is a widely and fundamentally investigated, but practically unexploited, field. Such fluids comprise mixtures of VES with HM or hydrophilic polymers. Systems comprising synthetic HM PAAm and VES are the most widely studied in the literature; at the same time, mixtures of HM natural polymers with VES are much less investigated. New developments in this field may be related to the cheaper and easier ways of the synthesis of HM polymers, including the use of hydrophobic groups other than alkyl chains, as well as the search for high-performance mixtures of VES with hydrophilic polymers.

Novel approaches in the studies of VES-based fluids offer promising ways for the improvement and optimization of their properties relevant to oil field applications and for the widening of their practical use in hydraulic fracturing technology.

Author Contributions: Conceptualization, A.V.S. and O.E.P.; writing—original draft preparation, A.V.S. and O.E.P.; writing—review and editing, A.V.S., A.A.O., and O.E.P.; supervision, O.E.P.; funding acquisition, O.E.P. All authors have read and agreed to the published version of the manuscript

Funding: This work is financially supported by the Russian Science Foundation (grant number 21-73-30013).

Institutional Review Board Statement: Not applicable.

Informed Consent Statement: Not applicable.

Conflicts of Interest: The authors declare no conflict of interest.

Abbreviations

Abbreviation	Expansion
CMC	critical micelle cocentration
CMHPG	carboxymethyl hydroxypropyl guar
CNF	cellulose nanofibrils
CTAB	cetyltrimethylammonium bromide
EDAA	N-erucamidopropyl-N,N-dimethyl-N-allylammonium bromide
EHAC	erucyl bis-(hydroxyethyl)methylammonium chloride
EOR	enhanced oil recovery
HM PAA	hydrophobically modified polyacrylic acid
HM PAAm	hydrophobically modified polyacrylamide
HM HPG	hydrophobically modified hydroxypropyl guar
HPGNaSal	hydroxypropyl guarsodium salicylate
NP	nanoparticle
PSS	sodium polystyrenesulfonate
SDS	sodium dodecyl sulfate
SOMT	sodium oleoyl methyl taurate
TMPDA	N,N,N′,N′-tetramethyl-1,3-propanediamine
VES	viscoelastic surfactant
WLM	wormlike micelle

References

1. Alvarado, V.; Manrique, E. Enhanced oil recovery: An update review. *Energies* **2010**, *3*, 1529–1575. [CrossRef]
2. Economides, M.J.; Nolte, K.G. *Reservoir Stimulation*, 3rd ed.; Wiley: New York, NY, USA; Chinchester, UK, 2000.
3. Osiptsov, A.A. Fluid mechanics of hydraulic fracturing: A review. *J. Pet. Sci. Eng.* **2017**, *156*, 513–535. [CrossRef]
4. Kalam, S.; Afagwu, C.; Al Jaberi, J.; Siddig, O.M.; Tariq, Z.; Mahmoud, M.; Abdulraheem, A. A review on non-aqueous fracturing techniques in unconventional reservoirs. *J. Nat. Gas Sci. Eng.* **2021**, *95*, 104223. [CrossRef]
5. Fink, J. *Petroleum Engineer's Guide to Oil Field Chemicals and Fluids*; Elsevier: Amsterdam, The Netherlands, 2012.
6. Barati, R.; Liang, J.-T. A review of fracturing fluid systems used for hydraulic fracturing of oil and gas wells. *J. Appl. Polym. Sci.* **2014**, *131*, 40735. [CrossRef]
7. Montgomery, C. Fracturing Fluids. In *Effective and Sustainable Hydraulic Fracturing*; InTech Open: Rijeka, Croatia, 2013; Chapter 1.
8. Al-Muntasheri, G.A. A critical review of hydraulic-fracturing fluids for moderate- to ultralow-permeability formations over the last decade. *SPE Prod. Oper.* **2014**, *29*, 243–260. [CrossRef]
9. Cao, X.; Shi, Y.; Li, W.; Zeng, P.; Zheng, Z.; Feng, Y.; Yin, H. Comparative studies on hydraulic fracturing fluids for high-temperature and high-salt oil reservoirs: Synthetic polymer versus guar gum. *ACS Omega* **2021**, *6*, 25421–25429. [CrossRef]
10. Xia, S.; Zhang, L.; Davletshin, A.; Li, Z.; You, J.; Tan, S. Application of polysaccharide biopolymer in petroleum recovery. *Polymers* **2020**, *12*, 1860. [CrossRef]
11. Harris, P.C. Chemistry and rheology of borate-crosslinked fluids at temperatures to 300F. *J. Pet. Technol.* **1993**, *45*, 264–269. [CrossRef]
12. Holtsclaw, J.; Funkhouser, G.P. A Crosslinkable synthetic-polymer system for high-temperature hydraulic-fracturing applications. *SPE Drill. Complet.* **2010**, *25*, 555–563. [CrossRef]
13. Cadix, A.; Wilson, J.; Carouhy, T.; Harrisson, S.; Guichon, H. A new class of associative polymer for hydraulic fracturing applications. In Proceedings of the SPE European Formation Damage Conference and Exhibition, Budapest, Hungary, 3 June 2015. [CrossRef]
14. Wei, B.; Romero-Zerón, L.; Rodrigue, D. Mechanical properties and flow behavior of polymers for enhanced oil recovery. *J. Macromol. Sci. Part B* **2014**, *53*, 625–644. [CrossRef]
15. Al-Muntasheri, G.A.; Li, L.; Liang, F.; Gomaa, A.M. Concepts in cleanup of fracturing fluids used in conventional reservoirs: A literature review. *SPE Prod. Oper.* **2018**, *33*, 196–213. [CrossRef]
16. Li, Q.; Xing, H.; Liu, J.; Liu, X. A review on hydraulic fracturing of unconventional reservoir. *Petroleum* **2015**, *1*, 8–15. [CrossRef]
17. Kang, W.; Mushi, S.J.; Yang, H.; Wang, P.; Hou, X. Development of smart viscoelastic surfactants and its applications in fracturing fluid: A review. *J. Pet. Sci. Eng.* **2020**, *190*, 107107. [CrossRef]
18. Pal, N.; Verma, A. Applications of surfactants as fracturing fluids: Chemical design, practice, and future prospects in oilfield stimulation operations. In *Surfactants in Upstream E & P*; Solling, T., Kamal, M.S., Hussain, S.M.S., Eds.; Springer International Publishing: Cham, Switzerland, 2021; pp. 331–355.
19. Sullivan, P.; Nelson, E.B.; Anderson, V.; Hudges, T. *Giant Micelles: Properties and Applications*; CRC Press: Boca Raton, FL, USA, 2002; pp. 453–472.
20. Samuel, M.M.; Card, R.J.; Nelson, E.B.; Brown, J.E.; Vinod, P.S.; Temple, H.L.; Qu, Q.; Fu, D.K. Polymer-free fluid for fracturing applications. *SPE Drill. Complet.* **1999**, *14*, 240–246. [CrossRef]
21. Chase, B.; Chmilowski, W.; Marcinew, R.; Mitchell, C.; Dang, Y.; Krauss, D.; Nelson, E.; Lantz, T.; Parham, C.; Plummer, J. Clear fracturing fluids for increased well productivity. *Oilfield Rev.* **1997**, *9*, 20–33.
22. Samuel, M.; Polson, D.; Graham, D.; Kordziel, W.; Waite, T.; Waters, G.; Vinod, P.; Fu, D.; Downey, R. Viscoelastic surfactant fracturing fluids: Applications in low permeability reservoirs. In Proceedings of the SPE Rocky Mountain Regional/Low-Permeability Reservoirs Symposium and Exhibition, Denver, CO, USA, 12–15 March 2000. [CrossRef]
23. Yang, X.; Mao, J.; Chen, Z.; Chen, Y.; Zhao, J. Clean fracturing fluids for tight reservoirs: Opportunities with viscoelastic surfactant. *Energy Sources Part A Recover. Util. Environ. Eff.* **2018**, *41*, 1446–1459. [CrossRef]
24. Hull, K.; Sayed, M.; Al-Muntasheri, G.A. Recent advances in viscoelastic surfactants for improved production from hydrocarbon reservoirs. *SPE J.* **2016**, *21*, 1340–1357. [CrossRef]
25. Dreiss, C.A. Wormlike micelles: Where do we stand? Recent developments, linear rheology and scattering techniques. *Soft Matter* **2007**, *3*, 956–970. [CrossRef]
26. Mitchell, D.J.; Ninham, B.W. Micelles, vesicles and microemulsions. *J. Chem. Soc. Faraday Trans. 2* **1981**, *77*, 601–629. [CrossRef]
27. Magid, L.J. The surfactant–polyelectrolyte analogy. *J. Phys. Chem. B* **1998**, *102*, 4064–4074. [CrossRef]
28. Rehage, H.; Hoffmann, H. Viscoelastic surfactant solutions: Model systems for rheological research. *Mol. Phys.* **1991**, *74*, 933–973. [CrossRef]
29. Wang, J.; Feng, Y.; Agrawal, N.R.; Raghavan, S.R. Wormlike micelles versus water-soluble polymers as rheology-modifiers: Similarities and differences. *Phys. Chem. Chem. Phys.* **2017**, *19*, 24458–24466. [CrossRef] [PubMed]
30. Raghavan, S.R.; Kaler, E.W. Highly viscoelastic wormlike micellar solutions formed by cationic surfactants with long unsaturated tails. *Langmuir* **2001**, *17*, 300–306. [CrossRef]
31. Croce, V.; Cosgrove, T.; Maitland, G.; Hughes, T.; Karlsson, G. Rheology, cryogenic transmission electron spectroscopy, and small-angle neutron scattering of highly viscoelastic wormlike micellar solutions. *Langmuir* **2003**, *19*, 8536–8541. [CrossRef]

32. Croce, V.; Cosgrove, T.; Dreiss, C.A.; Maitland, G.; Hughes, T. Mixed spherical and wormlike micelles: A contrast-matching study by small-angle neutron scattering. *Langmuir* **2004**, *20*, 9978–9982. [CrossRef] [PubMed]
33. Raghavan, S.R.; Fritz, A.G.; Kaler, E.W. Wormlike micelles formed by synergistic self-assembly in mixtures of anionic and cationic surfactants. *Langmuir* **2002**, *18*, 3797–3803. [CrossRef]
34. Ali, A.A.; Makhloufi, R. Effect of organic salts on micellar growth and structure studied by rheology. *Colloid Polym. Sci.* **1999**, *277*, 270–275. [CrossRef]
35. Hartmann, V.; Cressely, R. Linear and non linear rheology of a wormlike micellar system in presence of sodium tosylate. *Rheol. Acta* **1998**, *37*, 115–121. [CrossRef]
36. Hassan, P.A.; Raghavan, A.S.R.; Kaler, E.W. Microstructural changes in SDS micelles induced by hydrotropic salt. *Langmuir* **2002**, *18*, 2543–2548. [CrossRef]
37. Flood, C.; Dreiss, C.A.; Croce, V.; Cosgrove, T.; Karlsson, G. Wormlike micelles mediated by polyelectrolyte. *Langmuir* **2005**, *21*, 7646–7652. [CrossRef]
38. Kumar, R.; Kalur, G.C.; Ziserman, L.; Danino, D.; Raghavan, S. Wormlike micelles of a C22-tailed zwitterionic betaine surfactant: From viscoelastic solutions to elastic gels. *Langmuir* **2007**, *23*, 12849–12856. [CrossRef]
39. Chu, Z.; Feng, Y.; Su, X.; Han, Y. Wormlike micelles and solution properties of a C22-tailed amidosulfobetaine surfactant. *Langmuir* **2010**, *26*, 7783–7791. [CrossRef]
40. Bai, Y.; Liu, S.; Liang, G.; Liu, Y.; Chen, Y.; Bao, Y.; Shen, Y. Wormlike micelles properties and oil displacement efficiency of a salt-tolerant C22-tailed amidosulfobetaine surfactant. *Energy Explor. Exploit.* **2021**, *39*, 1057–1075. [CrossRef]
41. Imanishi, K.; Einaga, Y. Wormlike micelles of polyoxyethylene alkyl ether mixtures C10E5 + C14E5 and C14E5 + C14E7: Hydrophobic and hydrophilic chain length dependence of the micellar characteristics. *J. Phys. Chem. B* **2007**, *111*, 62–73. [CrossRef]
42. Wang, J.; Zhang, Y.; Chu, Z.; Feng, Y. Wormlike micelles formed by ultra-long-chain nonionic surfactant. *Colloid Polym. Sci.* **2021**, *299*, 1295–1304. [CrossRef]
43. Sato, T.; Acharya, D.P.; Kaneko, M.; Aramaki, K.; Singh, Y.; Ishitobi, M.; Kunieda, H. Oil-induced structural change of wormlike micelles in sugar surfactant systems. *J. Dispers. Sci. Technol.* **2006**, *27*, 611–616. [CrossRef]
44. Acharya, D.P.; Kunieda, H. Formation of viscoelastic wormlike micellar solutions in mixed nonionic surfactant systems. *J. Phys. Chem. B* **2003**, *107*, 10168–10175. [CrossRef]
45. Cates, M.E.; Candau, S.J. Statics and dynamics of worm-like surfactant micelles. *J. Phys. Condens. Matter* **1990**, *2*, 6869–6892. [CrossRef]
46. McCoy, T.; King, J.P.; Moore, J.E.; Kelleppan, V.T.; Sokolova, A.; de Campo, L.; Manohar, M.; Darwish, T.; Tabor, R.F. The effects of small molecule organic additives on the self-assembly and rheology of betaine wormlike micellar fluids. *J. Colloid Interface Sci.* **2019**, *534*, 518–532. [CrossRef]
47. Fieber, W.; Scheklaukov, A.; Kunz, W.; Pleines, M.; Benczédi, D.; Zemb, T. Towards a general understanding of the effects of hydrophobic additives on the viscosity of surfactant solutions. *J. Mol. Liq.* **2021**, *329*, 115523. [CrossRef]
48. Shibaev, A.V.; Tamm, M.V.; Molchanov, V.S.; Rogachev, A.V.; Kuklin, A.I.; Dormidontova, E.E.; Philippova, O.E. How a viscoelastic solution of wormlike micelles transforms into a microemulsion upon absorption of hydrocarbon: New insight. *Langmuir* **2014**, *30*, 3705–3714. [CrossRef]
49. Shibaev, A.V.; Aleshina, A.L.; Arkharova, N.A.; Orekhov, A.S.; Kuklin, A.I.; Philippova, O.E. Disruption of cationic/anionic viscoelastic surfactant micellar networks by hydrocarbon as a basis of enhanced fracturing fluids clean-up. *Nanomaterials* **2020**, *10*, 2353. [CrossRef]
50. Kamal, M.S. A review of gemini surfactants: Potential application in enhanced oil recovery. *J. Surfactants Deterg.* **2015**, *19*, 223–236. [CrossRef]
51. Zana, R. Dimeric and oligomeric surfactants. Behavior at interfaces and in aqueous solution: A review. *Adv. Colloid Interface Sci.* **2002**, *97*, 205–253. [CrossRef]
52. In, M.; Bec, V.; Aguerre-Chariol, O.; Zana, R. Quaternary ammonium bromide surfactant oligomers in aqueous solution: Self-association and microstructure. *Langmuir* **2000**, *16*, 141–148. [CrossRef]
53. Sun, Y.; Han, F.; Zhou, Y.; Xu, B. Synthesis and properties of quaternary ammonium gemini surfactants with hydroxyethyl at head groups. *J. Surfactants Deterg.* **2019**, *22*, 675–681. [CrossRef]
54. Maiti, P.K.; Lansac, Y.; Glaser, M.A.; Clark, N.; Rouault, Y. Self-assembly in surfactant oligomers: A coarse-grained description through molecular dynamics simulations. *Langmuir* **2002**, *18*, 1908–1918. [CrossRef]
55. Wu, H.; Xu, J.; He, X.; Zhao, Y.; Wen, H. Mesoscopic simulation of self-assembly in surfactant oligomers by dissipative particle dynamics. *Colloids Surf. A Physicochem. Eng. Asp.* **2006**, *290*, 239–246. [CrossRef]
56. Li, H.; Yang, H.; Yan, Y.; Wang, Q.; He, P. Synthesis and solution properties of cationic gemini surfactants with long unsaturated tails. *Surf. Sci.* **2010**, *604*, 1173–1178. [CrossRef]
57. Mao, J.; Yang, X.; Wang, D.; Li, Y.; Zhao, J. A novel gemini viscoelastic surfactant (VES) for fracturing fluids with good temperature stability. *RSC Adv.* **2016**, *6*, 88426–88432. [CrossRef]
58. Zhang, W.; Mao, J.; Yang, X.; Zhang, H.; Zhang, Z.; Yang, B.; Zhang, Y.; Zhao, J. Study of a novel gemini viscoelastic surfactant with high performance in clean fracturing fluid application. *Polymers* **2018**, *10*, 1215. [CrossRef] [PubMed]

59. Zhao, J.; Fan, J.; Mao, J.; Yang, X.; Zhang, H.; Zhang, W. High performance clean fracturing fluid using a new tri-cationic surfactant. *Polymers* **2018**, *10*, 535. [CrossRef] [PubMed]
60. Yang, C.; Hu, Z.; Song, Z.; Bai, J.; Zhang, Y.; Luo, J.; Du, Y.; Jiang, Q. Self-assembly properties of ultra-long-chain gemini surfactant with high performance in a fracturing fluid application. *J. Appl. Polym. Sci.* **2017**, *134*, 44602. [CrossRef]
61. Yang, C.; Song, Z.; Zhao, J.; Hu, Z.; Zhang, Y.; Jiang, Q. Self-assembly properties of ultra-long-chain gemini surfactants bearing multiple amide groups with high performance in fracturing fluid application. *Colloids Surf. A Physicochem. Eng. Asp.* **2017**, *523*, 62–70. [CrossRef]
62. Zhang, W.; Mao, J.; Yang, X.; Zhang, H.; Zhao, J.; Tian, J.; Lin, C.; Mao, J. Development of a sulfonic gemini zwitterionic viscoelastic surfactant with high salt tolerance for seawater-based clean fracturing fluid. *Chem. Eng. Sci.* **2019**, *207*, 688–701. [CrossRef]
63. Zhang, Y.; Mao, J.; Zhao, J.; Zhang, W.; Liao, Z.; Xu, T.; Du, A.; Zhang, Z.; Yang, X.; Ni, Y. Preparation of a novel sulfonic gemini zwitterionic viscoelastic surfactant with superior heat and salt resistance using a rigid-soft combined strategy. *J. Mol. Liq.* **2020**, *318*, 114057. [CrossRef]
64. Huang, F.; Pu, C.; Lu, L.; Pei, Z.; Gu, X.; Lin, S.; Wu, F.; Liu, J. Gemini surfactant with unsaturated long tails for viscoelastic surfactant (VES) fracturing fluid used in tight reservoirs. *ACS Omega* **2021**, *6*, 1593–1602. [CrossRef]
65. Mao, J.; Zhang, H.; Zhang, W.; Fan, J.; Zhang, C.; Zhao, J. Dissymmetric beauty: A novel design of heterogemini viscoelastic surfactant for the clean fracturing fluid. *J. Ind. Eng. Chem.* **2018**, *60*, 133–142. [CrossRef]
66. Hussain, S.M.S.; Kamal, M.S.; Solling, T.; Murtaza, M.; Fogang, L.T. Surface and thermal properties of synthesized cationic poly(ethylene oxide) gemini surfactants: The role of the spacer. *RSC Adv.* **2019**, *9*, 30154–30163. [CrossRef]
67. Tariq, Z.; Kamal, M.S.; Mahmoud, M.; Hussain, S.M.S.; Hussaini, S.R. Novel gemini surfactant as a clay stabilizing additive in fracturing fluids for unconventional tight sandstones: Mechanism and performance. *J. Pet. Sci. Eng.* **2020**, *195*, 107917. [CrossRef]
68. Tariq, Z.; Kamal, M.S.; Mahmoud, M.; Hussain, S.M.S.; Abdulraheem, A.; Zhou, X. Polyoxyethylene quaternary ammonium gemini surfactants as a completion fluid additive to mitigate formation damage. *SPE Drill. Complet.* **2020**, *35*, 696–706. [CrossRef]
69. Tariq, Z.; Kamal, M.S.; Mahmoud, M.; Murtaza, M.; Abdulraheem, A.; Zhou, X. Dicationic surfactants as an additive in fracturing fluids to mitigate clay swelling: A petrophysical and rock mechanical assessment. *ACS Omega* **2021**, *6*, 15867–15877. [CrossRef]
70. Shrestha, R.G.; Shrestha, L.K.; Aramaki, K. Formation of wormlike micelle in a mixed amino-acid based anionic surfactant and cationic surfactant systems. *J. Colloid Interface Sci.* **2007**, *311*, 276–284. [CrossRef]
71. Różańska, S. Rheology of wormlike micelles in mixed solutions of cocoamidopropyl betaine and sodium dodecylbenzenesulfonate. *Colloid. Sur. A Physicochem. Eng. Aspect.* **2015**, *482*, 394–402. [CrossRef]
72. Lu, H.; Yuan, M.; Fang, B.; Wang, J.; Guo, Y. Wormlike micelles in mixed amino acid-based anionic surfactant and zwitterionic surfactant systems. *J. Surfactants Deterg.* **2015**, *18*, 589–596. [CrossRef]
73. Shrestha, R.G.; Shrestha, L.; Aramaki, K. Wormlike micelles in mixed amino acid-based anionic/nonionic surfactant systems. *J. Colloid Interface Sci.* **2008**, *322*, 596–604. [CrossRef]
74. Wang, J.; Luo, X.; Chu, Z.; Feng, Y. Effect of residual chemicals on wormlike micelles assembled from a C22-tailed cationic surfactant. *J. Colloid Interface Sci.* **2019**, *553*, 91–98. [CrossRef]
75. Croce, V.; Cosgrove, T.; Dreiss, C.A.; Maitland, G.; Hughes, T.; Karlsson, G. Impacting the length of wormlike micelles using mixed surfactant systems. *Langmuir* **2004**, *20*, 7984–7990. [CrossRef]
76. Shibaev, A.V.; Philippova, O.E. Viscoelastic properties of new mixed wormlike micelles formed by a fatty acid salt and alkylpyridinium surfactant. *Nanosyst. Phys. Chem. Math.* **2017**, *8*, 732–739. [CrossRef]
77. Shibaev, A.V.; Ospennikov, A.; Kuklin, A.I.; Arkharova, N.A.; Orekhov, A.S.; Philippova, O.E. Structure, rheological and responsive properties of a new mixed viscoelastic surfactant system. *Colloids Surf. A Physicochem. Eng. Asp.* **2020**, *586*, 124284. [CrossRef]
78. Koehler, R.D.; Raghavan, S.R.; Kaler, E.W. Microstructure and dynamics of wormlike micellar solutions formed by mixing cationic and anionic surfactants. *J. Phys. Chem. B* **2000**, *104*, 11035–11044. [CrossRef]
79. Schubert, B.A.; Kaler, E.W.; Wagner, N.J. The microstructure and rheology of mixed cationic/anionic wormlike micelles. *Langmuir* **2003**, *19*, 4079–4089. [CrossRef]
80. Geng, F.; Yu, L.; Cao, Q.; Li, Z.; Zheng, L.; Xiao, J.; Chen, H.; Cao, Z. Rheological properties of wormlike micelles formed by cetyltrimethylammonium bromide and sodium laurate. *J. Dispers. Sci. Technol.* **2009**, *30*, 92–99. [CrossRef]
81. Koshy, P.; Verma, G.; Aswal, V.K.; Venkatesh, M.; Hassan, P.A. Viscoelastic fluids originated from enhanced solubility of sodium laurate in cetyl trimethyl ammonium bromide micelles through cooperative self-assembly. *J. Phys. Chem. B* **2010**, *114*, 10462–10470. [CrossRef]
82. Koshy, P.; Aswal, V.K.; Venkatesh, M.; Hassan, P.A. Swelling and elongation of tetradecyltrimethylammonium bromide micelles induced by anionic sodium laurate. *Soft Matter* **2011**, *7*, 4778–4786. [CrossRef]
83. Hao, L.-S.; Hu, P.; Nan, Y.-Q. Salt effect on the rheological properties of the aqueous mixed cationic and anionic surfactant systems. *Colloids Surf. A Physicochem. Eng. Asp.* **2010**, *361*, 187–195. [CrossRef]
84. Zhai, Z.; Yan, X.; Xu, J.; Song, Z.; Shang, S.; Rao, X. Phase behavior and aggregation in a catanionic system dominated by an anionic surfactant containing a large rigid group. *Chem. A Eur. J.* **2018**, *24*, 9033–9040. [CrossRef]
85. Bhattacharjee, J.; Aswal, V.K.; Hassan, P.A.; Pamu, R.; Narayanan, J.; Bellare, J. Structural evolution in catanionic mixtures of cetylpyridinium chloride and sodium deoxycholate. *Soft Matter* **2012**, *8*, 10130–10140. [CrossRef]

86. Zhou, M.; Li, S.; Zhang, Z.; Luo, G.; Zhao, J. Synthesis of oligomer betaine surfactant (DDTPA) and rheological properties of wormlike micellar solution system. *J. Taiwan Inst. Chem. Eng.* **2016**, *66*, 1–11. [CrossRef]
87. Baruah, A.; Shekhawat, D.S.; Pathak, A.K.; Ojha, K. Experimental investigation of rheological properties in zwitterionic-anionic mixed-surfactant based fracturing fluids. *J. Pet. Sci. Eng.* **2016**, *146*, 340–349. [CrossRef]
88. Pei, X.; Xu, Z.; Song, B.; Cui, Z.; Zhao, J. Wormlike micelles formed in catanionic systems dominated by cationic gemini surfactant: Synergistic effect with high efficiency. *Colloids Surf. A Physicochem. Eng. Asp.* **2014**, *443*, 508–514. [CrossRef]
89. Zhang, Y.; Mao, J.; Zhao, J.; Liao, Z.; Xu, T.; Mao, J.; Sun, H.; Zheng, L.; Ni, Y. Synergy between different sulfobetaine-type zwitterionic gemini surfactants: Surface tension and rheological properties. *J. Mol. Liq.* **2021**, *332*, 115141. [CrossRef]
90. Chieng, Z.H.; Mohyaldinn, M.E.; Hassan, A.M.; Bruining, H. Experimental investigation and performance evaluation of modified viscoelastic surfactant (VES) as a new thickening fracturing fluid. *Polymers* **2020**, *12*, 1470. [CrossRef]
91. Al-Muntasheri, G.A.; Liang, F.; Hull, K. Nanoparticle-enhanced hydraulic-fracturing fluids: A review. *SPE Prod. Oper.* **2017**, *32*, 186–195. [CrossRef]
92. Philippova, O.E.; Molchanov, V.S. Enhanced rheological properties and performance of viscoelastic surfactant fluids with embedded nanoparticles. *Curr. Opin. Colloid Interface Sci.* **2019**, *43*, 52–62. [CrossRef]
93. Philippova, O.E. Self-assembled networks formed by wormlike micelles and nanoparticles. In *Wormlike Micelles;* Royal Society of Chemistry (RSC): Cambridge, UK, 2017; Volume 6, Chapter 5; pp. 103–120.
94. Wu, H.; Zhou, Q.; Xu, D.; Sun, R.; Zhang, P.; Bai, B.; Kang, W. SiO_2 nanoparticle-assisted low-concentration viscoelastic cationic surfactant fracturing fluid. *J. Mol. Liq.* **2018**, *266*, 864–869. [CrossRef]
95. Sambasivam, A.; Dhakal, S.; Sureshkumar, R. Structure and rheology of self-assembled aqueous suspensions of nanoparticles and wormlike micelles. *Mol. Simul.* **2018**, *44*, 485–493. [CrossRef]
96. Helgeson, M.; Hodgdon, T.K.; Kaler, E.W.; Wagner, N.; Vethamuthu, M.; Ananthapadmanabhan, K.P. Formation and rheology of viscoelastic "double networks" in wormlike micelle−nanoparticle mixtures. *Langmuir* **2010**, *26*, 8049–8060. [CrossRef]
97. Jódar-Reyes, A.B.; Leermakers, F.A.M. Can linear micelles bridge between two surfaces? *J. Phys. Chem. B* **2006**, *110*, 18415–18423. [CrossRef]
98. Fan, Q.; Li, W.; Zhang, Y.; Fan, W.; Li, X.; Dong, J. Nanoparticles induced micellar growth in sodium oleate wormlike micelles solutions. *Colloid Polym. Sci.* **2015**, *293*, 2507–2513. [CrossRef]
99. Luo, M.; Jia, Z.; Sun, H.; Liao, L.; Wen, Q. Rheological behavior and microstructure of an anionic surfactant micelle solution with pyroelectric nanoparticle. *Colloids Surf. A Physicochem. Eng. Asp.* **2012**, *395*, 267–275. [CrossRef]
100. Nettesheim, F.; Liberatore, M.W.; Hodgdon, T.K.; Wagner, N.J.; Kaler, E.W.; Vethamuthu, M. Influence of nanoparticle addition on the properties of wormlike micellar solutions. *Langmuir* **2008**, *24*, 7718–7726. [CrossRef]
101. Huang, T.; Crews, J.B. Nanotechnology applications in viscoelastic surfactant stimulation fluids. *SPE Prod. Oper.* **2008**, *23*, 512–517. [CrossRef]
102. Crews, J.B.; Huang, T. Performance enhancements of viscoelastic surfactant stimulation fluids with nanoparticles. In Proceedings of the Society of Petroleum Engineers Europec/EAGE Conference and Exhibition, Rome, Italy, 9 June 2008. [CrossRef]
103. Huang, T.; Crews, J.B.; Agrawal, G. Nanoparticle pseudocrosslinked micellar fluids: Optimal solution for fluid-loss control with internal breaking. In Proceedings of the Society of Petroleum Engineers International Symposium and Exhibiton on Formation Damage Control, Lafayette, LA, USA, 10–12 February 2010. [CrossRef]
104. Pletneva, V.A.; Molchanov, V.S.; Philippova, O.E. Viscoelasticity of smart fluids based on wormlike surfactant micelles and oppositely charged magnetic particles. *Langmuir* **2015**, *31*, 110–119. [CrossRef]
105. Ismagilov, I.F.; Kuryashov, D.A.; Idrisov, A.R.; Bashkirtseva, N.Y.; Zakharova, L.Y.; Zakharov, S.V.; Alieva, M.R.; Kashapova, N.E. Supramolecular system based on cylindrical micelles of anionic surfactant and silica nanoparticles. *Colloids Surf. A Physicochem. Eng. Asp.* **2016**, *507*, 255–260. [CrossRef]
106. Dai, C.; Zhang, Y.; Gao, M.; Li, Y.; Lv, W.; Wang, X.; Wu, Y.; Zhao, M. The study of a novel nanoparticle-enhanced wormlike micellar system. *Nanoscale Res. Lett.* **2017**, *12*, 431. [CrossRef]
107. Zhao, M.; Zhang, Y.; Zou, C.; Dai, C.; Gao, M.; Li, Y.; Lv, W.; Jiang, J.; Wu, Y. Can more nanoparticles induce larger viscosities of nanoparticle-enhanced wormlike micellar system (NEWMS)? *Materials* **2017**, *10*, 1096. [CrossRef]
108. Qin, W.; Yue, L.; Liang, G.; Jiang, G.; Yang, J.; Liu, Y. Effect of multi-walled carbon nanotubes on linear viscoelastic behavior and microstructure of zwitterionic wormlike micelle at high temperature. *Chem. Eng. Res. Des.* **2017**, *123*, 14–22. [CrossRef]
109. Molchanov, V.S.; Pletneva, V.A.; Kuklin, A.I.; Philippova, O.E. Nanocomposite composed of charged wormlike micelles and magnetic particles. *J. Phys. Conf. Ser.* **2017**, *848*, 12013. [CrossRef]
110. Zhang, Y.; Dai, C.L.; Qian, Y.; Fan, X.; Jiang, J.; Wu, Y.; Wu, X.; Huang, Y.; Zhao, M. Rheological properties and formation dynamic filtration damage evaluation of a novel nanoparticle-enhanced VES fracturing system constructed with wormlike micelles. *Colloids Surf. A Physicochem. Eng. Asp.* **2018**, *553*, 244–252. [CrossRef]
111. Zhao, M.; Gao, Z.; Dai, C.; Sun, X.; Zhang, Y.; Yang, X.; Wu, Y. Effect of silica nanoparticles on wormlike micelles with different entanglement degrees. *J. Surfactants Deterg.* **2019**, *22*, 587–595. [CrossRef]
112. Xiong, J.; Fang, B.; Lu, Y.; Qiu, X.; Ming, H.; Li, K.; Zhai, W.; Wang, L.; Liu, Y.; Cao, L. Rheology and high-temperature stability of novel viscoelastic gemini micelle solutions. *J. Dispers. Sci. Technol.* **2018**, *39*, 1324–1327. [CrossRef]
113. Molchanov, V.S.; Pletneva, V.A.; Klepikov, I.A.; Razumovskaya, I.V.; Philippova, O.E. Soft magnetic nanocomposites based on adaptive matrix of wormlike surfactant micelles. *RSC Adv.* **2018**, *8*, 11589–11597. [CrossRef]

114. Omeiza, A.A.; Samsuri, A.B. Viscoelastic surfactants application in hydraulic fracturing, it's set back and mitigation-an overview. *ARPN J. Eng. Appl. Sci.* **2014**, *9*, 25–29.
115. Luo, M.; Jia, Z.; Sun, H.; Wen, Q. Performance of nano-TiO2 modified MES viscoelastic micelle solution. *Acta Petrol. Sin.* **2012**, *28*, 456–462. [CrossRef]
116. Nagarajan, R. Association of nonionic polymers with micelles, bilayers, and microemulsions. *J. Chem. Phys.* **1989**, *90*, 1980–1994. [CrossRef]
117. Brackman, J.C.; Engberts, J.B.F.N. Polymer-induced breakdown of rodlike micelles. A striking transition of a non-Newtonian to a Newtonian fluid. *J. Am. Chem. Soc.* **1990**, *112*, 872–873. [CrossRef]
118. Brackman, J.C.; Engberts, J.B.F.N. Influence of polymers on the micellization of cetyltrimethylammonium salts. *Langmuir* **1991**, *7*, 2097–2102. [CrossRef]
119. Lin, Z.; Eads, C.D. Polymer-induced structural transitions in oleate solutions: Microscopy, rheology, and nuclear magnetic resonance studies. *Langmuir* **1997**, *13*, 2647–2654. [CrossRef]
120. Li, X.; Lin, Z.; Cai, J.; Scriven, L.E.; Davis, H.T. Polymer-induced microstructural transitions in surfactant solutions. *J. Phys. Chem.* **1995**, *99*, 10865–10878. [CrossRef]
121. Nakamura, K.; Shikata, T. Hybrid threadlike micelle formation between a surfactant and polyelectrolyte. *Macromolecules* **2003**, *36*, 9698–9700. [CrossRef]
122. Oikonomou, E.; Bokias, G.; Kallitsis, J.K.; Iliopoulos, I. Formation of hybrid wormlike micelles upon mixing cetyl trimethylammonium bromide with poly (methyl methacrylate-co-sodium styrene sulfonate) copolymers in aqueous solution. *Langmuir* **2011**, *27*, 5054–5061. [CrossRef]
123. Nakamura, K.; Yamanaka, K.; Shikata, T. Hybrid threadlike micelle formation between a surfactant and polymer in aqueous solution. *Langmuir* **2003**, *19*, 8654–8660. [CrossRef]
124. Kwiatkowski, A.L.; Sharma, H.; Molchanov, V.S.; Orekhov, A.S.; Vasiliev, A.L.; Dormidontova, E.E.; Philippova, O.E. Wormlike surfactant micelles with embedded polymer chains. *Macromolecules* **2017**, *50*, 7299–7308. [CrossRef]
125. Kwiatkowski, A.L.; Molchanov, V.S.; Sharma, H.; Kuklin, A.I.; Dormidontova, E.E.; Philippova, O.E. Growth of wormlike micelles of surfactant induced by embedded polymer: Role of polymer chain length. *Soft Matter* **2018**, *14*, 4792–4804. [CrossRef]
126. Kwiatkowski, A.L.; Molchanov, V.S.; Kuklin, A.I.; Philippova, O.E. Opposite effect of salt on branched wormlike surfactant micelles with and without embedded polymer. *J. Mol. Liq.* **2020**, *311*, 113301. [CrossRef]
127. Peiffer, D. Hydrophobically associating polymers and their interactions with rod-like micelles. *Polymer* **1990**, *31*, 2353–2360. [CrossRef]
128. Shashkina, J.A.; Philippova, O.E.; Zaroslov, Y.D.; Khokhlov, A.R.; Pryakhina, T.A.; Blagodatskikh, I.V. Rheology of viscoelastic solutions of cationic surfactant. Effect of added associating polymer. *Langmuir* **2005**, *21*, 1524–1530. [CrossRef]
129. Gouveia, L.M.; Müller, A.J. The effect of NaCl addition on the rheological behavior of cetyltrimethylammonium p-toluenesulfonate (CTAT) aqueous solutions and their mixtures with hydrophobically modified polyacrylamide aqueous solutions. *Rheol. Acta* **2009**, *48*, 163–175. [CrossRef]
130. Philippova, O.E.; Khokhlov, A.R. Smart polymers for oil production. *Pet. Chem.* **2010**, *50*, 266–270. [CrossRef]
131. Pletneva, V.A.; Molchanov, V.S.; Philippova, O. Effect of polymer on rheological behavior of heated solutions of potassium oleate cylindrical micelles. *Colloid J.* **2010**, *72*, 716–722. [CrossRef]
132. Mei, Y.; Han, Y.; Zhou, H.; Yao, L.; Jiang, B. Study on synergistic effect between wormlike micelles and hydrophobically modified poly (acrylic acid) in salt solution. *J. Dispers. Sci. Technol.* **2013**, *34*, 651–656. [CrossRef]
133. Zhou, H.; Han, Y.; Mei, Y.; Wei, Y.; Wang, H. Study on response behaviors of mixed solution of polyelectrolytes and worms under shear. *J. Polym. Res.* **2014**, *21*, 351. [CrossRef]
134. Mei, Y.; Han, Y.; Wang, H.; Xie, L.; Zhou, H. Electrostatic effect on synergism of wormlike micelles and hydrophobically modified polyacrylic acid. *J. Surfactants Deterg.* **2013**, *17*, 323–330. [CrossRef]
135. Yongqiang, W.; Yixiu, H.; Hong, Z.; Ke, W.; Yongjun, M.; Hang, W. Investigation on the interaction between hydrophobically modified polyacrylic acid and wormlike micelles under shear. *J. Solut. Chem.* **2015**, *44*, 1177–1190. [CrossRef]
136. Molchanov, V.S.; Philippova, O.E.; Khokhlov, A.R.; Kovalev, Y.A.; Kuklin, A. Self-assembled networks highly responsive to hydrocarbons. *Langmuir* **2007**, *23*, 105–111. [CrossRef]
137. Molchanov, V.S.; Philippova, O.E. Dominant role of wormlike micelles in temperature-responsive viscoelastic properties of their mixtures with polymeric chains. *J. Colloid Interface Sci.* **2013**, *394*, 353–359. [CrossRef]
138. Li, X.; Sarsenbekuly, B.; Yang, H.; Huang, Z.; Jiang, H.; Kang, X.; Li, M.; Kang, W.; Luo, P. Rheological behavior of a wormlike micelle and an amphiphilic polymer combination for enhanced oil recovery. *Phys. Fluids* **2020**, *32*, 073105. [CrossRef]
139. Yang, J.; Cui, W.; Guan, B.; Lu, Y.; Qiu, X.; Yang, Z.; Qin, W. Supramolecular fluid of associative polymer and viscoelastic surfactant for hydraulic fracturing. *SPE Prod. Oper.* **2016**, *31*, 318–324. [CrossRef]
140. Jiang, Q.; Jiang, G.; Wang, C.; Zhu, Q.; Yang, L.; Wang, L.; Zhang, X.; Chong, L. A new high-temperature shear-tolerant supramolecular viscoelastic fracturing fluid. In Proceedings of the Paper Presented at the IADC/SPE Asia Pacific Drilling Technology Conference, Singapore, 22–24 August 2016.
141. Jiang, G.; Jiang, Q.; Sun, Y.; Liu, P.; Zhang, Z.; Ni, X.; Yang, L.; Wang, C. Supramolecular-structure-associating weak gel of wormlike micelles of erucoylamidopropyl hydroxy sulfobetaine and hydrophobically modified polymers. *Energy Fuels* **2017**, *31*, 4780–4790. [CrossRef]

142. Hill, A.; Candau, F.; Selb, J. Properties of hydrophobically associating polyacrylamides: Influence of the method of synthesis *Macromolecules* **1993**, *26*, 4521–4532. [CrossRef]
143. Candau, F.; Selb, J. Hydrophobically-modified polyacrylamides prepared by micellar polymerization. *Adv. Colloid Interface Sci.* **1999**, *79*, 149–172. [CrossRef]
144. Chen, F.; Wu, Y.; Wang, M.; Zha, R. Self-assembly networks of wormlike micelles and hydrophobically modified polyacrylamide with high performance in fracturing fluid application. *Colloid Polym. Sci.* **2014**, *293*, 687–697. [CrossRef]
145. Pu, W.-F.; Du, D.-J.; Liu, R. Preparation and evaluation of supramolecular fracturing fluid of hydrophobically associative polymer and viscoelastic surfactant. *J. Pet. Sci. Eng.* **2018**, *167*, 568–576. [CrossRef]
146. Xiong, C.; Peng, K.; Tang, X.; Ye, Z.; Shi, Y.; Yang, H. CO2-responsive self-healable hydrogels based on hydrophobically-modified polymers bridged by wormlike micelles. *RSC Adv.* **2017**, *7*, 34669–34675. [CrossRef]
147. Yoshida, T.; Taribagil, R.; Hillmyer, M.A.; Lodge, T.P. Viscoelastic synergy in aqueous mixtures of wormlike micelles and model amphiphilic triblock copolymers. *Macromolecules* **2007**, *40*, 1615–1623. [CrossRef]
148. Ramos, L.; Ligoure, C. Structure of a new type of transient network: Entangled wormlike micelles bridged by telechelic polymers. *Macromolecules* **2007**, *40*, 1248–1251. [CrossRef]
149. Nakaya–Yaegashi, K.; Ramos, L.; Tabuteau, H.; Ligoure, C. Linear viscoelasticity of entangled wormlike micelles bridged by telechelic polymers: An experimental model for a double transient network. *J. Rheol.* **2008**, *52*, 359. [CrossRef]
150. Tabuteau, H.; Ramos, L.; Nakaya-Yaegashi, K.; Imai, M.; Ligoure, C. Nonlinear rheology of surfactant wormlike micelles bridged by telechelic polymers. *Langmuir* **2009**, *25*, 2467–2472. [CrossRef]
151. Couillet, I.; Hughes, T.; Maitland, G.; Candau, F. Synergistic effects in aqueous solutions of mixed wormlike micelles and hydrophobically modified polymers. *Macromolecules* **2005**, *38*, 5271–5282. [CrossRef]
152. Lee, J.-H.; Gustin, J.P.; Chen, T.; Payne, G.F.; Raghavan, S. Vesicle—biopolymer gels: Networks of surfactant vesicles connected by associating biopolymers. *Langmuir* **2005**, *21*, 26–33. [CrossRef]
153. Rojas, M.; Müller, A.; Sáez, A. Synergistic effects in flows of mixtures of wormlike micelles and hydroxyethyl celluloses with or without hydrophobic modifications. *J. Colloid Interface Sci.* **2008**, *322*, 65–72. [CrossRef]
154. Zoratto, N.; Grillo, I.; Matricardi, P.; Dreiss, C.A. Supramolecular gels of cholesterol-modified gellan gum with disc-like and worm-like micelles. *J. Colloid Interface Sci.* **2019**, *556*, 301–312. [CrossRef]
155. Afifi, H.; da Silva, M.A.; Nouvel, C.; Six, J.-L.; Ligoure, C.; Dreiss, C.A. Associative networks of cholesterol-modified dextran with short and long micelles. *Soft Matter* **2011**, *7*, 4888–4899. [CrossRef]
156. Huang, F.; Pu, C.; Gu, X.; Ye, Z.; Khan, N.; An, J.; Wu, F.; Liu, J. Study of a low-damage efficient-imbibition fracturing fluid without flowback used for low-pressure tight reservoirs. *Energy* **2021**, *222*, 119941. [CrossRef]
157. Shibaev, A.V.; Mityuk, D.Y.; Muravlev, D.A.; Philippova, O.E. Viscoelastic solutions of wormlike micelles of a cationic surfactant and a stiff-chain anionic polyelectrolyte. *Polym. Sci. Ser. A* **2019**, *61*, 765–772. [CrossRef]
158. Shibaev, A.; Abrashitova, K.A.; Kuklin, A.; Orekhov, A.S.; Vasiliev, A.L.; Iliopoulos, I.; Philippova, O.E. Viscoelastic synergy and microstructure formation in aqueous mixtures of nonionic hydrophilic polymer and charged wormlike surfactant micelles. *Macromolecules* **2017**, *50*, 339–348. [CrossRef]
159. Shibaev, A.V.; Makarov, A.V.; Kuklin, A.I.; Iliopoulos, I.; Philippova, O.E. Role of charge of micellar worms in modulating structure and rheological properties of their mixtures with nonionic polymer. *Macromolecules* **2018**, *51*, 213–221. [CrossRef]
160. Roland, S.; Miquelard-Garnier, G.; Shibaev, A.V.; Aleshina, A.L.; Chennevière, A.; Matsarskaia, O.; Sollogoub, C.; Philippova, O.E.; Iliopoulos, I. Dual transient networks of polymer and micellar chains: Structure and viscoelastic synergy. *Polymers* **2021**, *13*, 4255. [CrossRef]
161. Shibaev, A.V.; Kuklin, A.I.; Torocheshnikov, V.N.; Orekhov, A.S.; Roland, S.; Miquelard-Garnier, G.; Matsarskaia, O.; Iliopoulos, I.; Philippova, O.E. Double dynamic hydrogels formed by wormlike surfactant micelles and cross-linked polymer. *J. Colloid Interface Sci.* **2022**. [CrossRef]
162. Gao, Z.; Dai, C.; Sun, X.; Huang, Y.; Gao, M.; Zhao, M. Investigation of cellulose nanofiber enhanced viscoelastic fracturing fluid system: Increasing viscoelasticity and reducing filtration. *Colloids Surf. A Physicochem. Eng. Asp.* **2019**, *582*, 123938. [CrossRef]
163. Yang, Y.; Zhang, H.; Wang, H.; Zhang, J.; Guo, Y.; Wei, B.; Wen, Y. Pseudo-interpenetrating network viscoelastic surfactant fracturing fluid formed by surface-modified cellulose nanofibril and wormlike micelles. *J. Pet. Sci. Eng.* **2022**, *208*, 109608. [CrossRef]

MDPI
St. Alban-Anlage 66
4052 Basel
Switzerland
Tel. +41 61 683 77 34
Fax +41 61 302 89 18
www.mdpi.com

Gels Editorial Office
E-mail: gels@mdpi.com
www.mdpi.com/journal/gels

www.ingramcontent.com/pod-product-compliance
Lightning Source LLC
LaVergne TN
LVHW070959170726
843515LV00006B/1028

* 9 7 8 3 0 3 6 5 5 1 9 6 8 *